NUMERICAL ANALYSIS

Numerical Analysis

KAISER S. KUNZ

Research Physicist
Schlumberger Well Surveying Corporation
Ridgefield Research Laboratory

McGRAW-HILL BOOK COMPANY, INC.

New York   Toronto   London

1957

NUMERICAL ANALYSIS

Library of Congress Catalog Card Number 56–13395

v

35630

THE MAPLE PRESS COMPANY, YORK, PA.

To my Mother and Father

PREFACE

This book has grown out of a set of lecture notes for a graduate course in numerical analysis which I gave at the Computation Laboratory of Harvard University during the years 1947 to 1949. Professor Howard Aiken of that laboratory foresaw the great need that would be felt throughout the country for men trained in the design and use of electronic digital computers and initiated at that early date a set of courses to meet this need. I was privileged during that time to devote full time to the preparation of the course on numerical methods.

It was felt that such a course should cover those topics most directly needed for an understanding of the methods used in the numerical solution of differential equations, both ordinary and partial, and in the solution of integral equations. Thus in this book, as in the lectures, considerable time is devoted to finite-difference tables and notation, to numerical differentiation and integration, but peripheral subjects such as the smoothing of experimental data, least-squares approximation, and harmonic analysis are omitted. These omissions made it possible to include in one book all the background material needed for obtaining numerical solutions to advanced problems in applied mathematics.

While the main reason for writing this book has been to acquaint the student with the best procedures available for obtaining numerical solutions to problems arising in applied mathematics, methods of doubtful practicality have often been included to broaden the student's understanding of the essential unity of all finite-difference methods. Since the goal sought is largely a pragmatic one, only a modest attempt has been made to preserve mathematical rigor and free use has been made of symbolic methods and heuristic arguments.

Although no attempt has been made to exclude methods suitable only for desk-type calculators, or even those for pencil and paper computations, more stress has been given to those procedures that can be most readily programmed for an electronic digital computer. Usually, however, a method that has merit for hand computation needs only minor modifications to adapt it to automatic machines.

While a knowledge of calculus and differential equations is presupposed, the level of mathematics required is rather modest, and except in Chap.

10, where a rudimentary knowledge of matrices is assumed, all results are obtained from first principles.

The book reflects the initial division of the lectures into a semester course covering Chaps. 1 to 9, including only material applying to a single independent variable, and another course covering the remaining six chapters, in which two or more independent variables are involved. The first leads in a very natural way to the treatment of numerical solution of ordinary differential equations in Chaps. 8 and 9, and the second leads by a much shorter route to a discussion of partial differential equations in Chaps. 12 to 14 and of integral equations in Chap. 15.

Methods for finding numerically the roots of polynomial and transcendental equations are covered in Chaps. 1 and 2. Finite-difference tables and notation and the theory of interpolation are treated in Chaps. 3 to 5. It is pointed out in these chapters that an interpolating polynomial can be conveniently designated by the highest order difference one can form using the points of fit between this polynomial and the given function and that the accuracy of the interpolation is determined by the interpolating polynomial used.

The treatment of summation of series and methods for improving the rate of convergence of an infinite series as given in Chap. 6 is much more complete than that usually given in a text on numerical analysis.

Chapter 7 on numerical differentiation and integration is unusual in that the lozenge-diagram technique previously used for interpolation is shown to be equally applicable to finding differentiation and integration formulas. Also, the simple method developed here for obtaining a new quadrature formula from a known one is, I believe, a new technique.

The numerical methods of solving ordinary differential equations are treated next. Methods of starting a solution are covered in Chap. 8 and methods for continuing the solution in Chap. 9. In these chapters considerable attention is given to the accumulation of errors in a computation.

The second part of the book, treating problems of more than one independent variable, starts off with Chap. 10 on the solution of simultaneous equations and Chap. 11 on multivariate interpolation. Some reduction in the material covered in a course could be made by omitting parts of the latter chapter.

Included in Chaps. 12 to 14 is a rather complete treatment of finite-difference methods for solving partial differential equations. The general technique of obtaining a difference equation from a differential equation is treated in Chap. 12. Elliptic equations are then covered in Chap. 13 and hyperbolic and parabolic equations in Chap. 14.

The last chapter is devoted to the numerical solution of integral equations, which are now of considerable importance in many fields of applied mathematics.

In Appendix A is given a short but fairly complete treatment of the general analysis of error in numerical computations. More specific applications of this analysis are given throughout the text as an integral part of the discussion.

I wish to acknowledge the influence of Professor Howard Aiken and the staff of Harvard Computation Laboratory on the initial formulation of the material and the stimulation afforded me by a great many of my former students. This acknowledgment should also include the influence of J. B. Scarborough, J. F. Steffensen, and E. Whittaker and G. Robinson through their pioneer works on the subject.

It is a pleasure to be able to thank John Baker, Jr., William Burkhardt, Frank Johnson, Jean Lebel, James Moran, William Polland, Francis Segesman, Gerald Simard, Sidney Soloway, and Jay Tittman for reading various chapters of the manuscript.

Besides those who helped with the actual writing, there are a great many others I would like to thank for having contributed, by their friendship and encouragement in difficult times, to the completion of this book.

Finally, I wish to thank my wife, Ruth Barbara Kunz, for her untiring support. Most assuredly, without her help this book would never have been written.

KAISER S. KUNZ

CONTENTS

CHAPTER 1

THE REAL ROOTS OF AN EQUATION

One of the oldest problems in mathematics is that of finding the roots of an equation. This has given rise to a whole branch of mathematics known as the *theory of equations*. Only a part of the material written on this subject, however, is directed toward finding numerical values of the roots, which is our main concern.

This chapter deals with those methods which are applicable to finding the real roots of the equation

$$f(x) = 0 \tag{1.1}$$

where $f(x)$ is any piecewise continuous function of x having numerical coefficients, whether a polynomial or transcendental function. Special methods that apply only to polynomial equations will be dealt with separately in Chap. 2. Although $f(x)$ is thus apparently quite general, it is here restricted by the assumption that the labor of obtaining the numerical value of $f(x)$ for any desired value of x is not excessive. Special consideration must be given any problem in which this requirement is not satisfied.

The finding of the real roots of $f(x)$ can in general be divided into two parts. The first part has as its goal the finding of an approximate value of the root. The second part makes use of this approximate knowledge of the root to obtain the root out to the desired number of significant figures.†

1.1. Finding an Approximate Value of a Real Root. In some cases the physical data giving rise to an equation will serve to fix a root as lying between two fairly narrow limits, and this fact may eliminate the necessity of using any special device for obtaining an approximate root. Sometimes, also, one may have a good guess as to the approximate value of the root; but in general it will be necessary to use one of the following methods.

---

† For a definition of the term *significant figures* see Sec. A.5. See also J. B. Scarborough, "Numerical Mathematical Analysis," 2d ed., p. 2, Johns Hopkins Press, Baltimore, 1950.

Graphical Methods. Generally speaking, the best method of finding the approximate value of a root is to plot the function

$$y = f(x) \tag{1.2}$$

and determine approximately the points at which this plot crosses the x axis. At these points $y = 0$, and hence the corresponding values of x, by Eq. (1.2), satisfy Eq. (1.1) and are therefore real roots of that equation. In some cases it is preferable to write the equation in the form

$$f_1(x) = f_2(x) \tag{1.3}$$

in which case we plot the two functions

$$y_1 = f_1(x) \quad \text{and} \quad y_2 = f_2(x) \tag{1.4}$$

The abscissas of their points of intersection obviously satisfy Eq. (1.3) and hence are the real roots of this equation.

In plotting the functions above, it is generally best to determine first as much of the following information as can be easily found:

1. The behavior of the function as $x \to -\infty$ and as $x \to +\infty$.
2. The value of $f(x)$ for those values of x which permit a rapid determination of $f(x)$. Usually $x = 0$ and $x = \pm 1$ are satisfactory for this purpose.
3. The values of x at which $f(x)$ becomes infinite.
4. The intercepts of the function on the x and y axes. This applies to Eqs. (1.4).

Usually enough of the above information is available to permit one to sketch in roughly the required function and thereby to determine the approximate location of each real root. If sufficient accuracy is achieved to permit one to say that one, and only one, root lies in a given region, then that root is said to be *isolated*. Additional points on the curve near the roots will serve to increase the accuracy with which the roots are known or will serve to isolate two or more adjacent roots.

Example 1. Find the approximate value of all the real roots of the equation

$$\sin x = \frac{x+1}{x-1} \tag{1.5}$$

The right- and left-hand members of this equation are plotted in Fig. 1.1. Their points of intersection I, II, III, etc., are all to the left of the y axis; thus there are no positive roots. Although the number of roots is infinite, the roots are isolated, since we see that only one root occurs in each of the intervals 0 to $-\pi/2$, $-\pi/2$ to $-3\pi/2$, $-3\pi/2$ to $-5\pi/2$, ..., $-(2k-1)\pi/2$ to $-(2k+1)\pi/2$, ... In fact the approximate values of the roots, the abscissas of the points of intersection, are seen to be $x = -0.4;\ -3.8,\ -5.5;\ -10.4,\ -11.5;\ -(11\pi/2 - \epsilon_3),\ -(11\pi/2 + \epsilon_3');$ ...; $-[(4k-1)\pi/2 - \epsilon_k],\ -[(4k-1)\pi/2 + \epsilon_k'];$ ..., where $\epsilon_k \to 0$ and $\epsilon_k' \to 0$ as $k \to \infty$.

The curves given in Fig. 1.1 can be sketched by observing that:

1. $\sin x$ is oscillatory and intersects the x axis at $n\pi$, where n takes on all integral values.

2. $\sin x$ achieves one-half its maximum absolute value at the points $n\pi \pm \pi/6$ and its full value 1 at the points $n\pi + \pi/2$.

3. $f(x) = (x + 1)/(x - 1)$ approaches $+\infty$ as $x \to 1$ from the right and approaches $-\infty$ as $x \to 1$ from the left.

4. $f(x) \to 1$ as $x \to \pm\infty$.

5. $f(0) = -1, f(-1) = 0, f(-2) = \frac{1}{3}, f(2) = 3,$ and $f(-9) = 0.8.$

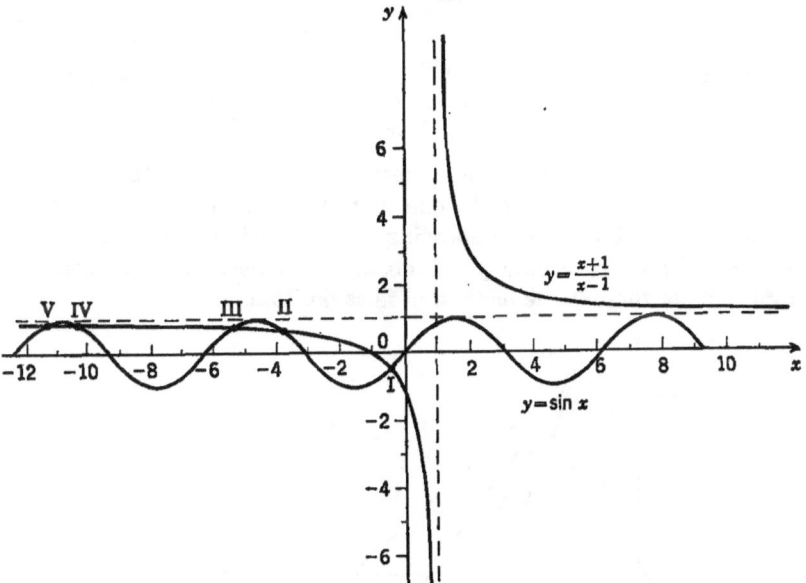

FIG. 1.1. Graphical solution of the equation $\sin x = (x + 1)/(x - 1)$.

Analytical Methods. There are two common analytical methods for finding the approximate value of the root of an equation. One is to find a simpler equation that has a root approximately equal to the required root of the given equation. This can often be done by neglecting a term known to be small. The second method makes use of the following theorem:

THEOREM I. If $f(x)$ is a real function that is continuous between $x = a$ and $x = b$, where a and b are real numbers, and if $f(a)$ and $f(b)$ are of opposite sign, then there is at least one real root between a and b.

The use of these two methods will be illustrated by the following examples:

Example 2. Find the approximate value of a real root of

$$f(x) = x^2 - \log x - 10 = 0 \qquad (1.6)$$

Since $0 < \log x < 1$ for $1 < x < 10$, one can obtain two approximate equations for a root in this range of x by replacing $\log x$ first by 0 then by 1. Thus

$$x_1^2 - 10 = 0$$
$$x_2^2 - 1 - 10 = 0$$

and, ordinarily, we should expect the root of Eq. (1.6) to lie between x_1 and x_2. This gives us a real root between 3.16 and 3.32. To check this result we apply Theorem I. Since $f(3.16) = -0.5141$ and $f(3.32) = 0.5013$ are of opposite sign and $f(x)$ is continuous for $x > 0$, there must be a root between these limits.

Example 3. Locate roughly the smallest positive real root of

$$f(x) = xe^x - 2 = 0 \tag{1.7}$$

Now $f(0) = -2$, and $f(1) = e - 2 = 0.718$; therefore there is at least one root between $x = 0$ and $x = 1$. Moreover $f(0.5) = -1.1756$, $f(0.8) = -0.2196$, and $f(0.9) = 0.2136$; therefore the root lies between $x = 0.8$ and $x = 0.9$.

1.2. Method of False Position (Regula Falsi). Having found a rough approximation to a real root by one of the methods suggested in Sec. 1.1, we have several ways of improving the approximation to obtain the required degree of accuracy. Probably the oldest and most generally applicable of these is the method of false position.

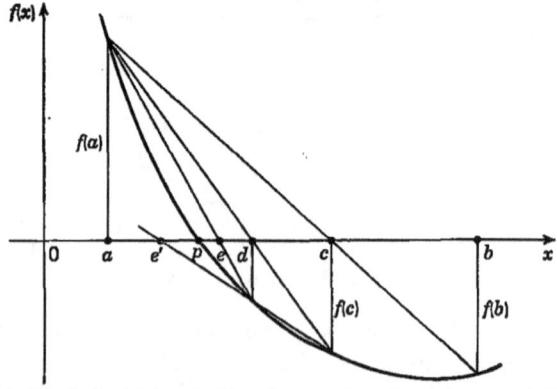

Fɪɢ. 1.2. Method of false position for locating the real roots of $f(x) = 0$.

Theorem I states that if $f(a)$ and $f(b)$ are of opposite signs, the plot of $f(x)$ crosses the x axis between $x = a$ and $x = b$ and that, therefore, a root $x = p$ lies between these limits. The method of false position goes further and predicts that the plot of $f(x)$ between these limits will be, roughly, a straight line; hence the root $x = p$ (see Fig. 1.2) will be approximately given by $x = c$. From the geometry of that figure

$$c = a + \frac{f(a)}{f(a) - f(b)} (b - a) \tag{1.8}$$

Since $f(a)$ and $f(b)$ are of opposite signs, $f(c)$ must be opposite in sign to one of them; therefore it is possible to apply the approximation again for a still better prediction of the value of p. In Fig. 1.2, $f(c)$ is opposite in sign to $f(a)$; therefore a better approximation

$$d = a + \frac{f(a)}{f(a) - f(c)} (c - a) \qquad (1.9)$$

to the root $x = p$ is obtained by replacing b in Eq. (1.8) by c.

Although one could proceed to replace c in Eq. (1.9) by d to get a still closer approximation e, it is better, ordinarily, to replace the more remote point a by d. This is preferable even though $f(c)$ and $f(d)$ are of the same sign and the predicted value $x = e'$ of the root lies outside of the interval $c < x < d.$† At the next step, however, one uses e' and d, which are ordinarily on opposite sides of the root and which serve, therefore, as upper and lower limits to the value of the root.

As the points approach the root and get closer and closer together, the curve becomes more nearly a straight line. This means that the rule gives best results when used to improve the accuracy of the roots once they are known approximately. Nevertheless, it can be used together with Theorem I to help locate a root roughly if due consideration is given to the possibility of erroneous predictions when the interval a to b is too large.

The method of false position is very simple in principle, as it merely replaces the plot of $f(x)$ between any two points a and b by its chord. Thus it represents nothing more than inverse interpolation, assuming linear variation, for that value of x which corresponds to $f(x) = 0$. The simplicity and complete generality of this method make it a very powerful tool for computation. It is readily adapted to an electronic calculator.‡

Example 4. Determine the smallest positive real root of

$$f(x) = xe^x - 2 = 0 \qquad (1.10)$$

to seven significant figures.

From Example 3 we know that the root lies between 0.8 and 0.9, since

$$f(0.8) = -0.2196$$

and $\qquad\qquad\qquad\qquad f(0.9) = 0.2136$

Therefore a better approximation, by the method of false position, is

$$x_1 = 0.8 + \frac{0.2196}{0.2196 - 0.2136} 0.1 = 0.851$$

† On an electronic computer one needs to ensure that this step will be disregarded if e' falls outside the range a to d.

‡ See Staff of the Computation Laboratory, "A Manual of Operation for the Automatic Sequence Controlled Calculator," p. 179, Harvard University Press, Cambridge, Mass., 1946.

From a table† for e^x

$$f(0.851) = -0.00697$$

Using the points 0.851 and 0.900,

$$x_3 = 0.851 + \frac{0.00697}{0.00697 - 0.2136} \, 0.049 = 0.8526$$

and
$$f(0.8526) = -0.00002\,396$$

Since the tables used give only four decimal places in the argument, and since the root lies between 0.8526 and 0.8527, we shall finish our calculation by applying the method of false position to these values:

$$f(0.8526) = -0.00002\,396$$
$$f(0.8527) = 0.00041\,067$$
$$x_4 = 0.8526 + \frac{2,396}{43,463} \, 0.0001 = 0.85260\,55$$

1.3. The Method of Iteration. As seen before, the equation $f(x) = 0$ can be written in the form

$$g(x) = h(x) \tag{1.11}$$

where there is, of course, wide latitude in the choice of $g(x)$ and $h(x)$. Moreover, since the real roots of the original equation are given by the abscissas of the points of intersection of the plots of the functions $g(x)$ and $h(x)$, it is convenient to replace Eq. (1.11) by the set of simultaneous equations

$$\begin{aligned} y &= g(x) \\ y &= h(x) \end{aligned} \tag{1.12}$$

If one can solve explicitly for x in the second equation, these can be written

$$\begin{aligned} y &= g(x) \\ x &= H(y) \end{aligned} \tag{1.13}$$

Suppose x_0 is any initial guess at the root of $f(x)$; then the *iterates* $x_1, x_2, x_3, \ldots, x_n$ and $y_0, y_1, y_2, y_3, \ldots, y_n$ may be defined by the formulas

$$\begin{aligned}
& & y_0 &= g(x_0) & \\
x_1 &= H(y_0) & y_1 &= g(x_1) & \\
x_2 &= H(y_1) & y_2 &= g(x_2) & \\
& \cdots \cdots & & \cdots \cdots & \tag{1.14} \\
x_i &= H(y_{i-1}) & y_i &= g(x_i) & \\
& \cdots \cdots & & \cdots \cdots & \\
x_n &= H(y_{n-1}) & y_n &= g(x_n) & \text{etc.}
\end{aligned}$$

† Tables of the Exponential Function e^x, *Natl. Bur. Standards Appl. Math. Ser.* 14, 1951.

Now if the absolute value of the slope of $g(x)$ is less than that of $h(x)$ at their intersection, i.e., if

$$\left| \frac{dg(x)}{dx} \right|_{x=p} < \left| \frac{dh(x)}{dx} \right|_{x=p} \tag{1.15}$$

where p is the desired root, then, for a sufficiently close guess x_0, $x_n \to p$ as $n \to \infty$. If the slope of $g(x)$ is much smaller than $h(x)$, the convergence of x_n to the root is rapid, and the method is a practical way of determining the root.

The nature of the iterative process and its speed of convergence can best be shown graphically (see Figs. 1.3 and 1.4). The slopes of $g(x)$ and $h(x)$ are of the same sign in Fig. 1.3 and of opposite sign in Fig. 1.4.

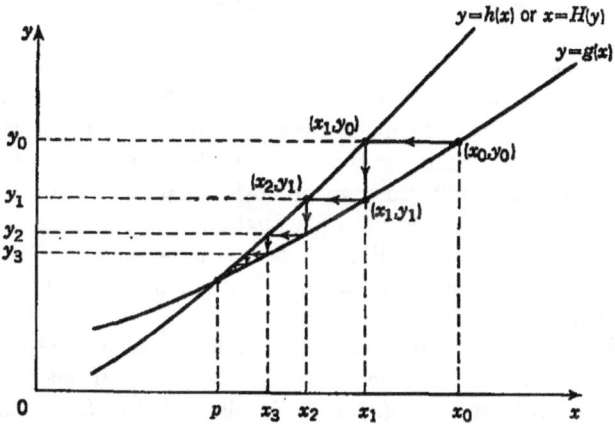

FIG. 1.3. Convergence of the iterates x_0, x_1, x_2, etc., to the root p when $g(x)$ and $h(x)$ both have positive slopes.

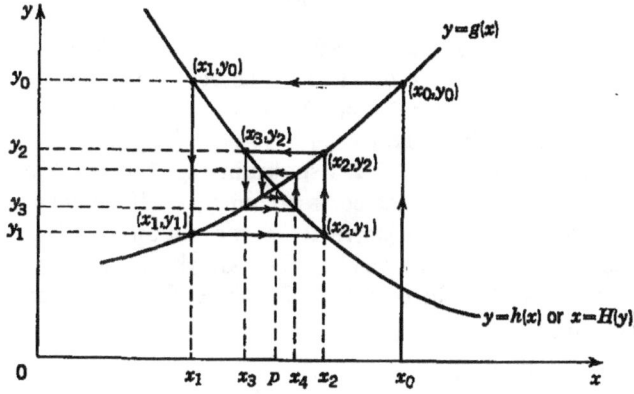

FIG. 1.4. Convergence of the iterates x_0, x_1, x_2, etc., to the root p when $g(x)$ and $h(x)$ have opposite slopes.

In both cases, the absolute values of the slopes are much more nearly equal than they should be for practical application of the method.

From Eqs. (1.14) it is at once apparent that the points (x_0,y_0), (x_1,y_1), (x_2,y_2), . . . , (x_i,y_i), . . . satisfy the equation $y = g(x)$ and hence lie on the graph of this equation. Likewise the points (x_1,y_0), (x_2,y_1), . . . , (x_i,y_{i-1}), . . . lie on the plot of the equation $x = H(y)$, which is also the plot of the equivalent equation $y = h(x)$. The repetition of two processes is involved in constructing these diagrams: (1) knowing any x_i, one can determine y_i by drawing a vertical line $x = x_i$ until it intersects the graph of $y = g(x)$; (2) knowing any y_i, one can determine x_{i+1} by drawing a horizontal line $y = y_i$ until it intersects the curve $y = h(x)$. Thus, starting with x_0, one obtains by the first process y_0 and then from the second process x_1. Repetition of this procedure gives y_1 and x_2, then y_2 and x_3, etc.

When $g(x)$ and $h(x)$ have opposite slopes, the intersection is approached in a rectangular spiral, as shown in Fig. 1.4, and the iterates x_0, x_1, x_2, . . . are alternately larger and smaller than p. Thus at all times we have an upper and a lower bound for the root. If, however, the slopes of $g(x)$ and $h(x)$ are either both positive or both negative, the intersection is approached along a stepped path like that shown in Fig. 1.3. Thus the iterates x_0, x_1, x_2, . . . are all larger or all smaller than p. One can, of course, choose an x_0 below p and an x_0' above p. Then at each stage in the calculation $x_i < p < x_i'$.

One always takes the inverse of that function which has the greater slope at the intersection, namely, the function $h(x)$. If one took instead the inverse of the function $g(x)$, then, in place of Eqs. (1.13), one would have the equations

$$x = G(y)$$
$$y = h(x) \tag{1.16}$$

The curves representing these equations are, of course, the same as those representing the equations of (1.13). Therefore the first equation is satisfied by the points (x_0,y_0), (x_1,y_1), (x_2,y_2), . . . , as before, and the second equation by the points (x_1,y_0), (x_2,y_1), (x_3,y_2), ... Thus, if we choose for the first guess the x_3 shown in Fig. 1.3 or 1.4 and use the second equation, we obtain the corresponding y_2. If we then substitute y_2 in the first equation and solve for x, we obtain x_2. On two more repetitions of this procedure we should obtain the value x_0. It is clear, therefore, that by this choice we get farther and farther from the root as we proceed.

Example 5. Find the root of

$$\sin x = \frac{x + 1}{x - 1} \tag{1.17}$$

whose approximate value is -5.5, to five significant figures.

The approximate value -5.5 was obtained from Example 1. The accompanying figure (Fig. 1.1) shows that the slope of $(x + 1)/(x - 1)$ is much smaller than, and of opposite sign to, that of sin x; hence the method of iteration should work well and give us a rectangular spiral approaching the intersection III. The equations we shall use for the iteration are

$$y = \frac{x + 1}{x - 1}$$

$$x = \arcsin y$$

Since $x_0 = -5.5$ [see Eqs. (1.14)],

$$y_0 = \frac{x_0 + 1}{x_0 - 1} = 0.692308$$

$$x_1 = \arcsin y_0 = -5.518\dagger$$

$$y_1 = \frac{x_1 + 1}{x_1 - 1} = 0.693157$$

$$x_2 = -5.5173$$

$$y_2 = 0.693124$$

$$x_3 = -5.5174$$

This is the answer correct to five significant figures. Any errors in the calculation would be automatically corrected in the subsequent iterations.

1.4. Special Iterative Forms. The weakness of the iterative method is the difficulty of being able to write $f(x) = 0$ in the form

$$y = g(x)$$
$$y = h(x)$$
$$(1.18)$$

and at the same time to make sure that the first equation has a much smaller slope at the intersection and that the second equation is readily solvable for x.

One can overcome these difficulties by establishing standard methods of forming the iterative equations (1.18). The shortcoming of such a procedure is, however, that one cannot make use of the possibilities of any one particular $f(x)$ but must treat it as an arbitrary function.

Now $f(x) = 0$ can be written in the form

$$g(x) = x \qquad (1.19)$$

by letting

$$g(x) = f(x) + x \qquad (1.20)$$

Therefore one can use for the iteration the equations

$$y = g(x)$$
$$y = x$$
$$(1.21)$$

The second equation is certainly solvable for x, and the slope of its plot is $+1$ everywhere. If $g(x)$ turns out to have a very small, or zero, slope at the intersection, these equations will serve admirably for iteration equations.

† Arcsin y_0 is multiple-valued, but one selects that value nearest -5.5.

Let p be the desired root of $f(x)$; then the slope of $g(x)$ at the intersection is

$$g'(p) = f'(p) + 1$$

Thus one would like most for $f'(p)$ to be equal to -1, in order that the slope of $g(x)$ at p be equal to zero. This cannot be done directly, but one can accomplish essentially the same result as follows. In place of $f(x) = 0$, write

$$F(x) = \frac{-f(x)}{f'(x)} = 0 \tag{1.22}$$

This function has the same root p as $f(x) = 0$, provided $f'(p)$ is not zero, and moreover

$$F'(p) = -1 + \frac{f(p)f''(p)}{[f'(p)]^2} = -1 \tag{1.23}$$

Therefore if one deals with $F(x)$ instead of $f(x)$, one obtains the iteration equations

$$\begin{aligned} y &= N(x) \\ x &= y \end{aligned} \tag{1.24}$$

where

$$N(x) = F(x) + x = x - \frac{f(x)}{f'(x)} \tag{1.25}$$

These equations are well suited for iteration, provided $f'(p)$ is not too small and is readily obtained, since $N'(p)$, the slope of the graph of the

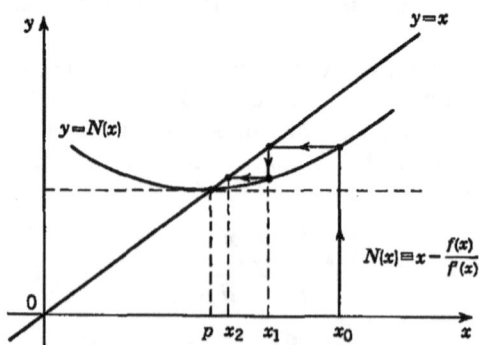

FIG. 1.5. The iterative equivalent of the Newton-Raphson formula.

first equation at the intersection, is zero and that of the second equation is unity (see Fig. 1.5).

1.5. The Newton-Raphson Method. Writing the Eqs. (1.24) in terms of the iterates, we have

$$\begin{aligned} y_{i+1} &= N(x_i) \\ x_{i+1} &= y_{i+1} \end{aligned}$$

Since the y_i are identical with the x_i, the y_i can be eliminated. In which case

$$x_{i+1} = N(x_i) = x_i - \frac{f(x_i)}{f'(x_i)} \tag{1.26}$$

This is the *Newton-Raphson* formula. Having obtained this formula as a development of the iterative method, it is informative to reinterpret Eq. (1.26) in terms of the graph of $f(x)$ (see Fig. 1.6).

In Fig. 1.6, x_{i+1} is determined by drawing a tangent to the curve at x_i and extending it until it intersects the x axis at the point $(x_{i+1},0)$. The slope of the tangent is given by

$$f'(x_i) = \frac{f(x_i)}{x_i - x_{i+1}}$$

Solving for x_{i+1}, we obtain Eq. (1.26); thus the successive iterates of that equation are represented graphically by Fig. 1.6.

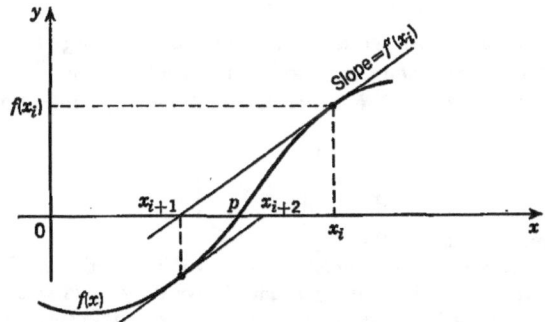

Fig. 1.6. Interpretation of the Newton-Raphson formula in terms of the plot of $f(x)$.

To determine the rate of convergence of the iterates in the Newton-Raphson method, we expand $f(x)$ about one of the iterates x_i using Taylor's series with a remainder

$$f(x) = f(x_i) + (x - x_i)f'(x_i) + \tfrac{1}{2}(x - x_i)^2 f''(\xi)$$

where ξ lies between x and x_i. Substituting $x = p$ and solving for p in the second term on the right-hand side, we have

$$p = x_i - \frac{f(x_i)}{f'(x_i)} - (p - x_i)^2 \frac{f''(\xi)}{2f'(x_i)}$$

therefore, by Eq. (1.26),

$$x_{i+1} - p = (x_i - p)^2 \frac{f''(\xi)}{2f'(x_i)} \tag{1.27}$$

Since $x_j - p$ is the error in the jth iterate to the root, Eq. (1.27) states that each iteration squares the error and then multiplies it by the factor $f''(\xi)/2f'(x_i)$, where ξ lies somewhere between x_i and the root p.

Note that the sign of the error in x_{i+1} is independent of the sign of the error in x_i. It is positive if $f''(x)f'(x) > 0$ in the neighborhood of p and negative if $f''(x)f'(x) < 0$. Thus, the iterates approach the root from one particular side that is independent of the initial iterate x_0.

Once one is close enough to the root so that the factor $f''/2f'$ in Eq. (1.27) is roughly constant from one iteration to the next, one may successfully employ the following rule, based on Eq. (1.27), for the rate of convergence of the iterative process:

If the root is known to lie in a range R for which an upper bound M can be established for the absolute value of $f''(\xi)/2f'(x_j)$, for any x_j in R, and if the ith iterate x_i is known to approximate the root to n decimal places, then x_{i+1} approximates the root to at least m decimal places, where m is the largest integer in $2n - \log M$. In this rule x_i is said to approximate p to n decimal places if n is the largest integer in $-\log |x_i - p|$. Proof of the rule is obtained by taking the logarithm to the base 10 of the absolute values of each side of Eq. (1.27).

It is clear from the above analysis that the convergence will be poor if $f''/2f'$ is large in the neighborhood of the root. This will usually happen if $f(x)$ does not cross the x axis at a sufficiently steep angle.

Example 6. Find to seven significant figures the root of

$$f(x) = \sin x - \frac{x+1}{x-1} = 0$$

whose approximate value is -0.4.

From Fig. 1.1 we see that this root is the abscissa of intersection I. Since the slopes of the two functions plotted are nearly equal, the convergence of an iterative process involving these two functions would be extremely slow. However, $f(x)$, which is the difference of these two functions, passes through zero at a steep angle; therefore the Newton-Raphson method should probably work well. By this method

$$f'(x) = \cos x + \frac{2}{(x-1)^2}$$

$$x_0 = -0.4$$

x	$f(x)$	$f'(x)$	$f(x)/f'(x)$
-0.4	0.039153	1.94147	0.020107
-0.4201	0.00050	1.90478	0.0002625

$$x_1 = x_0 - \frac{f(x_0)}{f'(x_0)} = -0.4 - 0.0201$$

$$x_2 = x_1 - \frac{f(x_1)}{f'(x_1)} = -0.42036\ 25$$

The correct value for the root is $-0.42036\ 24$. Thus two iterations give us the root to seven significant figures. Since the number of correct decimal places roughly doubles each iteration, one infers that M is of the order of 10 or less. By evaluating the second derivative one finds that M is about 5.

1.6. Iterative Formulas for the Square Root, the Reciprocal, and the Reciprocal Square Root. In numerical analysis it often happens that one reverses the steps taken in an analytical solution. Consider the three equations

$$x^2 - N = 0 \tag{1.28}$$

$$\frac{1}{x} - N = 0 \tag{1.29}$$

$$\frac{1}{x^2} - N = 0 \tag{1.30}$$

The analytical solutions of these equations are, respectively, $x = \sqrt{N}$, $x = 1/N$, and $x = 1/\sqrt{N}$, i.e., the square root, the reciprocal, and the reciprocal square root of N. When, however, it becomes necessary to find numerical values for these functions of N, it is often easier to do so indirectly by solving the appropriate equation above by the Newton-Raphson method.

Square Roots. The square root of N is obtained by solving Eq. (1.28). One lets $f(x) = x^2 - N$ and applies Newton-Raphson formula (1.26), which here becomes

$$x_{i+1} = x_i - \frac{x_i^2 - N}{2x_i}$$

or

$$x_{i+1} = \tfrac{1}{2}\left(x_i + \frac{N}{x_i}\right) \tag{1.31}$$

This is the required iteration formula for the square root.

Let $p = \sqrt{N}$, and write

$$x_i = p\frac{1 + \epsilon_i}{1 - \epsilon_i} = p\left(1 + \frac{2\epsilon_i}{1 - \epsilon_i}\right) \tag{1.32}$$

then $2\epsilon_i/(1 - \epsilon_i)$ is the relative error (see Appendix A) in using x_i as the square root of N. Since ϵ_i is generally quite small compared to 1, one can think of ϵ_i as being approximately half the relative error in x_i. By substituting from Eq. (1.32) into Eq. (1.31) one finds that

$$x_{i+1} = p\frac{1 + \epsilon_{i+1}}{1 - \epsilon_{i+1}} = \frac{1}{2}\left(\frac{1 + \epsilon_i}{1 - \epsilon_i} + \frac{1 - \epsilon_i}{1 + \epsilon_i}\right)p = p\frac{1 + \epsilon_i^2}{1 - \epsilon_i^2}$$

therefore

$$\epsilon_{i+1} = \epsilon_i^2 \tag{1.33}$$

Starting with any x_0 such that $|\epsilon_0| < 1$, one obtains a series of epsilons $\epsilon_0, \epsilon_1, \epsilon_2, \ldots, \epsilon_i, \ldots$ that approach zero as $i \to \infty$. If x_0 and p are real, by Eq. (1.33) all but the first of these will be positive; hence, by Eq. (1.32) x_i will approach p from above. Since ϵ_i is very nearly one-half the relative error in x_i, when one is close to the root, the effect of a single iteration is to square the relative error and then to divide it by 2.

This means that one slightly more than doubles the number of significant figures in the square root on each iteration.

Suppose one allows both N and x_0 to be complex. Since none of the above equations will be changed, one sees that by Eq. (1.33) x_i will approach $p = \sqrt{N}$ as i approaches ∞, provided x_0 is chosen so that ϵ_0 is less than 1 in absolute value. On the other hand, if ϵ_0 is greater than 1 in absolute value, ϵ_i will approach infinity, and hence, by Eq. (1.32), x_i will approach $-p$, the other square root of N. Only if $|\epsilon_0| = 1$ will the process fail to converge to one of the square roots. This occurs when x_0 lies on the perpendicular bisector of the line joining the two roots. Thus for a real N, the iterates x_i converge to p if the real part of x_0 is positive and to $-p$ if the real part is negative. It fails to converge if x_0 is purely imaginary.

Example 7. Obtain the $\sqrt{30}$ to nine significant figures.
Since $\sqrt{25} = 5$ and $\sqrt{36} = 6$, we shall take $x_0 = 5.5$. By Eq. (1.32)

$$x_1 = \tfrac{1}{2}\left(5.5 + \frac{30}{5.5}\right) = 5.47727\ 273$$

$$x_2 = \tfrac{1}{2}\left(x_1 + \frac{30}{x_1}\right) = 5.47722\ 557$$

The latter is correct to nine significant figures. This can be proved by noting that the root lies between x_2 and $30/x_2$.

In hand computation one would want to drop some of the meaningless figures in x_1, but with a desk-type calculator it is just as simple to leave them as to figure which should be dropped. One usually compromises, by dropping some but retaining a few more than the minimum needed, in order to dispense with any analysis.

Reciprocals. The reciprocal of any number N is obtained by applying the Newton-Raphson formula to Eq. (1.29). Since $f(x) = 1/x - N$, $f'(x) = -1/x^2$, and therefore in place of Eq. (1.26) one has

$$x_{i+1} = x_i + (x_i - x_i^2 N)$$
or
$$x_{i+1} = x_i(2 - x_i N) \tag{1.34}$$

Let $p = 1/N$ and write

$$x_i = p(1 - \epsilon_i) \tag{1.35}$$

then by Eq. (1.34)

$$x_{i+1} = p(1 - \epsilon_{i+1}) = p(1 - \epsilon_i)(1 + \epsilon_i) = p(1 - \epsilon_i^2)$$
Thus again
$$\epsilon_{i+1} = \epsilon_i^2$$

and ϵ_i will approach zero and x_i approach p as i approaches infinity, provided $|\epsilon_0| < 1$. For $|\epsilon_0| > 1$, ϵ_i approaches infinity and hence x_i does also. For $|\epsilon_0| < 1$, by Eq. (1.35), x_0 lies inside the circle in the complex x plane whose center is at $x = p$ and whose radius is p.

For N and p real, convergence occurs for any real x_0 in the range $0 < x_0 < 2/N$.

In electronic-calculator construction it is sometimes found inexpedient to build a dividing unit. This arises partly from the amount of equipment needed to perform the indirect operations required for conventional division. Not only is this equipment rather costly, but its presence also makes it more difficult to discover and locate faulty operation in the machine as a whole. By storing a few approximate reciprocals to be used as x_0 and using the iterative form above one can eliminate the need of a dividing unit. One simply multiplies by the reciprocal. Each iteration for the reciprocal requires two multiplications and one addition. The over-all time required is, generally speaking, not much different from what it would be if the division were done by a separate unit.

Example 8. Obtain the reciprocal of 3 by iteration using as the first guess $x_0 = 0.3$.
This is a very simple example but illustrates well the doubling of significant figures on each iteration. Equation (1.34) becomes

$$x_{i+1} = x_i(2 - 3x_i)$$

thus

$$x_1 = (0.3)(1.1) = 0.33$$
$$x_2 = (0.33)(1.01) = 0.3333$$
$$x_3 = (0.3333)(1.0001) = 0.33333\ 333 \qquad \text{etc.}$$

Reciprocal Square Roots. Letting $f(x) = 1/x^2 - N$ and therefore $f'(x) = -2/x^3$, one obtains the iteration formula

$$x_{i+1} = \tfrac{1}{2}x_i(3 - x_i^2 N) \tag{1.36}$$

for $1/\sqrt{N}$.

Having the reciprocal square root $N^{-\frac{1}{2}}$, one may obtain any negative power or half power of N by multiplication alone. By multiplying by N one obtains the square root $N^{\frac{1}{2}}$, and from this any positive power or half power of N.

Equation (1.36) has a distinct advantage over Eq. (1.31) as a method for finding the square root, since it does not require division. The divisions by 2 can be replaced by multiplications by 0.5.

PROBLEMS

1. Find by graphical construction the approximate values of the real roots of the equation $\tan x = 4x$.
2. Determine the number of real roots of each of the following equations:

(a) $\dfrac{\sin x}{x} = x^2$ (b) $\sin x = x^3 + 1$

(c) $\dfrac{x}{x^2 + 1} = e^{-x}$ (d) $\ln x = e^x$

3. Using the method shown in Example 2, find an approximate root of $e^x = \ln x + 64$.
4. Find an approximate root of $e^{-x} \sin x + 25x - 1 = 0$.

5. Show that a real root of the equation $\tan 2\pi x - 6 \tan \pi x + 1 = 0$ exists in the range $0 \leq x \leq \frac{1}{6}$.

6. Use the method of false position to determine the root in Prob. 5 to six decimal places.

7. Use the method of iteration to find a real root of the equation in Prob. 3 to eight significant figures.

8. Find a root of the equation $J_0(x) = \frac{1}{2}$ to five decimal places.

9. Show that, if Eq. (1.31) for the square root is applied twice, one obtains as an approximation for the square root of $N = AB$ the formula

$$\sqrt{N} \cong \frac{S}{4} + \frac{N}{S}$$

where S stands for the sum of the factors A and B. Show that these factors should be chosen as nearly equal as possible since the relative error in this formula is approximately

$$\frac{1}{8}\left(\frac{A-B}{A+B}\right)^4$$

10. Determine for what region of the complex plane, representing the possible initial guesses x_0 for the square root of N, the iterates x_i of formula (1.31) converge to one of the two square roots $\sqrt{N}$ and $-\sqrt{N}$.

11. Find to five decimal places the real roots of the cubic $t^3 - 4t^2 - 6t + 4 = 0$ lying between zero and one.

CHAPTER 2

ROOTS OF POLYNOMIAL EQUATIONS

The methods used in Chap. 1 for finding the real zeros of a general function $f(x)$ having numerical coefficients can, of course, be used when $f(x)$ is restricted to a polynomial in x with numerical coefficients. However, a great deal of general information is available concerning the roots of polynomials, and hence special methods can be employed in their solution. Moreover, we shall include in this chapter methods for finding the complex roots of polynomial equations.

2.1. General Theorems Relating to the Roots of Polynomials. Most of the theorems given in this section can be found in elementary discussions of the theory of equations, and hence the proofs are largely omitted.†

Consider the general polynomial equation,‡

$$P_n(x) = a_n x^n + a_{n-1} x^{n-1} + \cdots + a_1 x^1 + a_0 = 0 \qquad (2.1)$$

If

$$P_n(x) = (x - p)^m Q(x) \qquad (2.2)$$

where $Q(x)$ is a polynomial and m is any positive integer, and if $Q(p) \neq 0$ or ∞, then p is a root of $P_n(x)$ of *multiplicity m*.

THEOREM I. Provided a root of multiplicity m is counted as m roots, a polynomial $P_n(x)$ of degree n has exactly n roots.

THEOREM II (*Remainder Theorem*). If a polynomial $P_n(x)$ is divided by $x - \alpha$ until a constant remainder is obtained, then this remainder equals $P_n(\alpha)$.

Proof: After dividing $P_n(x)$ by $x - \alpha$, a quotient $q(x)$ and a remainder R, independent of x, are obtained; thus

$$P_n(x) = (x - \alpha)q(x) + R \qquad (2.3)$$

and hence

$$P_n(\alpha) = R \qquad (2.4)$$

† See, for example, J. B. Rosenbach and E. A. Whitman, "College Algebra," 3d ed., chap. XIV, Ginn & Company, Boston, 1949; H. T. Davis, "College Algebra," chap. XIII, Prentice-Hall, Inc., Englewood Cliffs, N.J., 1941.

‡ Unless otherwise specified, it will be assumed that the coefficients a_0, a_1, . . . , a_n are real numbers.

THEOREM III (*Factor Theorem*). If $P_n(p) = 0$, then $x - p$ is a factor of $P_n(x)$, and vice versa.

Proof: If $P_n(p) = 0$, then by Eq. (2.4) R is zero, and by Eq. (2.3) $P_n(x) = (x - p)q(x)$; therefore $x - p$ is a factor of $P_n(x)$. The converse is an immediate consequence of Theorem II.

THEOREM IV. The coefficients of a polynomial

$$Q_n(x) = b_0 x^n + b_1 x^{n-1} + b_2 x^{n-2} + \cdots + b_{n-1} x + b_n$$

are related to the n roots $p_1, p_2, p_3, \ldots, p_n$ by the following equations:

$$\frac{b_1}{b_0} = -\sum_i p_i = -(p_1 + p_2 + \cdots + p_n)$$

$$\frac{b_2}{b_0} = \sum_{i<j} p_i p_j = p_1 p_2 + p_1 p_3 + \cdots + p_1 p_n + p_2 p_3 + \cdots$$

$$+ p_2 p_n + \cdots + p_{n-1} p_n$$

$$\frac{b_3}{b_0} = -\sum_{i<j<k} p_i p_j p_k = -(p_1 p_2 p_3 + p_1 p_2 p_4 + \cdots + p_1 p_2 p_n$$

$$+ p_1 p_3 p_4 + \cdots + p_1 p_3 p_n + \cdots + p_{n-2} p_{n-1} p_n)$$

$$\cdots \cdots \cdots \cdots \cdots \cdots \cdots \cdots \cdots$$

$$\frac{b_m}{b_0} = (-1)^m \sum_{i<j<\cdots<z} p_i p_j p_k \cdots p_z \quad \text{(m factors)}$$

$$\cdots \cdots \cdots \cdots \cdots \cdots \cdots \cdots \cdots$$

$$\frac{b_n}{b_0} = (-1)^n p_1 p_2 p_3 \cdots p_n$$

where the summations are taken over all combinations of the subscripts satisfying the inequalities.

THEOREM V. If a rational number b/c, a fraction in its lowest terms, is a root of Eq. (2.1) where the coefficients a_i are all integers, then b is a factor of a_0, and c is a factor of a_n.[†]

THEOREM VI. If the coefficients a_i of $P_n(x)$ are real and $a + bi$, where a and b are real, is a root of the equation $P_n(x) = 0$, then $a - bi$ is also a root.[‡]

THEOREM VII. (*Descartes's Rule of Signs*). The number of positive roots of $P_n(x) = 0$, where the coefficients a_i are real, cannot exceed the number of variations in sign of the coefficients in the polynomial $P_n(x)$, and the number of negative roots cannot exceed the number of variations in sign of the coefficients in $P_n(-x)$. Moreover, in each case, if the

[†] Rosenbach and Whitman, *op. cit.*, p. 293.
[‡] As a consequence of this theorem, any polynomial of odd degree with real coefficient has at least one real root.

number of roots is less than the number of variations in sign, it must be smaller by some multiple of 2.

THEOREM VIII. The coefficients of an equation

$$Q_n(x) = A_n x^n + A_{n-1} x^{n-1} + \cdots + A_1 x + A_0 = 0 \qquad (2.5)$$

whose roots are less by a quantity h than those of $P_n(x) = 0$ [see Eq. (2.1)] may be obtained by dividing $P_n(x)$ by $x - h$, the resulting quotient by $x - h$, the next quotient by $x - h$, etc.† The remainders after each division are, respectively, the coefficients A_0, A_1, A_2, ..., A_n.

Proof: The roots of

$$Q_n(x - h) = A_n(x - h)^n + A_{n-1}(x - h)^{n-1} + \cdots + A_1(x - h) + A_0$$
$$= 0$$

are larger than the roots of Eq. (2.5) by the quantity h, and hence the latter has the same roots as $P_n(x)$. Therefore one can require that

$$Q_n(x - h) = P_n(x)$$

Successive division of $P_n(x)$ by $x - h$ is therefore equivalent to successive division of $Q_n(x - h)$ by $x - h$, and this, of course, yields as successive remainders the numbers A_0, A_1, A_2, ..., A_n.

2.2. Synthetic Division. Let us change the designation of coefficients in $P_n(x)$ and let

$$P_n(x) = c_0 x^n + c_1 x^{n-1} + c_2 x^{n-2} + \cdots + c_{n-1} x + c_n \qquad (2.6)$$

For a given set of coefficients c_0, c_1, ..., c_n there can be defined the following polynomials:

$$
\begin{aligned}
P_0(x) &= c_0 \\
P_1(x) &= x P_0(x) + c_1 \\
P_2(x) &= x P_1(x) + c_2 \\
&\cdots\cdots\cdots\cdots\cdots \\
P_{n-1}(x) &= x P_{n-2}(x) + c_{n-1} \\
P_n(x) &= x P_{n-1}(x) + c_n
\end{aligned}
\qquad (2.7)
$$

Multiplying the first equation by x^n, the second by x^{n-1}, the third by x^{n-2}, etc., and adding, we obtain at once Eq. (2.6). Therefore $P_n(x)$ of Eqs. (2.7) is identical with the $P_n(x)$ of Eq. (2.6).

Suppose we wish to evaluate $P_n(x_0)$. It is generally more convenient to calculate in order, from Eqs. (2.7), the values of $P_0(x_0)$, $P_1(x_0)$, $P_2(x_0)$,

† This should ordinarily be done by synthetic division. For an explanation of synthetic division see Sec. 2.2.

$\ldots$, $P_n(x_0)$ than to find all n powers of x_0 and substitute in Eq. (2.6). The evaluation of the P_i is best done as follows:

$$
\begin{array}{cccccccc}
c_0 & c_1 & c_2 & c_3 & \cdots & c_{n-1} & c_n & \underline{|x_0} \\
& x_0 P_0 & x_0 P_1 & x_0 P_2 & \cdots & x_0 P_{n-2} & x_0 P_{n-1} & \\
\hline
P_0 & P_1 & P_2 & P_3 & \cdots & P_{n-1} & P_n(x_0) &
\end{array}
\tag{2.8}
$$

The coefficients are written, in order, in the first row, zero coefficients being supplied where a power of x is missing, and the value of the argument is shown at the extreme right. The P_i are calculated in order by adding the two numbers above them, one of which is c_i and the other x_0 times P_{i-1}, the latter number having been previously calculated using the already known value of P_{i-1}.

The process shown in (2.8) is called *synthetic division* since, from another point of view, it can be looked upon as a shorthand division of $P_n(x)$ by $x - x_0$. In this case $P_n(x_0)$ is the remainder, and

$$
P_0(x_0)x^{n-1} + P_1(x_0)x^{n-2} + \cdots + P_{n-1}(x_0)
$$

is the quotient.†

Example 1. Determine $f(3)$, where

$$
f(x) = 2x^5 - 13x^3 + 5x^2 - 11x - 6
$$

By synthetic division,

$$
\begin{array}{ccccccc}
2 & 0 & -13 & 5 & -11 & -6 & \underline{|3} \\
& 6 & 18 & 15 & 60 & 147 & \\
\hline
2 & 6 & 5 & 20 & 49 & 141 & = f(3)
\end{array}
$$

2.3. Derivatives of a Polynomial by Synthetic Division. If the Eqs. (2.7) are differentiated with respect to x, we obtain the equations

$$
\begin{aligned}
P_0'(x) &= 0 \\
P_1'(x) &= P_0(x) \\
P_2'(x) &= xP_1'(x) + P_1(x) \\
P_3'(x) &= xP_2'(x) + P_2(x) \\
&\cdots\cdots\cdots\cdots\cdots \\
P_n'(x) &= xP_{n-1}'(x) + P_{n-1}(x)
\end{aligned}
\tag{2.9}
$$

$P_n'(x)$ is, of course, the derivative of the original polynomial. Comparing these equations with Eqs. (2.7), we see that $P_1'(x)$, $P_2'(x)$, $\ldots$, $P_n'(x)$ are obtained from $P_0(x)$, $P_1(x)$, $\ldots$, $P_{n-1}(x)$ in exactly the same way that $P_0(x)$, $P_1(x)$, $\ldots$, $P_{n-1}(x)$ are obtained from c_0, c_1, $\ldots$,

† For a discussion of synthetic division from this point of view, see Rosenbach and Whitman, *op. cit.*, p. 282.

c_{n-1}. Hence diagram (2.8) can be extended to give the derivative of each of the $P_i(x)$; thus

$$
\begin{array}{ccccccc}
c_0 & c_1 & c_2 & c_3 & \cdots & c_{n-1} & c_n & \underline{|x_0} \\
& x_0 P_0 & x_0 P_1 & x_0 P_2 & \cdots & x_0 P_{n-2} & x_0 P_{n-1} & \\
\hline
P_0 & P_1 & P_2 & P_3 & \cdots & P_{n-1} & P_n(x_0) & \\
& x P_1' & x P_2' & x P_3' & \cdots & x P_{n-1}' & & \\
\hline
P_1' & P_2' & P_3' & P_4' & \cdots & P_n'(x_0) & &
\end{array}
\qquad (2.10)
$$

Differentiating Eqs. (2.9), one obtains equations of the form

$$P_i^{(2)}(x) = x P_{i-1}^{(2)}(x) + 2 P_{i-1}'(x)$$

or

$$\tfrac{1}{2} P_i^{(2)}(x) = x[\tfrac{1}{2} P_{i-1}^{(2)}(x)] + P_{i-1}'(x)$$

where

$$P_n^{(s)}(x) = \frac{d^s}{dx^s} P_n(x)$$

Thus the next application of the process in (2.10) using the $P_i'(x_0)$ yields the second derivatives divided by 2.

It can be seen that by repeating the process a sufficient number of times one obtains in succession as end terms on the right

$$P_n(x_0),\; P_n'(x_0),\; \frac{1}{2!} P_n^{(2)}(x_0),\; \frac{1}{3!} P_n^{(3)}(x_0),\; \ldots,\; \frac{1}{n!} P_n^{(n)}(x_0)$$

Example 2. Given the equation

$$P(x) = x^3 - 3x^2 + 4x - 5 = 0 \qquad (2.11)$$

write an equation with these roots each diminished by 2.

If the roots of Eq. (2.11) are p_1, p_2, and p_3, the roots of

$$P(x + h) = x^3 + B(h)x^2 + C(h)x + D(h) = 0 \qquad (2.12)$$

are $p_1 - h$, $p_2 - h$, and $p_3 - h$, since $P((p_i - h) + h) = P(p_i) = 0$, by Eq. (2.11). Therefore if $h = 2$, Eq. (2.12) is the required equation.

Now by Taylor's series

$$P(x + h) = P(h) + x[P^{(1)}(h)] + x^2 \left[\frac{1}{2!} P^{(2)}(h)\right] + x^3 \left[\frac{1}{3!} P^{(3)}(h)\right] \qquad (2.13)$$

and therefore the coefficients $B(h)$, $C(h)$, and $D(h)$ are just the end terms in a continuation of the synthetic division shown in (2.10). Thus

$$
\begin{array}{rrrr}
1 & -3 & 4 & -5 \qquad \underline{|2} \\
& 2 & -2 & 4 \\
\hline
1 & -1 & 2 & -1 = P(2) \\
& 2 & 2 & \\
\hline
1 & 1 & 4 = P^{(1)}(2) & \\
& 2 & & \\
\hline
1 & 3 = \dfrac{1}{2!} P^{(2)}(2) & & \\
\hline
1 = \dfrac{1}{3!} P^{(3)}(2) & & &
\end{array}
$$

and the required equation is

$$P(x+2) = x^3 + 3x^2 + 4x - 1 = 0$$

2.4. Birge-Vieta Method. In the Newton-Raphson method any close approximation x_i to a root of $P_n(x) = 0$ is improved by the formula

$$x_{i+1} = x_i - \frac{P_n(x_i)}{P'_n(x_i)} \tag{2.14}$$

This same procedure is followed in the Birge-Vieta method, but the values of $P_n(x_i)$ and $P'_n(x_i)$ are calculated by synthetic division, as shown in diagram (2.10). An example will illustrate the procedure.†

Example 3. Find a real root of the polynomial

$$P(x) = x^5 - 6.28427\,31x^4 + 23.714994x + 3 = 0 \tag{2.15}$$

to eight significant figures.

By direct substitution,

$$P(0) = 3 \qquad P(1) = 21.430721 \qquad P(2) = -18.118382$$

therefore a real root, p, lies between $x = 1$ and $x = 2$. By synthetic division, inserting zeros for the coefficients of x^3 and x^2 (see Table 2.1), $P(1) = 21.430721$ (which checks the above result), and $P'(1) = 3.577902$. Thus with $x_0 = 1$ as the first guess, the Newton-Raphson formula gives

$$x_1 = 1 - \frac{21.430721}{3.577902} \cong -5$$

Since the root, p, is known to lie in the interval $1 < x < 2$, this estimate is completely faulty. Now the slope is positive at $x = 1$, but from the values of $P(1)$ and $P(2)$ it must be negative somewhere in the interval; hence it is obvious that the plot of $P(x)$ reaches a maximum somewhere beyond $x = 1$ and then turns downward to cross the x axis before $x = 2$.

To gain further information let us try $x = 1.7$. From Table 2.1, $P(1.7) = 4.192461$, and $P'(1.7) = -58.023041$; thus the slope is negative, as one would expect, and the Newton-Raphson formula gives the reasonable estimate

$$x_1 = 1.7 + \frac{4.192461}{58.023041} = 1.7723$$

From the table,

$$f(x_1) = 0.51415\,39 \qquad \text{and} \qquad f'(x_1) = -66.88924\,41$$

so that as an improved estimate one has

$$x_2 = 1.7723 + 0.00769\,866 = 1.77998\,66$$

Again by the synthetic division shown in Table 2.1,

$$f(x_2) = -0.00370\,94 \qquad \text{and} \qquad f'(x_2) = -67.85648\,93$$

† See also *The Birge-Vieta Method of Finding Real Roots of Rational Integral Functions, Marchant Methods,* MM-225, Marchant Calculating Machine Co., Oakland, Calif., 1942.

and hence $\quad x_3 = 1.77998\,66 - 0.00005\,47 = 1.77993\,19$
By the table,

$$f(x_3) = -0.00000\,05$$

and, since $f'(x_3) \cong -68$, this value is good to all eight figures.

TABLE 2.1. SYNTHETIC DIVISION EMPLOYED IN EXAMPLE 3

1	−6.2842731	0	0	23.714994	3.000000	⌊1
	1	−5.2842731	−5.2842731	−5.284273	18.430721	
1	−5.2842731	−5.2842731	−5.2842731	18.430721	21.430721	
	1	−4.2842731	−9.5685462	−14.852819		
1	−4.2842731	−9.5685462	−14.8528193	3.577902		
1	−6.2842731	0	0	23.7149943	3.000000	⌊1.7
	1.7	−7.7932643	−13.2485493	−22.5225333	1.192461	
1	−4.5842731	−7.7932643	−13.2485493	1.192461	4.192461	
	1.7	−4.9032643	−21.5840993	−59.215502		
1	−2.8842731	−12.6965286	−34.8326483	−58.023041		
1	−6.2842731	0	0	23.7149943	3.0000000	⌊1.7723
	1.7723	−7.9965699	−14.1723209	−25.1176045	−2.4858461	
1	−4.5119731	−7.9965699	−14.1723209	−1.4026102	0.5141539	
	1.7723	−4.8555226	−22.7777627	−65.4866339		
1	−2.7396731	−12.8520925	−36.9500836	−66.8892441		
1	−6.2842731	0	0	23.7149943	3.0000000	⌊1.7799866
	1.7799866	−8.0175696	−14.2711665	−25.4024842	−3.0037094	
1	−4.5042865	−8.0175696	−14.2711665	−1.6874899	−0.0037094	
	1.7799866	−4.8492173	−22.9027084	−66.1689994		
1	−2.7242999	−12.8667869	−37.1738749	−67.8564893		
1	−6.2842731	0	0	23.7149943	3.0000000	⌊1.77999319
	1.7799319	−8.0174206	−14.2704627	−25.4004523	−3.0000005	
1	−4.5043412	−8.0174206	−14.2704627	−1.6854580	−0.0000005	

2.5. Horner's Method.†

When one must use pencil and paper to calculate the roots, the divisions of large numbers required in the Birge-Vieta method are very time-consuming. The Horner method was devised to reduce such pencil work; however, when used with a desk-type calculator, it is much longer than the foregoing method.

The following example will illustrate the method.

Example 4. Find the roots of

$$P(x) = x^3 - 3x^2 + 4x - 5 = 0 \tag{2.16}$$

to six significant figures.

Since there are three variations in sign of the coefficients of $P(x)$ and no variation of sign for $P(-x)$, by Descartes's rule there are one or three positive real roots and no negative roots.

The only possible rational roots by Theorem V are ± 1 and ± 5. Since there are no negative roots, -1 and -5 are eliminated; and on actual substitution 1 and 5 are also eliminated. Thus the real root (or roots) is irrational.

Since $P(0) = -5$, $P(1) = -3$, $P(2) = -1$, and $P(3) = 7$, a real root lies between 2 and 3. Suppose the root is $p = 2 + p_1$. An equation with the root p_1 can be

† E. Whittaker and G. Robinson, "The Calculus of Observations," 4th ed., p. 100, Blackie & Son, Ltd., Glasgow, 1944; Rosenbach and Whitman, *op. cit.*, p. 315.

obtained from Eq. (2.16) by reducing the roots of that equation by 2 by synthetic division (see Example 2). The required equation from that example is

$$f_1(x) = x^3 + 3x^2 + 4x - 1 = 0 \qquad (2.17)$$

Since p_1 is a root of Eq. (2.17) and $p_1 < 1$, a guess, x_1, at the value of p_1 can be obtained by neglecting the first and second terms in Eq. (2.17). This gives $4x_1 - 1 = 0$, or $x_1 = 0.25$.†

By the synthetic division shown below, $f_1(0.3)$ and $f_1(0.2)$ are found to have opposite signs; therefore p_1 lies between 0.2 and 0.3.

1	3	4	−1	|0.3
	0.3	0.99	1.497	
1	3.3	4.99	0.497 = $f_1(0.3)$	
1	3	4	−1	|0.2
	0.2	0.64	0.928	
1	3.2	4.64	−0.072 = $f_1(0.2)$	
	0.2	0.68		
1	3.4	5.32		
	0.2			
1	3.6			

At this stage we know that $p = 2.2 + p_2$, where p_2 is a root of the equation obtained by reducing the roots of Eq. (2.17) by 0.2. The coefficients of this equation are the end terms in the synthetic division above involving 0.2. Thus p_2 is a root of

$$f_2(x) = x^3 + 3.6x^2 + 5.32x - 0.072 = 0 \qquad (2.18)$$

Again neglecting the first two terms (this time the error in doing so is much less), one obtains as an approximate value of p_2

$$x_2 = \frac{0.072}{5.32} \cong \frac{0.07}{5} = 0.014$$

Thus p_2 probably lies between 0.01 and 0.02.‡

1	3.6	5.32	−0.072	|0.02	
	0.02	0.0724	0.107848		
1	3.62	5.3924	0.035848		
1	3.6	5.32	−0.072	|0.01	
	0.01	0.0361	0.053561		(2.19)
1	3.61	5.3561	−0.018439		
	0.01	0.0362			
1	3.62	5.3923			
	0.01				
1	3.63				

† This is equivalent to applying the Newton-Raphson iteration formula to Eq. (2.17) with the initial guess $x = 0$ for p_1.

‡ Usually decimals are eliminated in Horner's method by writing equations with roots multiplied by some power of 10. This is of doubtful advantage and only complicates the presentation of the principles of the method.

By the same reasoning as employed above, $p = 2.21 + p_3$, where p_3 is the root of

$$f_3(x) = x^3 + 3.63x^2 + 5.3923x - 0.018439 = 0$$

The approximate value x_3 of p_3 is

$$x_3 = \frac{0.018439}{5.3923} \cong \frac{0.018}{5} = 0.0036\dagger$$

1	3.63	5.3923	-0.018439		0.004
	0.004	0.014536	0.021627344	
1	3.634	5.406836	0.003188344	
1	3.63	5.3923	-0.018439		0.003
	0.003	0.010899	0.016209597	
1	3.633	5.403199	-0.002229403	
	0.003	0.010908		
1	3.636	5.414107		
	0.003			
1	3.639			

Thus by the synthetic division above, $p = 2.213 + p_4$, where p_4 is a root of

$$x^3 + 3.639x^2 + 5.414107x - 0.00222\,9403 = 0$$

The approximate value of p_4 is

$$x_4 = \frac{0.00222\,9403}{5.414107} \cong \frac{0.00223}{5.41} = 0.00041$$

therefore the root out to six figures is very close to

$$p = 2.21341$$

One can check this root by synthetic division using the original equation. Note that, while accurate division is not required in Horner's method, accurate multiplication is, since subsequent calculations depend on the accuracy of the reduced equation. This necessitates many-digit numbers' being carried along in the synthetic divisions. For the same reason, any errors in the reduction will show up in the final answer. In the Newton-Raphson and Birge-Vieta methods any errors are automatically corrected in the subsequent steps. This property of a numerical method is very important.

Since we have one of the roots of (2.16), we can divide $P(x)$ synthetically by $x - 2.21341$ to obtain an equation of second degree:

1	-3	4	-5		2.21341
	2.21341	-1.741046	4.999991	
1	-0.78659	2.258954	-0.000009	

† An accurate division would, of course, yield 0.00341 95, the same value as that obtained when the Birge-Vieta method is applied to the synthetic division of (2.19). This would give $p = 2.21 + x_3 = 2.21342$, which is off one unit in the sixth place. Thus Horner's method eliminates the need of accurate division, but each step yields only one integer instead of several.

This equation has for coefficients the first three numbers in the third row; thus the other two roots of Eq. (2.16) are roots of the quadratic

$$x^2 - 0.78659x + 2.258954 = 0 \qquad (2.20)$$

which by the quadratic formula are

$$x = 0.393295 \pm 1.45061i$$

The three roots obtained can now be checked by Theorem IV.

The methods discussed so far are expedient only when real roots are desired. We shall next discuss several methods that are applicable to complex roots.

2.6. Root-squaring Method of Dandelin and Graeffe.† Besides being applicable to complex roots, Graeffe's method has the advantage of yielding, in many cases, all the roots of a polynomial directly without initial information concerning the roots. Its two main disadvantages are (1) the complexity of the method and (2) the need of a sufficiently accurate table of logarithms.

The underlying principle of the method is to isolate the roots of a polynomial by forming an equation in which the roots are a very high power of the roots of the original polynomial. For the purposes of discussion let this be the 256th power of the roots. If two roots differ by as little as 10 per cent, say a and $1.1a$, then the ratio of their 256th powers is

$$\frac{1.1^{256}a^{256}}{a^{256}} \sim 3 \times 10^{10}$$

Since the new equation has roots differing greatly in magnitude, it can often be solved rather simply for the magnitude of the new roots. By the use of logarithm tables the magnitude of the roots of the original equation can then be determined. The amplitude (or angle) of the root is determined, in turn, by making use of the original equation, e.g., direct substitution will fix the sign of the real roots.

2.7. The Root-squaring Process. Suppose we wish to form an equation whose roots are the negative of the square of the roots of the quartic

$$f(x) = b_0x^4 + b_1x^3 + b_2x^2 + b_3x + b_4 = 0 \qquad (2.21)$$

This can be done, as is proved below, by forming the even function of x

$$\begin{aligned}
F(-x^2) &= f(x)f(-x) \\
&= (b_0x^4 + b_1x^3 + b_2x^2 + b_3x + b_4)(b_0x^4 - b_1x^3 + b_2x^2 - b_3x + b_4) \\
&= b_0^2x^8 - (b_1^2 - 2b_0b_2)x^6 + (b_2^2 - 2b_1b_3 + 2b_0b_4)x^4 - (b_3^2 - 2b_2b_4)x^2 + b_4^2
\end{aligned}$$

† J. B. Scarborough, "Numerical Mathematical Analysis," 2d ed., chap. X, Johns Hopkins Press, Baltimore, 1950; Whittaker and Robinson, *op. cit.*, p. 106; R. E. Doherty and E. G. Keller, "Mathematics of Modern Engineering," vol. I, John Wiley & Sons, Inc., New York, 1936.

substituting $y = -x^2$, and setting the resulting polynomial equal to zero:

$$F(y) = b_0^2 y^4 + (b_1^2 - 2b_0 b_2)y^3 + (b_2^2 - 2b_1 b_3 + 2b_0 b_4)y^2$$
$$+ (b_3^2 - 2b_2 b_4)y + b_4^2 = 0 \quad (2.22)$$

Let the zeros of $f(x)$ be $x = p_1, p_2, p_3,$ and p_4; then the zeros of

$$F(-x^2) = f(x)f(-x)$$

are $x = \pm p_1, \pm p_2, \pm p_3,$ and $\pm p_4$, and since $y = -x^2$, the roots of Eq. (2.22) are $y = -p_1^2, -p_2^2, -p_3^2,$ and $-p_4^2$. Thus the latter is the required equation. Notice that it is of the same degree as the original equation.

This procedure can be generalized to a polynomial equation of any degree and can be performed schematically. Suppose, for example, that the given equation is

$$b_0 x^6 + b_1 x^5 + b_2 x^4 + b_3 x^3 + b_4 x^2 + b_5 x + b_6 = 0 \quad (2.23)$$

and the required equation is

$$B_0 x^6 + B_1 x^5 + B_2 x^4 + B_3 x^3 + B_4 x^2 + B_5 x + B_6 = 0 \quad (2.24)$$

One writes down the detached coefficients $b_0, b_1, \ldots, b_6$, as shown in (2.25), and under them their squares and the products indicated. The coefficients $B_0, B_1, \ldots, B_6$ are now determined by adding the quantities beneath the corresponding $b_0, b_1, \ldots, b_6$:

b_0	b_1	b_2	b_3	b_4	b_5	b_6
b_0^2	b_1^2	b_2^2	b_3^2	b_4^2	b_5^2	b_6^2
	$-2b_0 b_2$	$-2b_1 b_3$	$-2b_2 b_4$	$-2b_3 b_5$	$-2b_4 b_6$	
		$2b_0 b_4$	$2b_1 b_5$	$2b_2 b_6$		
			$-2b_0 b_6$			
B_0	B_1	B_2	B_3	B_4	B_5	B_6

$$(2.25)$$

Example 5. Write down an equation whose roots are the negative of the square of the roots of

$$x^5 - 4x^4 + 2x^3 - 6x + 5 = 0 \quad (2.26)$$

The coefficients of the required equation are found as in computational scheme (2.25) by working with detached coefficients of Eq. (2.26) as follows (note that a zero is inserted as the coefficient of x^2):

1	−4	2	0	−6	5
1	16	4	0	36	25
	−4	0	24	0	
		−12	−40		
1	12	−8	−16	36	25

The numbers in the last row being the new coefficients, the required equation is

$$x^5 + 12x^4 - 8x^3 - 16x^2 + 36x + 25 = 0 \quad (2.27)$$

Aside from a consideration of the labor involved, the root-squaring process can be repeated until any two roots differing initially in their absolute values can be separated as far as one chooses.

2.8. Roots All Real and Unequal in Magnitude. The root-squaring method is most easily applied to an equation having only real roots and these all of different magnitudes. Suppose that

$$P(x) = b_0 x^n + b_1 x^{n-1} + \cdots + b_{n-1} x + b_n = 0 \qquad (2.28)$$

has n real roots $p_1, p_2, \ldots, p_n$ and that

$$|p_1| > |p_2| > |p_3| > \cdots > |p_n| \qquad (2.29)$$

After the first root-squaring process, the root corresponding to p_k is $-p_k^2$; after the second, the corresponding root is $-p_k^4$; after the third, it is $-p_k^8$; and after the mth, it is $q_k = -p_k^{2m}$ (thus for $m = 8$ it is $-p_k^{256}$). Each root-squaring process increases greatly the inequalities of (2.29). When these inequalities are so great that, to the number of significant figures decided upon,†

$$1 + \frac{q_{i+1}(n-1)}{q_i} = 1 \qquad i = 1, 2, \ldots, n-1 \qquad (2.30)$$

the root-squaring process should be terminated.

Suppose that the final equation, after the m root squarings required, is

$$Q(x) = B_0 x^n + B_1 x^{n-1} + \cdots + B_{n-1} x + B_n = 0 \qquad (2.31)$$

and that its roots $q_1, q_2, \ldots, q_n$ satisfy Eqs. (2.30).

By Theorem IV,

$$\frac{B_1}{B_0} = -(q_1 + q_2 + q_3 + \cdots + q_n) = -q_1 \left(1 + \frac{q_2}{q_1} + \cdots + \frac{q_n}{q_1} \right)$$

therefore, by Eqs. (2.30), and to the approximation there indicated,

$$\frac{B_1}{B_0} = -q_1$$

Likewise, to the same order of approximation,

$$\frac{B_2}{B_0} = q_1(q_2 + q_3 + \cdots + q_n) + q_2(q_3 + \cdots + q_n) + \cdots + q_{n-1}(q_n)$$

$$= q_1 q_2 \left[\left(1 + \frac{q_3}{q_2} + \cdots + \frac{q_n}{q_2} \right) + \left(\frac{q_2}{q_1} + \cdots + \frac{q_n}{q_1} \right) \right.$$

$$\left. + \cdots + \frac{q_{n-1}}{q_1} \frac{q_n}{q_2} \right] \cong q_1 q_2$$

† This requires that $|q_i| > 2(n-1)10^k |q_{i+1}|$, where k is the number of significant figures employed in the calculations.

Continuing in this way, we have

$$B_1 \cong -q_1 B_0$$
$$B_2 \cong q_1 q_2 B_0$$
$$B_3 \cong -q_1 q_2 q_3 B_0$$
$$B_4 \cong q_1 q_2 q_3 q_4 B_0 \qquad (2.32)$$
$$\cdots\cdots\cdots\cdots$$
$$B_k \cong (-1)^k q_1 q_2 \cdots q_k B_0$$
$$\cdots\cdots\cdots\cdots\cdots\cdots$$
$$B_n \cong (-1)^n q_1 q_2 \cdots q_n B_0$$

From these equations, we have

$$B_0 q_1 + B_1 = 0$$
$$B_1 q_2 + B_2 = 0$$
$$\cdots\cdots\cdots \qquad (2.33)$$
$$B_{k-1} q_k + B_k = 0$$
$$\cdots\cdots\cdots$$
$$B_{n-1} q_n + B_n = 0$$

Since, by Eqs. (2.33),

$$|q_k| = \frac{|B_k|}{|B_{k-1}|} = |p_k|^{2^m}$$
$$\log |p_k| = 2^{-m}(\log |B_k| - \log |B_{k-1}|) \qquad (2.34)$$

from which one can determine $|p_k|$. Substitution in the original equation (2.28) will determine whether the roots are positive or negative.

Termination of the Root-squaring Process. As stated in the above discussion, the root squaring should be terminated when q_i is negligible compared to q_{i-1} and when, therefore, the coefficients of the equation are given in terms of the root by Eqs. (2.32). One needs, however, to be able to judge from the root-squaring process itself when these conditions are met.

Suppose that an additional root squaring is performed on $Q(x)$ [see Eq. (2.31)] to obtain

$$\bar{Q}(x) = \bar{B}_0 x^n + \bar{B}_1 x^{n-1} + \cdots + \bar{B}_{n-1} x + \bar{B}_n = 0 \qquad (2.35)$$

with the roots $\bar{q}_1, \bar{q}_2, \ldots, \bar{q}_n$, where by the nature of the process

$$\bar{q}_k = -q_k^2 \qquad (2.36)$$

Now the roots will be separated even farther than before. Therefore the approximations used in obtaining Eqs. (2.32) are certainly valid, and hence

$$\bar{B}_k = (-1)^k \bar{q}_1 \bar{q}_2 \cdots \bar{q}_k \bar{B}_0 \qquad (2.37)$$

Directly from the root-squaring process [see (2.25)] $\bar{B}_0 = B_0^2$. Substituting this in Eq. (2.37) together with the expressions for $\bar{q}_k$ from Eq. (2.36), one obtains the approximate relation

$$\bar{B}_k = q_1^2 q_2^2 \cdots q_k^2 B_0^2 = B_k^2 \qquad (2.38)$$

This will be true if the cross-product terms in the root-squaring process are negligible.

Thus the vanishing of all the cross-product terms in the root-squaring process can be used to indicate that the roots are sufficiently separated for the degree of accuracy required and that the process should be terminated.

Example 6. Find the roots of

$$x^3 - 2x^2 - 5x + 6 = 0 \qquad (2.39)$$

to eight significant figures.

By repeated root squaring [see (2.25)] one obtains the following table:

1	−2	−5	6	Given equation
1	4	25	36	
	10	24		
1	14	49	36	2d power
1	1.96^2	2.401^3	1.296^3	
	-0.98	-1.008		
1	0.98^2	1.393^3	1.296^3	4th power
1	0.9604^4	1.940449^6	1.679616^6	
	-0.2786	-0.254016		
1	0.6818^4	1.686433^6	1.679616^6	8th power
1	0.46485124^8	2.8440563^{12}	2.8211099^{12}	
	-0.03372866	-0.0229032		
1	0.43112258^8	2.8211531^{12}	2.8211099^{12}	16th power
1	1.8586668^{15}	0.79589048^{25}	0.79586611^{25}	
	-0.0056423	-0.00002432		
1	1.8530245^{15}	0.79589048^{25}	0.79586611^{25}	32nd power
1	3.4336998^{30}	0.63340294^{50}	0.63340286^{50}	
	-0.0000159			
1	3.4336839^{30}	0.63340294^{50}	0.63340286^{50}	64th power

The powers of 10 are indicated by an exponent on the number; thus 3.4336839^{30} means 3.4336839×10^{30}. The root-squaring process is terminated at the 64th power of the roots, since in any further squaring the cross-product terms would be negligible.

From Eqs. (2.33) the absolute values of the roots q_1, q_2, and q_3 of the final equation are

$$|q_1| = 3.4336839 \times 10^{39} = |p_1|^{64}$$

$$|q_2| = \frac{6.3340294}{3.4336839} \times 10^{19} = |p_2|^{64}$$

$$|q_3| = \frac{6.3340286}{6.3340294} = |p_3|^{64}$$

Using logarithms,

$$\log |p_1| = \tfrac{1}{64}(30.535760) = 0.47712\,125$$
$$\log |p_2| = \tfrac{1}{64}(19.801680 - 0.535760) = 0.30103\,000$$
$$\log |p_3| = \tfrac{1}{64}(-0.00000\,030) = -0.00000\,000$$

therefore

$$|p_1| = 3.00000\,00$$
$$|p_2| = 2.00000\,00$$
$$|p_3| = 1.00000\,00$$

To determine the sign we write the equation in the form

$$x^3 - 5x = 2x^2 - 6$$

and substitute only positive values for the roots. If the two sides of the equation are alike in sign, the root is positive; if they differ in sign, the root should have a negative sign.

x	$x^3 - 5x$	$2x^2 - 6$	Root
3	12	12	3
2	−2	2	−2
1	−4	−4	1

It is of interest to know how close we should have been to the roots if we had terminated the root squaring earlier. In the table below are shown the results obtained by terminating at any particular step, starting with the original equation.

Root squaring terminated at	p_1	p_2	p_3
Original equation...............	2.000	−2.500	1.2000
2d power.......................	3.742	−1.870	0.8571
4th power......................	3.146	−1.942	0.9821
8th power......................	3.014	−1.991	0.9996
16th power.....................	3.00028	−1.99981	0.999995
32nd power.....................	3.0000000	−2.0000000	1.00000000
64th power.....................	3.0000000	−2.0000000	1.00000000

2.9. Complex Roots and Roots Equal in Magnitude. For a polynomial with real coefficients the complex roots occur in pairs. Since such roots are equal in magnitude, no amount of squaring will separate them. The equation cannot, therefore, be completely broken down into the linear equations (2.33). However, if the various pairs of complex roots all differ in absolute value, the final equation obtained by the m root-squaring procedure can be broken down into linear equations for the real (and unequal) roots and quadratic equations for the pairs of complex roots.

As an example, suppose that the roots of the given equation are p_1, $x \pm iy$, p_2, p_3, where $|p_1| > |x \pm iy| > |p_2| > |p_3|$, and that the final equation after m root squarings is

$$B_0 x^5 + B_1 x^4 + B_2 x^3 + B_3 x^2 + B_4 x + B_5 = 0$$

Using an argument similar to that used to obtain the equations in (2.33), we can break this equation down into the equations

$$
\begin{aligned}
B_0 x + B_1 &= 0 \\
B_1 x^2 + B_2 x + B_3 &= 0 \\
B_3 x + B_4 &= 0 \\
B_4 x + B_5 &= 0
\end{aligned}
\qquad (2.40)
$$

and the roots of those equations are, respectively, $-p_1^N$; $-(x + iy)^N$ and $-(x - iy)^N$; $-p_2^N$; $-p_3^N$, where $N = 2^m$.

The presence of the complex pair of roots in the above example would be indicated in the root-squaring process by the nonvanishing of the product terms in the B_2 column and the frequent changes in sign of B_2, although the latter may not always occur.

If two or more complex pairs are equal in magnitude, quartic equations must be solved; and therefore the method is of little or no use.

The presence of two real roots of equal magnitude will likewise give rise to quadratic equations, such as found in Eqs. (2.40) for complex roots. The presence of such roots is marked by the nonvanishing of cross-product terms. These cross-product terms, in this case, approach a value equal to half the squared term. Again the method is of little or no use if three or more roots are equal in magnitude.

In general, the root-squaring method of Dandelin and Graeffe is rather unsatisfactory because of the complexity of rules for its proper application,[†] the uncertainty of obtaining an answer if too many roots are equal in magnitude, and the tendency for numerical errors to occur in the root-squaring process. These errors are not corrected by the subsequent calculations. For real roots, it is therefore best to use the method of false position or the Birge-Vieta method. However, for finding the complex roots of polynomials of the eighth order and higher, the root-squaring method has few successful competitors.[‡]

[†] See the references under footnote, p. 26.

[‡] This statement is based on the assumption that all real roots have previously been determined and the order of the polynomial reduced accordingly.

2.10. Complex Roots of a Quartic. One can obtain the complex roots of a quartic,

$$w^4 + Bw^3 + Cw^2 + Dw + E = 0 \tag{2.41}$$

in a very simple way by the following procedure.

Reduce all the roots of Eq. (2.41) by $-\frac{1}{4}B$ using synthetic division (see Example 2).

The new equation for $z = w + \frac{1}{4}B$,

$$z^4 + cz^2 + dz + e = 0 \tag{2.42}$$

does not contain a term in z^3. Substituting in this equation $z = x + iy$ and setting the real and imaginary parts each equal to zero, one obtains the two equations

$$y^4 - (6x^2 + c)y^2 + (x^4 + cx^2 + dx + e) = 0 \tag{2.43}$$
$$y[4xy^2 - (4x^3 + 2cx + d)] = 0$$

Since we are investigating the complex roots of (2.42), we assume that $y \neq 0$. If the first equation of (2.43) is multiplied by $16x^2$ and the last equation is used to substitute for $4xy^2$, we have

$$(4x^3 + 2cx + d)^2 - 4x(6x^2 + c)(4x^3 + 2cx + d)$$
$$+ 16x^2(x^4 + cx^2 + dx + e) = 0$$

or

$$64x^6 + 32cx^4 + 4(c^2 - 4e)x^2 - d^2 = 0 \tag{2.44}$$

Let

$$s = (2x)^2 \qquad \text{or} \qquad x = \frac{\pm \sqrt{s}}{2} \tag{2.45}$$

then Eq. (2.44) can be written

$$s^3 + 2cs^2 + (c^2 - 4e)s - d^2 = 0 \tag{2.46}$$

and the second equation of (2.43) requires that

$$y = \pm \frac{1}{2} \sqrt{s + 2c + \frac{d}{x}} \tag{2.47}$$

Since x must be real, only the positive real roots of (2.46) need be obtained. If we let

$$y_1 = \frac{1}{2} \sqrt{s + 2c + \frac{d}{|x|}}$$
$$y_2 = \frac{1}{2} \sqrt{s + 2c - \frac{d}{|x|}} \tag{2.48}$$

then the required roots of Eq. (2.42) are

$$\begin{aligned}|x| + iy_1\\ |x| - iy_1\\ -|x| + iy_2\\ -|x| - iy_2\end{aligned} \tag{2.49}$$

where $|x| = \sqrt{s}/2$ and s is the positive root of Eq. (2.46).† The roots of Eq. (2.41) can be found by adding $-B/4$ to each of the above roots.

2.11. Lin-Bairstow Method for Complex Roots. A general method for determining the complex roots of a polynomial equation

$$P_n(x) = a_0 x^n + a_1 x^{n-1} + a_2 x^{n-2} + \cdots + a_{n-1} x + a_n = 0 \quad (2.50)$$

involves finding a quadratic factor $x^2 + \alpha x + \beta$ of the polynomial by an iterative procedure. If $P_n(x)$ is divided by a trial factor $x^2 + rx + s$, one obtains as a quotient a polynomial $Q(x)$ of degree $n - 2$ and a remainder $Rx + S$. One may therefore write

$$\sum_{k=0}^{n} a_k x^{n-k} = (x^2 + rx + s) \sum_{k=0}^{n-2} b_k x^{n-k-2} + Rx + S \quad (2.51)$$

from which it follows that

$$
\begin{aligned}
a_0 &= b_0 \\
a_1 &= b_1 + rb_0 \\
a_2 &= b_2 + rb_1 + sb_0 \\
&\cdots\cdots\cdots\cdots \\
a_k &= b_k + rb_{k-1} + sb_{k-2} \\
&\cdots\cdots\cdots\cdots \\
a_{n-1} &= R + rb_{n-2} + sb_{n-3} \\
a_n &= S + sb_{n-2}
\end{aligned}
\quad (2.52)
$$

If one sets

$$
\begin{aligned}
b_{-1} &= b_{-2} = 0 \\
b_{n-1} &= R \\
b_n &= S - rR
\end{aligned}
\quad (2.53)
$$

one may write Eqs. (2.52) in the form

$$b_k = a_k - rb_{k-1} - sb_{k-2} \qquad k = 0, 1, 2, \ldots, n \quad (2.54)$$

The coefficients b_k of the polynomial $Q(x)$ and the coefficients R and S of the remainder are functions of r and s. We now attempt by an iterative process to solve the simultaneous equations

$$R(r,s) = 0 \qquad \text{and} \qquad S(r,s) = 0 \quad (2.55)$$

which, if satisfied by r^* and s^*, make $x^2 + r^* x + s^*$ a factor of the polynomial. To find r^* and s^* we assume that we have an r and s near these values so that

$$
\begin{aligned}
r^* &= r + \Delta r \\
s^* &= s + \Delta s
\end{aligned}
\quad (2.56)
$$

† Since Eq. (2.46) has at least one alteration in the sign of its coefficients, there must be, by Descartes's rule, at least one positive real root.

where Δr and Δs are small. Using Taylor's expansion for functions of two variables and neglecting terms of higher power than the first in these increments, we have

$$R(r,s) + \Delta r \frac{\partial R}{\partial r} + \Delta s \frac{\partial R}{\partial s} \cong R(r^*,s^*) = 0$$

$$S(r,s) + \Delta r \frac{\partial S}{\partial r} + \Delta s \frac{\partial S}{\partial s} \cong S(r^*,s^*) = 0$$

(2.57)

We next find the four partial derivatives in Eqs. (2.57) and solve these equations for Δr and Δs. Use of Eqs. (2.56) then yields an r^* and s^*, which, of course, because of the approximate nature of Eqs. (2.57), are now only an improved estimate of the roots of Eqs. (2.55).

To find the partial derivatives in Eqs. (2.57) we differentiate Eqs. (2.54); thus

$$\frac{\partial b_k}{\partial r} = -b_{k-1} - r \frac{\partial b_{k-1}}{\partial r} - s \frac{\partial b_{k-2}}{\partial r}$$

$$\frac{\partial b_k}{\partial s} = -b_{k-2} - r \frac{\partial b_{k-1}}{\partial s} - s \frac{\partial b_{k-2}}{\partial s}$$

(2.58)

Since from Eqs. (2.52) $b_0 = a_0$, it is not a function of r or s, and therefore from the above equations

$$\frac{\partial b_0}{\partial r} = 0 \qquad\qquad \frac{\partial b_1}{\partial s} = 0$$

$$\frac{\partial b_1}{\partial r} = -b_0 \qquad\qquad \frac{\partial b_2}{\partial s} = -b_0 - r \frac{\partial b_1}{\partial s} = -b_0$$

$$\frac{\partial b_2}{\partial r} = -b_1 - r \frac{\partial b_1}{\partial r} \qquad \frac{\partial b_3}{\partial s} = -b_1 - r \frac{\partial b_2}{\partial s} - s \frac{\partial b_1}{\partial s}$$

$$= -b_1 + rb_0 \qquad\qquad = -b_1 + rb_0$$

(2.59)

Thus for $k = 0, 1,$ and 2

$$\frac{\partial b_k}{\partial r} = \frac{\partial b_{k+1}}{\partial s} = -c_{k-1}$$

(2.60)

and by mathematical induction this equation holds for all k. For suppose that Eq. (2.60) holds for all k up to $m - 1$; then by Eqs. (2.58)

$$\frac{\partial b_m}{\partial r} = -b_{m-1} - r \frac{\partial b_{m-1}}{\partial r} - s \frac{\partial b_{m-2}}{\partial r} = -b_{m-1} - r \frac{\partial b_m}{\partial s} - s \frac{\partial b_{m-1}}{\partial s} = \frac{\partial b_{m+1}}{\partial s}$$

and thus it holds for m.

Making use of Eq. (2.60) one may write in place of the equations in (2.58) the single recurrence relation

$$c_k = b_k - rc_{k-1} - sc_{k-2}$$

(2.61)

In particular, from Eqs. (2.59) and (2.60) $c_{-1} = 0$ and $c_0 = b_0$. Thus the c's are obtained from the b's in exactly the same way the b's were obtained from the a's.

From Eqs. (2.53) and (2.60)

$$R = b_{n-1}$$
$$\frac{\partial R}{\partial r} = \frac{\partial b_{n-1}}{\partial r} = -c_{n-2}$$
$$\frac{\partial R}{\partial s} = \frac{\partial b_{n-1}}{\partial s} = -c_{n-3}$$
$$S = b_n + rb_{n-1} \tag{2.62}$$
$$\frac{\partial S}{\partial r} = \frac{\partial b_n}{\partial r} + b_{n-1} + r\frac{\partial b_{n-1}}{\partial r} = -c_{n-1} - rc_{n-2} + b_{n-1}$$
$$\frac{\partial S}{\partial s} = \frac{\partial b_n}{\partial s} + r\frac{\partial b_{n-1}}{\partial s} = -c_{n-2} - rc_{n-3}$$

and therefore the equations (2.57) for Δr and Δs may be written

$$c_{n-2}\,\Delta r + c_{n-3}\,\Delta s = b_{n-1}$$
$$(c_{n-1} - b_{n-1})\,\Delta r + c_{n-2}\,\Delta s = b_n \tag{2.63}$$

Note that c_n is not needed in these equations and that in place of c_{n-1} we may compute directly

$$\bar{c}_{n-1} = c_{n-1} - b_{n-1} = -rc_{n-2} - sc_{n-3} \tag{2.64}$$

the coefficient needed in Eqs. (2.63).

To summarize, one computes the b_k and c_k from the given coefficients a_k using recurrence relations (2.54), (2.61), and (2.64) and assuming that

$$a_{-1} = b_{-1} = c_{-1} = 0$$
$$a_0 = b_0 = c_0 \tag{2.65}$$

It is convenient for this purpose to arrange the coefficients in the following array:

$$
\begin{array}{ccc}
a_0 & b_0 & c_0 \\
a_1 & b_1 & c_1 \\
a_2 & b_2 & c_2 \\
\multicolumn{3}{c}{\cdots\cdots\cdots} \\
a_{n-2} & b_{n-2} & c_{n-2} \\
a_{n-1} & b_{n-1} & \bar{c}_{n-1} \\
a_n & b_n &
\end{array} \tag{2.66}
$$

Having these coefficients, one may write down the two simultaneous equations in (2.63) for Δr and Δs. These increments added on to r and s, respectively, give by Eqs. (2.56) improved estimates of coefficients r^* and s^* in a quadratic factor $x^2 + r^*x + s^*$ of the given polynomial.

When such a quadratic factor is found with sufficient accuracy, two roots of the given equation (2.50) are determined by setting

$$x^2 + r^*x + s^* = 0 \qquad (2.67)$$

Such a method permits one to compute a pair of complex roots of a polynomial having real coefficients by operating only with real numbers.

PROBLEMS

1. Determine the fourth-degree polynomial that has roots ± 1 and $\pm i$.

2. Prove Theorem V.

3. Prove Theorem VI.

4. Given the fourth-degree polynomial $P(x) = 3x^4 - 7x^2 + 10x - 5$, find $P(4)$, first by direct substitution, and then by synthetic division.

5. Find the derivative at $x = -2$ of the polynomial in Prob. 4 by synthetic division.

6. Find a fourth-degree polynomial that has all its zeros decreased by 2 relative to the polynomial in Prob. 4.

7. Determine the nature of all the zeros of the polynomial in Prob. 4.

8. Find a real zero of the polynomial $P(x) = x^5 - 0.346284x^4 + x^3 + 3.762848x + 10$. Correct to six decimal places by the Birge-Vieta method, and write out the reduced fourth-degree polynomial that arises when this zero is removed.

9. Solve Prob. 8 using Horner's method.

10. Find an equation whose roots are the negative of the square of the roots of $x^5 - 2x^4 + 10x^3 - 9x + 3 = 0$.

11. Find all the roots of the equation in Prob. 10 using the method of Dandelin and Graeffe.

12. Find all the roots of the quartic $x^4 - 6x^3 + 14x^2 - 24x + 40 = 0$ using the method of Sec. 2.10.

13. Show that the root-squaring method breaks down for the quartic $x^4 - x^3 + x^2 - x + 1 = 0$. What can therefore be said about the roots of this equation?

CHAPTER 3

FINITE-DIFFERENCE TABLES AND THE THEORY
OF INTERPOLATION

The underlying theory of numerical analysis is based on the calculus of finite differences, which in itself has an extensive literature. In this chapter, however, we shall make use only of that portion of the subject which applies rather directly to the construction and use of numerical tables.

3.1. Descending Differences. The *first descending difference* of a function $f(x)$, usually referred to simply as the first difference of $f(x)$, is indicated by the symbol $\Delta f(x)$ and is defined by the equation

$$\Delta f(x) = f(x + h) - f(x) \tag{3.1}$$

Since it is a function of the parameter h, it is sometimes indicated by $\Delta_h f(x)$.

Suppose that a function $y = f(x)$ is tabulated at $y_s = f(x_s)$, where $x_s = x_0 + sh$ and s takes on integral values. The interval between two successive values of the *argument* x is thus equal to h. (The numbers y_s are referred to as *tabulated values* or *entries*.) In such a table the *first difference* of y_s is given by the expression

$$\Delta y_s = y_{s+1} - y_s \tag{3.2}$$

The *second difference* $\Delta^2 y_s$ is obtained by taking the first difference of Δy_s; thus

$$\Delta^2 y_s = \Delta(\Delta y_s) = \Delta y_{s+1} - \Delta y_s \tag{3.3}$$

and similarly

$$\Delta^3 y_s = \Delta(\Delta^2 y_s) = \Delta^2 y_{s+1} - \Delta^2 y_s$$
$$\Delta^4 y_s = \Delta(\Delta^3 y_s) = \Delta^3 y_{s+1} - \Delta^3 y_s \tag{3.4}$$
. . . .
$$\Delta^k y_s = \Delta(\Delta^{k-1} y_s) = \Delta^{k-1} y_{s+1} - \Delta^{k-1} y_s \quad \text{etc.}$$

The kth difference of y_s can also be expressed, using Eq. (3.1), as

$$\Delta^k f(x_s) = \Delta^{k-1} f(x_s + h) - \Delta^{k-1} f(x_s) \tag{3.5}$$

These differences are conveniently obtained by forming Table 3.1. In this table the differences in any column are obtained by subtracting the two neighboring differences in the column to its left, the upper difference being subtracted from the lower.

It is convenient for some purposes to introduce the variable

$$u = \frac{x - x_0}{h} \qquad (3.6)$$

Then $f(x)$ can be written $f(x + uh) = g(u)$, and the entries are $y_s = g(s)$, where s assumes only integral values. In terms of $g(u)$,

$$\Delta^k y_s = \Delta^k g(s) = \Delta^{k-1} g(s + 1) - \Delta^{k-1} g(s) \qquad (3.7)$$

TABLE 3.1

x_0	y_0						
x_1	y_1	Δy_0	$\Delta^2 y_0$	$\Delta^3 y_0$			
x_2	y_2	Δy_1	$\Delta^2 y_1$	$\Delta^3 y_1$	$\Delta^4 y_0$	$\Delta^5 y_0$	
x_3	y_3	Δy_2	$\Delta^2 y_2$	$\Delta^3 y_2$	$\Delta^4 y_1$	$\Delta^5 y_1$	$\Delta^6 y_0$
.	.	Δy_3	.	.	.		
.	.	.	.	.			
.	.	.	.	.			
.	.	.	.				
x_{n-2}	y_{n-2}	Δy_{n-3}	$\Delta^2 y_{n-3}$	$\Delta^3 y_{n-4}$	$\Delta^4 y_{n-4}$		
x_{n-1}	y_{n-1}	Δy_{n-2}	$\Delta^2 y_{n-2}$	$\Delta^3 y_{n-3}$			
x_n	y_n	Δy_{n-1}					

For $h = 1$ and x_0 set equal to zero, $u = x$, and the functions $g(u)$ and $f(x)$ are identical. In general, however, $g(u)$ contains h and x_0 as parameters.

Note that by Eqs. (3.1) and (3.6)

$$\Delta u = 1 \qquad (3.8)$$

Example 1. Write down the difference table for $y = x^3$ where x takes on the integral values 0, 1, 2, 3, . . . , 8.†

TABLE 3.2

x	y	Δy	$\Delta^2 y$	$\Delta^3 y$	$\Delta^4 y$
0	0				
		1			
1	1		6		
		7		6	
2	8		12		0
		19		6	
3	27		18		0
		37		6	
4	64		24		0
		61		6	
5	125		30		0
		91		6	
6	216		36		0
		127		6	
7	343		42		
		169			
8	512				

† h is therefore equal to one.

Notice that all the third differences are equal and that consequently all fourth and higher differences are zero.

If one labels the first entry as y_0, one has a unique designation for each of the numbers occurring in the table; thus, for example,

$$\Delta y_5 = 91 \qquad \Delta^2 y_3 = 24 \qquad \text{and} \qquad \Delta^2 y_6 = 42$$

3.2. Leading Differences. The tabulated values y_s can be looked upon as zero-order differences. With this in mind, we refer to the first numbers in each difference column (in Table 3.2 the numbers 0, 1, 6, and 6) as *leading differences*. The leading 0th difference is called the *leading term* in the table. If the difference of some particular order is constant, one can construct the entire table by addition alone from a knowledge of the leading differences. For Table 3.2 this may be done as follows.

Since $\Delta^3 y$ is constant, each number in this column is 6. Moreover, since $\Delta^3 y$ is the difference between terms in the $\Delta^2 y$ column, the numbers in the latter column are obtained in succession by adding 6 to the preceding number in the column. This follows from Eqs. (3.4)

$$\Delta^2 y_{s+1} = \Delta^2 y_s + \Delta^3 y_s = \Delta^2 y_s + 6$$

After these second differences are determined, the first differences can be obtained from the relation

$$\Delta y_{s+1} = \Delta y_s + \Delta^2 y_s$$

which means that one obtains Δy_{s+1} from Δy_s by adding to it the second difference lying between them. Thus, knowing $\Delta y_3 = 37$, one finds $\Delta y_4 = 37 + 24 = 61$. In exactly the same way the values of y are obtained from the knowledge of values of Δy, since

$$y_{s+1} = y_s + \Delta y_s$$

Incidentally, one may check a difference table by the use of the equation

$$\sum_{s=n}^{m} \Delta^k y_s = \sum_{s=n}^{m} \Delta^{k-1} y_{s+1} - \sum_{s=n}^{m} \Delta^{k-1} y_s = \Delta^{k-1} y_{m+1} - \Delta^{k-1} y_n \qquad (3.9)$$

which expresses the sum of the differences in any difference column in terms of differences of the next lower order.

3.3. Other Examples of Difference Tables. It will be seen later that, in general, the set of values $y_0, y_1, y_2, \ldots, y_n$ obtained from any function $y = f(x)$ for equally spaced values of the argument will give rise to a difference table in which some order of difference is nearly constant, provided only that h is sufficiently small. Even experimentally determined functions are found to behave in this way, within the accuracy of the data obtained. In fact a difference table is a very good test of the accuracy of physical data. Thus in Table 3.5, although the data are given to seven significant figures, the variation in the second difference indicates clearly that the data are accurate to only five significant figures. Higher-order differences added to this table would, obviously, not be any more nearly constant than the second difference.

TABLE 3.3

x	$y = \log x$	Δy	$\Delta^2 y$	$\Delta^3 y$
1.00	0.0000 000			
		43 214		
1.01	0.0043 214		−426	
		42 788		8
1.02	0.0086 002		−418	
		42 370		9
1.03	0.0128 372		−409	
		41 961		8
1.04	0.0170 333		−401	
		41 560		7
1.05	0.0211 893		−394	
		41 166		
1.06	0.0253 059			

TABLE 3.4

	d = sun's declination	Δd	$\Delta^2 d$	$\Delta^3 d$
Feb. 1	−17°0′19″0			
		34′56″1		
Feb. 3	−16°25′22″9		1′8″0	
		36′4″1		−3″1
Feb. 5	−15°49′18″8		1′4″9	
		37′9″0		−3″2
Feb. 7	−15°12′9″8		1′1″7	
		38′10″7		−3″1
Feb. 9	−14°33′59″1		58″6	
		39′9″3		−3″1
Feb. 11	−13°54′49″8		55″5	
		40′4″8		
Feb. 13	−13°14′45″0			

It is often convenient to write the differences without regard to the decimal point, as is done in Table 3.3.

3.4. Errors in a Table. When one is constructing a table, it is usually highly important that the table be free of errors. No matter how careful one is, in any extended table there is likely to be an occasional error. The chances are good that such an error can be spotted by running a difference table of the entries y_s, provided only one or two errors occur in neighboring values of y.

Suppose that the interval h is chosen small enough so that some order difference of y is nearly constant when the y's are correctly calculated; this might, for instance, be the fourth difference. An error ϵ in one value of y, say y_4, will then have the effect shown in Table 3.6 (set $\eta = 0$). This effect is shown to grow as the order of difference increases. More-

TABLE 3.5

Temp., °C	Vapor pressure of water, mm	Δ	Δ^2
200	116 59.16		
		246.24	
201	119 05.40		3.80
		250.04	
202	121 55.44		3.04
		253.08	
203	124 08.52		4.56
		257.64	
204	126 66.16		5.32
		262.96	
205	129 29.12		5.32
		268.28	
206	131 97.40		2.28
		270.56	
207	134 67.96		

TABLE 3.6. EFFECT OF ERRORS ON A DIFFERENCE TABLE

x_0	y_0				
		Δy_0			
x_1	y_1		$\Delta^2 y_0$		
		Δy_1		$\Delta^3 y_0$	
x_2	y_2		$\Delta^2 y_1$		$\Delta^4 y_0 + \epsilon$
		Δy_2		$\Delta^3 y_1 + \epsilon$	
x_3	y_3		$\Delta^2 y_2 + \epsilon$		$\Delta^4 y_1 - 4\epsilon$
		$\Delta y_3 + \epsilon$		$\Delta^3 y_2 - 3\epsilon$	
x_4	$y_4 + \epsilon$		$\Delta^2 y_3 - 2\epsilon$		$\Delta^4 y_2 + 6\epsilon$
		$\Delta y_4 - \epsilon$		$\Delta^3 y_3 + 3\epsilon$	
x_5	y_5		$\Delta^2 y_4 + \epsilon$		$\Delta^4 y_3 - 4\epsilon$
		Δy_5		$\Delta^3 y_4 - \epsilon$	
x_6	y_6		$\Delta^2 y_5$		$\Delta^4 y_4 + \epsilon + \eta$
		Δy_6		$\Delta^3 y_5 + \eta$	
x_7	y_7		$\Delta^2 y_6 + \eta$		$\Delta^4 y_5 - 4\eta$
		$\Delta y_7 + \eta$		$\Delta^3 y_6 - 3\eta$	
x_8	$y_8 + \eta$		$\Delta^2 y_7 - 2\eta$		$\Delta^4 y_6 + 6\eta$
		$\Delta y_8 - \eta$		$\Delta^3 y_7 + 3\eta$	
x_9	y_9		$\Delta^2 y_8 + \eta$		
		Δy_9			
x_{10}	y_{10}				

over, the distribution of the error in any one difference column is in accordance with the binomial coefficients of that order with, however, alternation of sign. Thus the errors in the fourth differences due to an error ϵ in one of the entries are ϵ, -4ϵ, 6ϵ, -4ϵ, ϵ, in agreement, except for sign, with the binomial coefficients $\binom{4}{0}$, $\binom{4}{1}$, $\binom{4}{2}$, $\binom{4}{3}$, $\binom{4}{4}$.† Such a fluctuation in the difference in a normally constant-difference column can readily be detected. Moreover, since as can be seen from Table 3.6, the error is maximum directly opposite the incorrect value of y, it is possible to correct this value at once.

If more than one value of y is in error, as shown in Table 3.6, the fluctuation in the normally constant-difference column is just the sum of that due to each. It is more difficult, therefore, to correct the table than when only a single error is present. Nevertheless, one can sometimes correct several errors, one at a time, by keeping in mind the binomial distribution of the errors due to each incorrect entry.

Example 2. Find the error (or errors) in Table 3.7.

The first three values of the third difference, as well as the last value, are very close to 11.‡ Moreover, the average value of all the third differences is approximately equal to 11; therefore, it appears that the large fluctuations in the other values of $\Delta^3 y$ are probably due to an error in the table. If the fourth differences are taken, they represent the fluctuations in the third differences, which presumably are due almost entirely to the errors in the tabulated values of y. These latter, of course, include round-off errors. The fact that the average value of the fourth differences is approximately zero corroborates this conclusion.

Since the maximum absolute value of the fourth difference, and hence presumably the maximum error in $\Delta^4 y$, occurs opposite $x = 0.06$, it is the corresponding value of y that is probably in error. Suppose the error in this value is ϵ; in the extreme right-hand column of the table is shown the effect of this error on the fourth differences. By equating the terms in the two last columns one obtains for values of ϵ the values -8, -10, -10, -10, and -10, which indicates clearly that the error in 1.06183 55 is -10 in the last place, and hence the correct value is 1.06183 65. With this value inserted, the difference table becomes smooth. In particular the fourth differences become, respectively, -2, -1, 2, 1, -3, and 0, which are well within the range of the probable errors due to the initial round-off errors.

Suppose one has two separate tables with entries $y_s = f(x_s)$ and $z_s = g(x_s)$, $s = 1$, 2, . . . , n, and has built up a difference table in each case so that one has the various differences $\Delta^k y_s$ and $\Delta^k z_s$, $k = 1, 2, \ldots, m$. Then one may easily form the difference table for the sum $w_s = y_s + z_s$ by adding the corresponding differences.

---

† The expansion of the nth power of a binomial is given by

$$(a + b)^n = \binom{n}{0}a^n + \binom{n}{1}a^{n-1}b + \binom{n}{2}a^{n-2}b^2 + \cdots + \binom{n}{n}b^n$$

where
$$\binom{n}{k} = \frac{n!}{k!(n-k)!}$$

‡ The small fluctuation in these values is easily explained by the round-off errors in the tabulated values. Thus for $x = 0.02$, $y = 1.02020\ 134003 = 1.02020\ 13 +$ round-off error.

TABLE 3.7

x	$y = e^x$	Δy	$\Delta^2 y$	$\Delta^3 y$	$\Delta^4 y$	Effect of an error
0.00	1.00000 00					
		1005 02				
0.01	1.01005 02		1009			
		1015 11		12		
0.02	1.02020 13		1021		−2	
		1025 32		10		
0.03	1.03045 45		1031		−1	
		1035 63		9		
0.04	1.04081 08		1040		−8	ϵ
		1046 03		1		
0.05	1.05127 11		1041		41	−4ϵ
		1056 44		42		
0.06	1.06183 55		1083		−63	6ϵ
		1067 27		−21		
0.07	1.07250 82		1062		42	−4ϵ
		1077 89		21		
0.08	1.08328 71		1083		−10	ϵ
		1088 72		11		
0.09	1.09417 43		1094			
		1099 66				
0.10	1.10517 09					

This can be seen from the definition of a difference, since

$$\Delta w_s = (y_{s+1} + z_{s+1}) - (y_s + z_s) = \Delta y_s + \Delta z_s$$
$$\Delta^2 w_s = \Delta(y_{s+1} + z_{s+1}) - \Delta(y_s + z_s) = \Delta^2 y_s + \Delta^2 z_s$$

and by mathematical induction

$$\Delta^k w_s = \Delta^k y_s + \Delta^k z_s$$

This *linearity* of the differencing operation is a very helpful fact to remember.

Thus Table 3.6 represents the sum of the difference table of the y's and the difference table of the errors. The entries in the error table are all zero except at x_4 and x_8, where they are ϵ and η, respectively.

3.5. Symbolic Operators. Let us introduce the symbolic operator E, which, when applied to $f(x)$, increases its argument by h, the interval between successive values of x_s; thus

$$Ef(x) = f(x + h) \tag{3.10}$$

With this notation, the first difference, by Eq. (3.1), is

$$\Delta f(x) = Ef(x) - f(x) = (E - 1)f(x)$$

the operator $E - 1$ being defined by the latter of these equations. It is convenient to abbreviate the above by writing

$$\Delta = E - 1 \tag{3.11}$$

the *operand* $f(x)$ on each side being understood. Such an equation is called an *operator equation*, and it is valid provided the results obtained when each side is allowed to operate on a general function $f(x)$ are identical.

It can be shown that these symbolic operators behave in almost every respect like algebraic symbols. In particular they satisfy the distributive, associative, and commutative laws of algebra. Just as the operation of multiplying $f(x)$ by 1 is indicated by the operator 1 in Eq. (3.11), it is convenient to use k as the operator that multiplies $f(x)$ by the numerical value of k.

That the commutative law holds between k, Δ, and E is illustrated by the following examples. By definition

$$\Delta k f(x) = k f(x + h) - k f(x) = k[f(x + h) - f(x)] = k \Delta f(x)$$

or
$$\Delta k = k \Delta$$

Likewise
$$E k f(x) = k f(x + h) = k E f(x)$$

or
$$E k = k E$$

Also

$$E \Delta f(x) = E[f(x + h) - f(x)] = f(x + 2h) - f(x + h) = \Delta f(x + h)$$
$$= \Delta E f(x)$$

or
$$E \Delta = \Delta E$$

Similar investigations will convince the student of the distributive and associative laws and the exponential laws

$$\Delta^m \Delta^n = \Delta^{m+n}$$

and
$$E^m E^n = E^{m+n} \tag{3.12}$$

Example 3. Express the nth difference of $y_0 = f(x_0)$ in terms of the values of y_0, y_1, y_2, ..., y_n, where $y_s = f(x_0 + sh)$.

By using the operators Δ and E, one obtains the relations

$$\Delta^n y_0 = (E - 1)^n y_0 = [E^n - \binom{n}{1}E^{n-1} + \binom{n}{2}E^{n-2} - \cdots + (-1)^n]y_0$$
$$\Delta^n y_0 = E^n y_0 - \binom{n}{1}E^{n-1}y_0 + \binom{n}{2}E^{n-2}y_0 - \cdots + (-1)^n y_0$$

Now
$$E y_s = E f(x_0 + sh) = f(x_0 + (s + 1)h) = y_{s+1}$$

and hence by induction

$$E^k y_s = y_{s+k} \tag{3.13}$$

Therefore†
$$\Delta^n y_0 = y_n - \binom{n}{1}y_{n-1} + \binom{n}{2}y_{n-2} - \cdots + (-1)^n y_0 \tag{3.14}$$

Example 4. Express y_n in terms of y_0 and its successive differences Δy_0, $\Delta^2 y_0$, ..., $\Delta^n y_0$.

By Eq. (3.13), $y_n = E^n y_0$, and since $E^n = (1 + \Delta)^n$,

$$y_n = (1 + \Delta)^n y_0 = [1 + \binom{n}{1}\Delta + \binom{n}{2}\Delta^2 + \cdots + \Delta^n]y_0 = y_0 + \binom{n}{1}\Delta y_0 + \binom{n}{2}\Delta^2 y_0 + \cdots + \Delta^n y_0 \tag{3.15}$$

† Since $\binom{n}{k} = n!/[k!(n - k)!] = \binom{n}{n-k}$, one has in particular

$$\binom{n}{0} = 1 \qquad \binom{n}{1} = n \qquad \binom{n}{2} = \frac{n(n - 1)}{2}, \quad \ldots, \binom{n}{n-1} = n \qquad \text{and} \qquad \binom{n}{n} = 1$$

3.6. Differences of a Polynomial. By definition

$$\Delta x^n = (x + h)^n - x^n = \binom{n}{1}hx^{n-1} + \binom{n}{2}h^2x^{n-2} + \cdots + h^n \quad (3.16)$$

Therefore, by the distributive law for Δ

$$\Delta P_n(x) = \Delta(c_0x^n + c_1x^{n-1} + c_2x^{n-2} + \cdots + c_{n-1}x + c_n) = c_0(\Delta x^n)$$
$$+ c_1(\Delta x^{n-1}) + c_2(\Delta x^{n-2}) + \cdots + c_{n-1}\Delta x$$

which, by Eq. (3.16), is a polynomial of degree $n - 1$. The nth difference of $P_n(x)$ is therefore a polynomial of zero degree, i.e., a constant. All higher differences are consequently zero. To evaluate the constant $\Delta^n P_n(x)$ one notes that

$$\Delta^n P_n(x) = c_0 \Delta^n x^n + c_1 \Delta^n x^{n-1} + \cdots + c_{n-1}\Delta^n x = c_0 \Delta^n x^n \quad (3.17)$$

because, by the above argument,

$$\Delta^n x^{n-p} = 0 \qquad p > 0 \quad (3.18)$$

Using Eqs. (3.12) and (3.16),

$$\Delta^n x^n = \Delta^{n-1}(\Delta x^n) = \binom{n}{1}h \Delta^{n-1}x^{n-1} + \binom{n}{2}h^2 \Delta^{n-1}x^{n-2} + \cdots$$

Only the first term, by Eq. (3.18), differs from zero. Therefore, since $\binom{n}{1} = n$,

$$\Delta^n x^n = hn \Delta^{n-1}x^{n-1} = hn[h(n - 1) \Delta^{n-2}x^{n-2}]$$

and hence by mathematical induction

$$\Delta^n x^n = h^{n-1}[n(n - 1) \cdots 1] \Delta x = n!h^n \quad (3.19)$$

Applying this result to Eq. (3.17), one has

$$\Delta^n P_n(x) = c_0 n!h^n \quad (3.20)$$

For example, if $h = 1$

$$\Delta^3 x^3 = 3! = 6$$

which agrees with the value of the third difference found in Table 3.2.

3.7. Factorial Polynomials. Since the differences of x^n are not simply expressed, it is convenient to let $u = (x - x_0)/h$ and to introduce the *factorial polynomials*

$$\begin{aligned}
u^{[0]} &= 1 \\
u^{[1]} &= u \\
u^{[2]} &= u(u - 1) \\
u^{[3]} &= u(u - 1)(u - 2) \\
&\cdots\cdots\cdots\cdots\cdots \\
u^{[r]} &= u(u - 1)(u - 2) \cdots [u - (r - 1)] \qquad \text{etc.}
\end{aligned} \quad (3.21)$$

and the reciprocals of polynomials associated with them

$$u^{[-1]} = \frac{1}{u+1}$$

$$u^{[-2]} = \frac{1}{(u+1)(u+2)}$$

$$u^{[-3]} = \frac{1}{(u+1)(u+2)(u+3)} \tag{3.22}$$

$$\cdots\cdots\cdots\cdots\cdots\cdots$$

$$u^{[-r]} = \frac{1}{(u+1)(u+2)\cdots(u+r)} \qquad \text{etc.}$$

The superscripts are referred to as *pseudo exponents*, and the functions in Eqs. (3.21) and (3.22) are called the *pseudo powers* of u.

They are defined in such a way that

$$\Delta u^{[k]} = k u^{[k-1]}$$
$$\Delta u^{[-k]} = -k u^{[-(k+1)]} \tag{3.23}$$

in complete analogy to the rules for differentiating a power of u with respect to u. These rules can be verified as follows:

$$\Delta u^{[k]} = (u+1)^{[k]} - u^{[k]} = (u+1)u^{[k-1]} - [u - (k-1)]u^{[k-1]}$$
$$= k u^{[k-1]}$$

and

$$\Delta u^{[-k]} = (u+1)^{[-k]} - u^{[-k]} = (u+1)u^{[-(k+1)]} - (u+k+1)u^{[-(k+1)]}$$
$$= -k u^{[-(k+1)]}$$

It can be easily established, by inspection, that

$$u^{[n]}(u-n)^{[m]} = u^{[n+m]} \tag{3.24}$$

where n and m are any positive or negative integers. This is referred to as the *exponential law* for pseudo exponents. It can also be written

$$\frac{u^{[n]}}{u^{[m]}} = (u-m)^{[n-m]} \tag{3.25}$$

The factorial polynomials of Eqs. (3.21) have roots that are the successive positive integers starting with zero. Moreover, for m and n any two positive integers

$$m^{[n]} = \begin{cases} 0 & \text{when } m < n \\ n! & \text{when } m = n \\ \dfrac{m!}{(m-n)!} & \text{when } m > n \end{cases} \tag{3.26}$$

Example 5. Write the polynomial

$$P(u) = u^4 - 3u^2 + 2u + 6 \tag{3.27}$$

in the form

$$P(u) = C_0 u^{[4]} + C_1 u^{[3]} + C_2 u^{[2]} + C_3 u^{[1]} + C_4 \qquad (3.28)$$

From Eq. (3.28) and the definition of $u^{[n]}$

$$P(u) = u\big[(u-1)\{(u-2)[C_0(u-3) + C_1] + C_2\} + C_3\big] + C_4 \qquad (3.29)$$

Hence $P(u) = G_4(u)$, where

$$
\begin{aligned}
G_4(u) &= uG_3(u) + C_4 \\
G_3(u) &= (u-1)G_2(u) + C_3 \\
G_2(u) &= (u-2)G_1(u) + C_2 \\
G_1(u) &= (u-3)G_0(u) + C_1 \\
G_0(u) &= C_0
\end{aligned}
\qquad (3.30)
$$

It follows that the coefficients $C_4, C_3, \ldots, C_0$ are the successive remainders obtained on dividing $P(u)$ by u, the resulting quotient by $u-1$, that quotient by $u-2$, etc. Thus by synthetic division (see Chap. 2)

1	0	−3	2	6	$\big\vert 0$
	0	0	0	0	
1	0	−3	2	$6 = C_4$	$\big\vert 1$
	1	1	−2		
1	1	−2	$0 = C_3$		$\big\vert 2$
	2	6			
1	3	$4 = C_2$			$\big\vert 3$
	3				
$1 = C_0$	$6 = C_1$				

It is clear that the first two lines can be dispensed with. As seen from this table, the successive quotients are

$$
\begin{aligned}
G_3(u) &= u^3 - 3u + 2 \\
G_2(u) &= u^2 + u - 2 \\
G_1(u) &= u + 3 \\
G_0(u) &= 1
\end{aligned}
$$

Substituting the values of the C_k obtained from the above table in Eq. (3.28), one has

$$P(u) = u^{[4]} + 6u^{[3]} + 4u^{[2]} + 6$$

Example 6. Find the value of

$$P(u) = 2u^{[4]} + 3u^{[2]} + u^{[1]} - 7 \qquad \text{at } u = 5$$

Writing $P(u)$ in the form of Eq. (3.29),

$$P(u) = u\big[(u-1)\{(u-2)[2(u-3) + 0] + 3\} + 1\big] - 7$$

one has

$$
\begin{aligned}
G_0(u) &= 2 \\
G_1(u) &= (u-3)G_0(u) + 0 \\
G_2(u) &= (u-2)G_1(u) + 3 \\
G_3(u) &= (u-1)G_2(u) + 1 \\
G_4(u) &= uG_3(u) - 7 = P(u)
\end{aligned}
$$

For $u = 5$, $G_0(5) = 2$, $G_1(5) = (2)(2) = 4$, $G_2(5) = (3)(4) + 3 = 15$, $G_3(5) = (4)(15) + 1 = 61$, $G_4(5) = (5)(61) - 7 = 298 = P(5)$.

This calculation is best carried out using the following computational scheme, which is closely related to synthetic division:

$$
\begin{array}{c}
C_0 = G_0 \longrightarrow u - 3 \longrightarrow (u - 3)G_0 \\
\dfrac{C_1}{} \\
(u - 2)G_1 \longleftarrow u - 2 \longleftarrow G_1 \\
\dfrac{C_2}{G_2} \longrightarrow u - 1 \longrightarrow (u - 1)G_2 \qquad (3.31) \\
\dfrac{C_3}{} \\
uG_3 \longleftarrow u \longleftarrow G_3 \\
\dfrac{C_4}{G_4} = P(u)
\end{array}
$$

For the numerical values above, this becomes

$$
\begin{array}{c}
2 \longrightarrow 2 \longrightarrow 4 \\
0 \\
12 \longleftarrow 3 \longleftarrow \overline{4} \\
\overline{3} \\
\overline{15} \longrightarrow 4 \longrightarrow 60 \\
1 \\
305 \longleftarrow 5 \longleftarrow \overline{61} \\
\dfrac{-7}{298} = P(5)
\end{array}
$$

3.8. Expansion of a Polynomial Using Differences. Any polynomial of degree n in x

$$P_n(x) = a_n x^n + a_{n-1} x^{n-1} + \cdots + a_1 x + a_0 \qquad (3.32)$$

can be written as an nth-degree polynomial in $u = (x - x_0)/h$. This polynomial, as we have seen, can also be written

$$P_n(x) = d_n \frac{u^{[n]}}{n!} + d_{n-1} \frac{u^{[n-1]}}{(n-1)!} + \cdots + d_2 \frac{u^{[2]}}{2!} + d_1 u + d_0 \qquad (3.33)$$

by making use of the factorials of Sec. 3.7.

To determine the coefficients in Eq. (3.33) we take successive differences of $P_n(x)$; thus, by Eqs. (3.23),

$$\Delta P_n(x) = d_n \frac{u^{[n-1]}}{(n-1)!} + d_{n-1} \frac{u^{[n-2]}}{(n-2)!} + \cdots + d_1$$

$$\Delta^2 P_n(x) = d_n \frac{u^{[n-2]}}{(n-2)!} + d_{n-1} \frac{u^{[n-3]}}{(n-3)!} + \cdots + d_2$$

$$\cdots \cdots \cdots \cdots \cdots \cdots \cdots \cdots \cdots \cdots \qquad (3.34)$$

$$\Delta^r P_n(x) = d_n \frac{u^{[n-r]}}{(n-r)!} + d_{n-1} \frac{u^{[n-r-1]}}{(n-r-1)!} + \cdots + d_r$$

$$\cdots \cdots \cdots \cdots \cdots \cdots \cdots \cdots \cdots \cdots$$

$$\Delta^n P_n(x) = d_n$$

Letting $x = x_0$ and $u = (x - x_0)/h = 0$ in Eqs. (3.32) and (3.33), we have $d_0 = P_n(x_0)$, $d_1 = \Delta P_n(x_0)$, and, in general,

$$d_r = \Delta^r P_n(x_0) \qquad r = 0, 1, 2, \ldots, n \qquad (3.35)$$

Therefore

$$P_n(x) = P_n(x_0) + u\,\Delta P_n(x_0) + \frac{u^{[2]}}{2!}\Delta^2 P_n(x_0) + \cdots$$

$$+ \frac{u^{[n]}}{n!}\Delta^n P_n(x_0) = \sum_{r=0}^{n} \frac{u^{[r]}}{r!}\Delta^r P_n(x_0) \qquad (3.36)$$

The original nth-degree polynomial of Eq. (3.32) required for its specification the $n + 1$ coefficients a_k. In the form of Eq. (3.36) the polynomial requires a knowledge of the $n + 1$ differences $\Delta^k P_n(x_0)$.

Example 7. A third-degree polynomial $P(x)$ is passed through the points $(0, -1)$, $(1,1)$, $(2,1)$, and $(3, -2)$. Find its value at $x = 1.2$.

First Method. Let the third-degree polynomial be

$$P(x) = a_3 x^3 + a_2 x^2 + a_1 x + a_0$$

If it is to pass through the above points, this equation must be satisfied for the above four pairs of values for x and y; therefore

$$
\begin{aligned}
a_0 &= -1 \\
a_3 + a_2 + a_1 + a_0 &= 1 \\
8a_3 + 4a_2 + 2a_1 + a_0 &= 1 \\
27a_3 + 9a_2 + 3a_1 + a_0 &= -2
\end{aligned}
\qquad (3.37)
$$

Solving these four equations in four unknowns, we have $a_0 = -1$, $a_1 = \frac{8}{3}$, $a_2 = -\frac{1}{2}$, $a_3 = -\frac{1}{6}$ or

$$P(x) = -\tfrac{1}{6}(x^3 + 3x^2 - 16x + 6) \qquad (3.38)$$

Thus, at $x = 1.2$, by synthetic division

$$
\begin{array}{rrrr|l}
1 & 3 & -16 & 6 & \underline{1.2} \\
  & 1.2 & 5.04 & -13.152 & \\
\hline
1 & 4.2 & -10.96 & -7.152 &
\end{array}
$$

$$P(1.2) = \frac{7.152}{6} = 1.192$$

This method is not recommended, as the solution of a set of linear equations usually involves a considerable amount of computation.

Second Method. Tabulate the values of $P(x)$ given, and form all possible differences, as below:

x	$y = P(x)$	Δy	$\Delta^2 y$	$\Delta^3 y$
0	-1			
		2		
1	1		-2	
		0		-1
2	1		-3	
		-3		
3	-2			

If one sets $x_0 = 0$, the differences required in Eq. (3.36) are seen to be just the leading differences $-1, 2, -2, -1$ in this table. With these differences known, Eq. (3.36) permits us to write down the polynomial. Since $h = 1$ and $x_0 = 0$, $u = x$, and this polynomial is

$$P(x) = -1 + 2x^{[1]} - x^{[2]} - \tfrac{1}{6}x^{[3]} \tag{3.39}$$

therefore the required value of $P(x)$ at $x = 1.2$ is

$$P(1.2) = -1 + (2)(1.2) - (1.2)(0.2) - \tfrac{1}{6}(1.2)(0.2)(-0.8) = 1.192$$

The calculations in the last step can best be performed by using the computational form shown in Example 6.

3.9. Difference Triangles. Suppose one has a table of $y = f(x)$ for equally spaced values of the argument $x_s = x_0 + sh$, like that represented by Table 3.1, and one passes an mth-degree polynomial $P_m(x)$ through the $m + 1$ points (x_0,y_0), (x_1,y_1), . . . , (x_m,y_m). Then, since $P_m(x_r) = y_r$ for $r = 0, 1, 2, . . . , m$, the differences of $P_m(x)$ must of necessity be equal to the differences of y for all tabulated differences in the triangle whose vertices are at y_0, y_m, and $\Delta^m y_0$. This triangle

$$
\begin{array}{ll}
x_0 & y_0 \\
 & \quad \Delta y_0 \\
x_1 & y_1 \quad \cdot \\
 \cdot & \quad \cdot \quad \cdot \\
 \cdot & \quad \cdot \quad \cdot \quad \Delta^m y_0 \\
 \cdot & \quad \cdot \quad \cdot \\
x_{m-1} & y_{m-1} \quad \cdot \\
 & \quad \Delta y_{m-1} \\
x_m & y_m
\end{array}
\tag{3.40}
$$

will be referred to as the *difference triangle with vertex at* $\Delta^m y_0$.

If one speaks of the entries y_k as *zero-order differences*, one observes that the above difference triangle, for an mth-degree polynomial, contains $\tfrac{1}{2}(m + 1)(m + 2)$ differences; however, the specification of any $m + 1$ independent differences in this triangle determines the rest. Thus, for example, the requirement that $\Delta^r P_m(x_0) = \Delta^r y_0$ for $r = 0, 1, 2, . . . ,$ m ensures that the differences agree over the entire difference triangle. This must be true since, given these differences, one can complete the triangle.

3.10. Interpolating Polynomials. Just as a polynomial of the mth degree that passes through $m + 1$ consecutive pairs of values (x_k,y_k) in a table of $y = f(x)$ determines a common difference triangle belonging both to the polynomial and $f(x)$, the difference triangle with vertex at $\Delta^m y_s$ determines uniquely an mth-degree polynomial. This polynomial will be referred to as the *interpolating polynomial associated with* $\Delta^m y_s$. It agrees with $f(x)$ at the $m + 1$ values x_s, x_{s+1}, . . . , x_{s+m} required in the definition of $\Delta^m y_s$. The use of this polynomial as an approximation for $f(x)$ over the interval $x_s \leq x \leq x_{s+m}$ is usually referred to as *polynomial interpolation*. The interval $x_s \leq x \leq x_{s+m}$ covered by the base of the triangle will be termed the *interval of interpolation* for this particular polynomial. Use of the polynomial outside this interval is referred to as *extrapolation*.

As indicated above, the interpolating polynomial associated with $\Delta^m y_s$ may be expressed in terms of any $m + 1$ independent quantities in the difference triangle with vertex at $\Delta^m y_s$.

3.11. Approximating a Tabulated Function by a Polynomial. We wish to determine in this section how well one can predict the intermediate values of a function $f(x)$ from just the tabulated values $y_s = f(x_s)$. Consider first a table with entries $y_0, y_1, \ldots, y_n$ in which the mth differences of y are strictly constant for some m less than n. Comparison of this table of $f(x)$ with a table of the interpolating polynomial

$$P_m(x) = y_0 + u^{[1]} \Delta y_0 + \frac{u^{[2]}}{2!} \Delta^2 y_0 + \cdots + \frac{u^{[m]}}{m!} \Delta^m y_0 \quad (3.41)$$

for the same values $x_s = x_0 + sh$, $s = 0, 1, 2, \ldots, n$ of the arguments leads to the conclusion that they have the same leading differences† $y_0, \Delta y_0, \Delta^2 y_0, \ldots, \Delta^m y_0$ (see Sec. 3.9) and the same constant-difference column $\Delta^m y$. For a table having a constant mth-difference column the knowledge of the leading differences permits one to fill in the entire table uniquely (see Sec. 3.2). Therefore the above table of $P_m(x)$ agrees everywhere with the original table for $f(x)$.

This latter fact can also be established by writing down the interpolating polynomial

$$P_n(x) = \sum_{r=0}^{n} \frac{u^{[r]}}{r!} \Delta^r y_0$$

associated with the difference triangle with vertex at $\Delta^n y_0$. This difference triangle includes the entire table, and hence $y_s = P_n(x_s)$ for $s = 0, 1, 2, \ldots, n$. Since all differences above the mth are zero, $P_n(x)$ reduces to the mth-degree interpolating polynomial of Eq. (3.41); therefore $y_s = P_m(x_s)$ at all the x_s.

This does not mean, however, that $P_m(x)$ is necessarily the same function as $f(x)$, since they do not need to agree for intermediate values of x lying between the tabulated arguments x_s or for values outside the range of tabulation $x_0 \leq x \leq x_n$. In fact, all that can be said is that

† As stated in Sec. 3.2, it is convenient to speak of the tabulated values as *zero-order differences*.

$f(x) = P_m(x) + g(x)$, where $g(x)$ is some function having zeros at x_0, x_1, . . . , x_n. The following example will serve to illustrate this point.

Example 8. Tabulate the function

$$f(x) = x^7 - 21x^6 + 175x^5 - 735x^4 + 1{,}624x^3 - 1{,}764x^2 + 725x + 210 = 0 \quad (3.42)$$

at $x = 0, 1, 2, 3, 4, 5$, and 6. Then determine the interpolating polynomial $P_m(x)$, m less than, or equal to, 6, associated with this table.

One obtains by direct substitution in Eq. (3.42) the table

x	$y = f(x)$	Δy	$\Delta^2 y$
0	210		
		5	
1	215		0
		5	
2	220		0
		5	
3	225		0
		5	
4	230		0
		5	
5	235		0
		5	
6	240		

Computation of the table, however, is speeded by the use of synthetic division (see Chap. 2). Since the first difference is constant, $m = 1$ in Eq. (3.41), and the required interpolating polynomial is

$$P_1(x) = 210 + u^{[1]}\,\Delta y_0 = 5x + 210$$

It is clear that, while $P_1(x) = f(x)$ at $x = 0, 1, 2, . . . , 6$, it will differ markedly at other values of x. For instance, $P_1(0.5) = 212.5$, and $f(0.5) = 293.71$; $P_1(7) = 245$, and $f(7) = 5{,}285$. Nothing, however, in the table itself can be used to predict the failure of $P_1(x)$ to approximate the tabulated function $f(x)$.

3.12. Gregory-Newton Formula for Forward Interpolation. Although, as seen in Sec. 3.11, a table of values of a function $y = f(x)$ for equidistant values of x does not permit one to say anything about the values that y might have at points between these tabulated values, nevertheless, it can usually be presumed that the person making up a table has tabulated sufficient points so that a smooth curve drawn through them gives a faithful picture of the variation in y. We shall show later that any analytic function can be approximated to any given degree of accuracy by a polynomial, provided only that the range is sufficiently small (see Sec. 3.13). It is very likely, therefore, that in any actual table the observation that the mth difference is very nearly constant over a portion of the table is sufficient indication that the function differs only slightly from an mth-degree polynomial in this range.

Let $y = P_m(x)$ represent an mth-degree interpolating polynomial associated with the difference triangle with vertex at $\Delta^m y_0$. By Eq. (3.41) this polynomial may be expressed in terms of $u = (x - x_0)/h$ and the differences of y_0 as

$$y = y_0 + u^{[1]} \Delta y_0 + \frac{1}{2!} u^{[2]} \Delta^2 y_0 + \cdots + \frac{1}{m!} u^{[m]} \Delta^m y_0 \qquad (3.43)$$

where $\qquad u^{[k]} = u(u - 1)(u - 2) \cdots (u - k + 1)$

This is the *Gregory-Newton formula for forward interpolation*. It may be used to interpolate in a table of $y = f(x)$ for any value of x between x_0 and x_m, the interval of interpolation. In terms of u this interval is $0 \leq u \leq m$. Customarily, however, $0 \leq u \leq 1$, so that one is using the interpolating polynomial in the first panel, $x_0 \leq x \leq x_1$, of the above interval of interpolation. This property makes the formula suitable for interpolation near the beginning of a table; however, it also means that the approximation is not as good as when interpolation is near the middle of the interval of interpolation, as it is when one uses the central-difference formulas of Chap. 4.

The computational scheme (3.31) for evaluation of a polynomial in factorial form can be applied directly to the Gregory-Newton formula.

Example 9. Using Table 3.3, find the value of log 1.024. Choosing x_0 to be 1.02,

$$y_0 = 0.00860\ 02 \qquad \Delta y_0 = 0.00423\ 70$$

$$\frac{1}{2!} \Delta^2 y_0 = -0.00002\ 045 \qquad \frac{1}{3!} \Delta^3 y_0 = 0.00000\ 013$$

Since $h = 0.01$,

$$u = \frac{1.024 - 1.02}{0.01} = 0.4 \qquad u - 1 = -0.6 \qquad u - 2 = -1.6$$

and therefore, by Eq. (3.43) and the computational scheme (3.31),

0.00000 013	−1.6	−0.00000 021
		−0.00002 045
0.00001 240	−0.6	−0.00002 066
0.00423 70		
0.00424 940	0.4	0.00169 976
		0.00860 02
	$y =$	0.01030 00

This is the approximate value of $y = \log 1.024$.

3.13. Remainder Term in the Gregory-Newton Formula. As stated in Sec. 3.12, one must have more information about a function than its values at the discrete points given in a table in order to interpolate reliably for a value lying between these tabulated values. In some cases, of course, this additional information is merely the knowledge that the maker of the table considered that sufficient points were given so that, by drawing a smooth curve through the points, one could get an essentially correct picture of the graph of the function. In any case, no quantitative measure of the error involved in using the Gregory-Newton formula can be obtained without some knowledge of the function tabulated. When, in particular, the function is known, it is possible to write down an exact equation for the intermediate values of the function by adding a remainder term to the Gregory-Newton interpolation formula [see Eq. (3.43)].

For the following development it will be assumed that the function $f(x)$ possesses a continuous $(m + 1)$th-order derivative. Although for interpolation x lies in the interval of interpolation, $x_0 \leq x \leq x_m$, one may obtain a more general result that is also applicable to extrapolation (see Sec. 3.16) by permitting x to be either inside or outside this interval.

Since $P_m(x)$ has the same value as $f(x)$ at the tabulated points x_0, x_1, . . . , x_m (see Sec. 3.11),

$$f(x) = P_m(x) + g(x)$$

where $g(x)$ has the roots x_0, x_1, . . . , x_m; therefore

$$f(x) = P_m(x) + K(x) \frac{(x - x_0)(x - x_1) \cdots (x - x_m)}{(m + 1)!} \tag{3.44}$$

It remains to express $K(x)$ in terms of the function $f(x)$. This can be done as follows.

Consider the function

$$W(t) = f(t) - P_m(t) - K(x) \frac{(t - x_0)(t - x_1) \cdots (t - x_m)}{(m + 1)!} \tag{3.45}$$

By virtue of Eq. (3.44) this equation has at least the $m + 2$ real roots x, x_0, x_1, . . . , x_m. Therefore by *Rolle's*† *theorem* $W^{(1)}(t)$ has at least $m + 1$ real roots lying between the smallest and greatest of the above roots. If we restrict ourselves to interpolation, this interval is simply $x_0 \leq t \leq x_m$. Likewise, the second derivative $W^{(2)}(t)$ has at least m real roots in this same interval, and finally the $(m + 1)$th derivative $W^{(m+1)}(t)$ has at least one root $t = \xi$ in the interval x_0 to x_m.

Taking the $(m + 1)$th derivative of each side of Eq. (3.45),

$$W^{(m+1)}(t) = f^{(m+1)}(t) - K(x) \tag{3.46}$$

since $P_m^{(m+1)}(t) = 0$. Therefore, since ξ is a root of the left-hand side,

$$K(x) = f^{(m+1)}(\xi) \tag{3.47}$$

† See any standard advanced calculus text, e.g., Philip Franklin, "Methods of Advanced Calculus," p. 13, McGraw-Hill Book Company, Inc., New York, 1944.

where for interpolation $x_0 < \xi < x_m$. Putting this result back into Eq. (3.44),

$$f(x) = P_m(x) + f^{(m+1)}(\xi)\,\frac{(x - x_0)(x - x_1)\,\cdots\,(x - x_m)}{(m + 1)!}$$

Finally substituting for $P_m(x)$ from Eq. (3.41) and introducing u in the remainder term above,

$$f(x) = y_0 + u^{[1]}\,\Delta y_0 + \frac{1}{2!}\,u^{[2]}\,\Delta^2 y_0 + \cdots$$
$$+ \frac{1}{m!}\,u^{[m]}\,\Delta^m y_0 + \frac{h^{m+1}}{(m + 1)!}\,u^{[m+1]}f^{(m+1)}(\xi) \qquad (3.48)$$

where $x_0 < \xi < x_m$. This is the *Gregory-Newton interpolation formula with remainder term.*

Notice that for any value of m an interval h can be chosen small enough so that the remainder term, and hence the error, is less than any preassigned value, provided only that $f^{(m+1)}(x)$ is bounded in the interval $x_0 < x < x_m$. This is the theoretical justification for the use of polynomial interpolation. How small h must be taken to achieve a given accuracy is determined by this upper bound for the absolute value of $f^{(m+1)}(x)$ in the interval of interpolation.

It is clear from Eq. (3.48) that the accuracy with which the interpolating polynomial given by the Gregory-Newton formula approximates the function $f(x)$ is not necessarily improved, for a particular value of h, by increasing the order m of the polynomial; however, the higher the order, the more rapidly the remainder term, and hence the error, approaches zero as h is decreased.

3.14. Estimating the Numerical Error in Interpolation. In any interpolation using numerical values one must consider the following three classes of error:

1. *Class A error, or truncation error,* is due to replacing the tabulated function $f(x)$ by an interpolating polynomial $P_m(x)$. If $f(x)$ is a known function possessing a continuous $(m + 1)$th derivative, this error for an mth-degree polynomial is given by the remainder term in Eq. (3.48).

2. *Class B error* is due to the fact that the tabulated values of $f(x)$ are expressed by finite decimals and are therefore subject to round-off error. It is clear that this error, which does not include the error introduced in the evaluation of the polynomial, is independent of the way in which the polynomial is expressed. Consideration of this error will be left until Lagrange's expression for the interpolating polynomial is introduced in Chap. 5 (see Sec. 5.3).

3. *Class C error* is due to the necessity of rounding off the intermediate numbers needed in the actual evaluation of the interpolating polynomial of Eq. (3.43). The magnitude of this error is dependent on the exact

way round-off is introduced in the calculations, as discussed in Appendix A. In practice, one may reduce this error to a negligible amount compared to class B by carrying along one more decimal place in the calculations than that used for the tabulated values of $f(x)$.

As shown in Appendix A, the maximum over-all error is just the sum of the errors of the above three classes.

Example 10. Determine the maximum error in the interpolation of Example 9. Since $m = 3$, the class A error is given by the remainder term [Eq. (3.48)]

$$\frac{1}{4!} h^4 u^{[4]} f^{(4)}(\xi)$$

where $x_0 < \xi < x_4$. In Example 9

$$h = 10^{-2} \qquad u = 0.4 \qquad f(x) = \log x$$
$$x_0 = 1.02 \qquad \text{and} \qquad x_4 = 1.05$$

Therefore the error is

$$\frac{10^{-8}}{4!} (0.4)(-0.6)(-1.6)(-2.6) \left[\frac{d^4}{dx^4} \log x \right]_{x=\xi}$$

where $1.02 < \xi < 1.05$.

Since

$$\frac{d^4}{dx^4} \log x = \frac{-3!}{x^4} \log e$$

the class A error ϵ_A in the interpolation must be in the range

$$0.892 \times 10^{-8} < \epsilon_A < 1.002 \times 10^{-8}$$

Ordinarily, one would find only a rough upper limit for $|\epsilon_A|$.

The class B error ϵ_B will be discussed in Chap. 5. Its maximum absolute value turns out to be only slightly larger than the possible round-off error in the individual entries, namely, 5×10^{-8}. The class C error is relatively small since an extra decimal place was used in the calculations.

It is thus clear that the maximum over-all error in the interpolated value $y = 0.01030\ 00$ is less than one unit in the last decimal place.

3.15. Ascending Differences. The *first ascending difference* $\nabla f(x)$ of $f(x)$ is defined by the equation

$$\nabla f(x) = f(x) - f(x - h) \qquad (3.49)$$

Suppose one is dealing with a table of $y = f(x)$ having as entries

$$y_s = f(x_s) = f(x_0 + sh)$$

where s is any integer. Then the ascending differences of y are given by the equations

$$\nabla y_s = f(x_s) - f(x_s - h) = y_s - y_{s-1}$$
$$\nabla^2 y_s = \nabla(\nabla y_s) = \nabla y_s - \nabla y_{s-1}$$
$$\nabla^3 y_s = \nabla(\nabla^2 y_s) = \nabla^2 y_s - \nabla^2 y_{s-1} \qquad (3.50)$$
$$\cdots \cdots \cdots \cdots \cdots \cdots$$
$$\nabla^k y_s = \nabla(\nabla^{k-1} y_s) = \nabla^{k-1} y_s - \nabla^{k-1} y_{s-1} \qquad \text{etc.}$$

These differences may be arranged as in Table 3.8 below and calculated in the same way that the descending differences were in Table 3.1. In fact, all that is involved is a change of notation, and by an inspection of these tables one sees that†

$$\nabla^k y_s = \Delta^k y_{s-k} \tag{3.51}$$

TABLE 3.8. ASCENDING-DIFFERENCE TABLE

x_{-2}	y_{-2}						
x_{-1}	y_{-1}	∇y_{-1}	$\nabla^2 y_0$				
x_0	y_0	∇y_0	$\nabla^2 y_1$	$\nabla^3 y_1$	$\nabla^4 y_2$		
x_1	y_1	∇y_1	$\nabla^2 y_2$	$\nabla^3 y_2$	$\nabla^4 y_3$	$\nabla^5 y_3$	$\nabla^6 y_4$
x_2	y_2	∇y_2	$\nabla^2 y_3$	$\nabla^3 y_3$	$\nabla^4 y_4$	$\nabla^5 y_4$	
x_3	y_3	∇y_3	$\nabla^2 y_4$	$\nabla^3 y_4$			
x_4	y_4	∇y_4					

3.16. Gregory-Newton Formula for Backward Interpolation. Suppose one is interpolating near the end of a table; then some care must be exercised to ensure that the differences called for in the interpolation formula can be obtained from the table. Let y_0 designate the last entry in a table (see Table 3.9) and m be the order of the interpolation desired. Then in order to use formula (3.43) to interpolate between x_{-1} and x_0 it is necessary to use the differences of the tabulated value y_{-m}. This is due to the fact that $\Delta^m y_{-p}$ for $p < m$ does not occur in the table.

One may write Eq. (3.43), for this purpose, in the form

$$y = \bar{y}_0 + \bar{u}\,\Delta\bar{y}_0 + \frac{1}{2!}\,\bar{u}^{[2]}\,\Delta^2\bar{y}_0 + \cdots + \frac{1}{m!}\,\bar{u}^{[m]}\,\Delta^m\bar{y}_0$$

where $\bar{y}_0 = y_{-m}$ and $\bar{u}$ is measured from $\bar{x}_0 = x_{-m}$; however, since

$$\bar{u} = \frac{1}{h}(x - x_{-m}) = \frac{1}{h}(x - x_0) + \frac{1}{h}(x_0 - x_{-m}) = u + m$$

this equation may be written

$$y = y_{-m} + (u + m)\,\Delta y_{-m} + \frac{1}{2!}(u + m)^{[2]}\,\Delta^2 y_{-m} + \cdots$$
$$+ \frac{1}{m!}(u + m)^{[m]}\,\Delta^m y_{-m} \tag{3.52}$$

In either form it represents the interpolating polynomial associated with the difference triangle with vertex at $\Delta^m y_{-m}$. This polynomial passes through the points (x_{-m}, y_{-m}), . . . , (x_{-1}, y_{-1}), and (x_0, y_0). Although this equation would be satisfactory for interpolation, it is not generally employed; instead one usually prefers to express the above interpolating polynomial in terms of the differences $\Delta^k y_{-k}$.

† Scarborough uses Δ_k in place of ∇^k. The latter notation has the advantage of indicating clearly that ∇^k is kth power of the operator ∇ (the nabla operator).

TABLE 3.9

u	x	y						
.	.	.	.	.	.	.	.	.
.	.	.	.	.	.	.	.	.
-6	x_{-6}	y_{-6}		.	.	.	.	.
-5	x_{-5}	y_{-5}	Δy_{-6}	$\Delta^2 y_{-6}$	.	.	.	.
-4	x_{-4}	y_{-4}	Δy_{-5}	$\Delta^2 y_{-5}$	$\Delta^3 y_{-6}$	$\Delta^4 y_{-6}$	$\Delta^5 y_{-6}$	.
-3	x_{-3}	y_{-3}	Δy_{-4}	$\Delta^2 y_{-4}$	$\Delta^3 y_{-5}$	$\Delta^4 y_{-5}$	$\Delta^5 y_{-5}$	$\Delta^6 y_{-6}$
-2	x_{-2}	y_{-2}	Δy_{-3}	$\Delta^2 y_{-3}$	$\Delta^3 y_{-4}$	$\Delta^4 y_{-4}$		
-1	x_{-1}	y_{-1}	Δy_{-2}	$\Delta^2 y_{-2}$	$\Delta^3 y_{-3}$			
0	x_0	y_0	Δy_{-1}					

This can be done by tabulating the entries y_{-k} in reverse order and forming therefrom a new descending-difference table. The result is shown in Table 3.10, in which we note that along with an inversion of the difference table there is a change in sign in all odd-difference columns. One may now interpolate in this table by means of a Gregory-Newton formula employing $u' = -u$ and the leading differences y_0, $-\Delta y_{-1}$, $\Delta^2 y_{-2}$, $-\Delta^3 y_{-3}$, . . . , $(-1)^m \Delta^m y_{-m}$; therefore, by reference to Eq. (3.43),

$$y = y_0 + (-u)^{[1]}(-\Delta y_{-1}) + \frac{1}{2!}(-u)^{[2]}\Delta^2 y_{-2} + \frac{1}{3!}(-u)^{[3]}(-\Delta^3 y_{-3})$$

$$+ \cdots + \frac{1}{m!}(-u)^{[m]}(-1)^m \Delta^m y_{-m} \quad (3.53)$$

TABLE 3.10. THE EFFECT OF INVERTING TABULATED VALUES

$u' = -u$	x	y						
0	x_0	y_0						
1	x_{-1}	y_{-1}	$-\Delta y_{-1}$	$\Delta^2 y_{-2}$	$-\Delta^3 y_{-3}$			
2	x_{-2}	y_{-2}	$-\Delta y_{-2}$	$\Delta^2 y_{-3}$	$-\Delta^3 y_{-3}$	$\Delta^4 y_{-4}$	$-\Delta^5 y_{-5}$	$\Delta^6 y_{-6}$
3	x_{-3}	y_{-3}	$-\Delta y_{-3}$	$\Delta^2 y_{-4}$	$-\Delta^3 y_{-4}$	$\Delta^4 y_{-5}$	$-\Delta^5 y_{-6}$	
4	x_{-4}	y_{-4}	$-\Delta y_{-4}$	$\Delta^2 y_{-5}$	$-\Delta^3 y_{-5}$	$\Delta^4 y_{-6}$	.	
5	x_{-5}	y_{-5}	$-\Delta y_{-5}$	$\Delta^2 y_{-6}$	$-\Delta^3 y_{-6}$	.	.	.
6	x_{-6}	y_{-6}	$-\Delta y_{-6}$	.	.	.	.	.
.	.	.	.	.	.	.		
.	.	.	.	.	.	.		

Since by Eqs. (3.21)

$$(-u)^{[k]} = -u(-u-1)(-u-2) \cdots [-u-(k-1)]$$
$$= (-1)^k u(u+1)(u+2) \cdots (u+k-1) \quad (3.54)$$

$$y = y_0 + u\,\Delta y_{-1} + \frac{1}{2!}u(u+1)\,\Delta^2 y_{-2} + \frac{1}{3!}u(u+1)(u+2)\,\Delta^3 y_{-3}$$

$$+ \cdots + \frac{1}{m!}u(u+1) \cdots (u+m-1)\,\Delta^m y_{-m} \quad (3.55)$$

This is the *Gregory-Newton formula for backward interpolation* in descending-difference notation. With ascending differences and the notation

$$u^{[k]} = u(u + 1) \cdots (u + k - 1) \tag{3.56}$$

this formula becomes

$$y = y_0 + u^{[1]} \nabla y_0 + \frac{1}{2!} u^{[2]} \nabla^2 y_0 + \frac{1}{3!} u^{[3]} \nabla^3 y_0 + \cdots$$

$$+ \frac{1}{m!} u^{[m]} \nabla^m y_0 \tag{3.57}$$

If one lets $u^{[0]} = 1$, then Eq. (3.57) becomes

$$y = \sum_{k=0}^{m} \frac{1}{k!} u^{[k]} \nabla^k y_0 \tag{3.58}$$

From Eq. (3.56),

$$\nabla u^{[k]} = u^{[k]} - (u - 1)^{[k]} = (u + k - 1) u^{[k-1]} - (u - 1) u^{[k-1]}$$

or
$$\nabla u^{[k]} = k u^{[k-1]} \tag{3.59}$$

Applying this formula to the interpolating polynomial $y = P_m(x)$ given by Eq. (3.57), we may easily show that this polynomial has the same ascending differences at $x = x_0$ as $y = f(x)$. By this token, $P_m(x)$ is the interpolating polynomial associated with the difference triangle with vertex at $\nabla^m y_0$ (see Sec. 3.10), and therefore $y_s = P_m(x_s) = y_s$ for $-m \le s \le 0$.

The polynomial given by Eq. (3.52) is associated with the same difference triangle and is hence the same interpolating polynomial as that given by Eq. (3.57). By comparing Eq. (3.52) with Eq. (3.48) it is seen that

$$f(x) = y_{-m} + (u + m) \Delta y_{-m} + \frac{1}{2!} (u + m)^{[2]} \Delta^2 y_{-m} + \cdots$$

$$+ \frac{1}{m!} (u + m)^{[m]} \Delta^m y_{-m} + \frac{1}{(m + 1)!} (u + m)^{[m+1]} h^{m+1} f^{(m+1)}(\xi) \tag{3.60}$$

where $x_{-m} < \xi < x_0$. The same remainder as used in this equation therefore applies to Eq. (3.57), and since $(u + m)^{[m+1]} = u^{[m+1]}$, one may write

$$f(x) = y_0 + u^{[1]} \nabla y_0 + \frac{1}{2!} u^{[2]} \nabla^2 y_0 + \cdots + \frac{1}{m!} u^{[m]} \nabla^m y_0$$

$$+ \frac{1}{(m + 1)!} u^{[m+1]} h^{m+1} f^{(m+1)}(\xi) \tag{3.61}$$

where again $x_{-m} < \xi < x_0$.

Example 11. Using Table 3.4, find the sun's declination d on Feb. 12.

First Method. Let $y_0 = -13°14'45''.0$, and use the ascending differences

$$\nabla y_0 = 40'4''.8$$
$$\nabla^2 y_0 = 55''.5$$
$$\nabla^3 y_0 = -3''.1$$

Since Feb. 11 represents $u = -1$ and Feb. 13 represents $u = 0$, Feb. 12 corresponds to $u = -0.5$; and hence

$$u^{\{1\}} = -0.5$$
$$u^{\{2\}} = (-0.5)(0.5) = -0.25$$
$$u^{\{3\}} = (-0.5)(0.5)(1.5) = -0.375$$

Therefore by Eq. (3.57)

$$d = y = -13°14'45''.0 - (0.5)(40'4''.8) - \tfrac{1}{2}(0.25)(55''.5) + \tfrac{1}{6}(0.375)(3''.1)$$

or
$$d = -13°34'54''.1$$

Clearly a slight modification of the computational scheme used in Example 9 can be employed in these calculations.

Second Method. Let $y_0 = -15°12'9''.8$, and use the descending differences

$$\Delta y_0 = 38'10''.7$$
$$\Delta^2 y_0 = 58''.6$$
$$\Delta^3 y_0 = -3''.1$$

Since Feb. 7 represents $u = 0$, Feb. 12 corresponds to $u = 2.5$, and thus

$$u^{[1]} = 2.5$$
$$u^{[2]} = (2.5)(1.5) = 3.75$$
$$u^{[3]} = (2.5)(1.5)(0.5) = 1.875$$

Therefore by Eq. (3.43)

$$d = -15°12'9''.8 + (2.5)(38'10''.7) + \tfrac{1}{2}(3.75)(58''.6) - \tfrac{1}{6}(1.875)(3''.1) = -13°34'54''.1$$

The answers obtained by the two methods are thus the same. This is due to the fact that one uses the same interpolating polynomial, a third-degree polynomial passing through the last four points in the table.† The first method is to be preferred because of the ease with which one can increase the degree of the interpolating polynomial by adding on extra terms.

Example 12. Derive the interpolation formulas (3.43) and (3.57) by the use of symbolic operators.

Let $P_m(x)$ be an interpolating polynomial of degree m. Then by a natural generalization of Eq. (3.10)

$$P_m(x) = P_m(x_0 + uh) = E^u P_m(x_0) \tag{3.62}$$

and by Eq. (3.11)

$$P_m(x) = (1 + \Delta)^u P_m(x_0)$$

Using the binomial expansion and the fact that all differences above the mth are zero, one has

$$P_m(x) = \left(\sum_{r=0}^{\infty} \frac{u^{[r]}}{r!} \Delta^r \right) P_m(x_0) = \sum_{r=0}^{m} \frac{u^{[r]}}{r!} \Delta^r P_m(x_0) \tag{3.63}$$

† A table with entries $y_s = f(x_s)$ associates a y_s with each x_s, and one may interpret the pair of values (x_s, y_s) as a point in the xy plane.

If this interpolating polynomial is required to have the same leading differences as $f(x)$, i.e., $\Delta^k P_m(x_0) = \Delta^k y_0$, this equation reduces to the Gregory-Newton formula [Eq. (3.43)].

To obtain Eq. (3.57) one needs first to express E in terms of ∇. From the first equation of (3.50)

$$\nabla = 1 - E^{-1} \quad \text{or} \quad E^{-1} = 1 - \nabla \qquad (3.64)$$

Therefore by Eq. (3.62)

$$P_m(x) = (1 - \nabla)^{-u} P_m(x_0)$$

By the binomial expansion for negative fractional powers and the fact that differences above the mth are zero

$$P_m(x) = \left(\sum_{r=0}^{\infty} \frac{u^{[r]}}{r!} \nabla^r \right) P_m(x_0) = \sum_{r=0}^{m} \frac{u^{[r]}}{r!} \nabla^r P_m(x_0) \qquad (3.65)$$

If the ascending differences of $P_m(x_0)$ are the same as those for $f(x_0)$, i.e., if $\nabla^r P_m = \nabla^r y_0$, then this equation is the same as Eq. (3.57).

3.17. Extrapolation. It is sometimes necessary to find the value of a function outside the values of x tabulated. This is done by *extrapolation*, a process differing in theory only slightly from interpolation. In fact, Eqs. (3.43) and (3.57) can be used directly as extrapolation formulas.

Suppose that a table starts with the value y_0 corresponding to the argument x_0 and that one needs to find the value of y at $x = x_0 + uh$, where u is negative. Since Eq. (3.43) represents the interpolating polynomial passing through the points (x_0,y_0), (x_1,y_1), . . . , (x_m,y_m), it can be used to predict the value of y at such a point. Likewise, for points beyond a table Eq. (3.57) can be used with the appropriate positive value of u.

The justification for attributing a polynomial behavior to the function rests, as in interpolation, on auxiliary information about the function. Without some assurance that the values are tabulated for sufficiently small steps h in the argument, the value predicted by any such formula becomes merely a good guess. Even when one knows that the function is sufficiently smooth for a polynomial to fit the curve over the range x_0 to x_m, it is to be expected that the polynomial will deviate widely from the curve if one goes much outside this range. However, if the extrapolation is confined to a distance less than h outside the table, the error in the predicted value of a reasonably smooth function is apt to be nearly as small as in interpolation. Special care should be exercised, however, to make sure that the function is behaving normally near the last tabulated values.

If the function is known, the remainder terms in Eqs. (3.48) and (3.61) will furnish one with an upper and lower limit to the error. The argument ξ is now limited only to the range x to x_m for extrapolation at the beginning of a table and to range x_0 to x for extrapolation at the end of a table.

Whenever one wishes to extrapolate a distance greater than h beyond the limits of the table, one may find it helpful to first extend the entries in the table by extrapolation. Once this is done, the required value can be obtained by interpolating in the extended table. Clearly the greatest error is to be expected in the extrapolation for the new entries. Thus one can estimate the error to be expected by considering the error in these new values.

A semiquantitative estimate of the error introduced by extrapolating for the new entries can be obtained if mth differences are nearly constant and if one has reason to believe that no abrupt change is to be expected in the tabulated function in the range of the extrapolation. One simply equates the error to be expected to the error one would make in extrapolating for the end values from a knowledge of earlier values. The distance of extrapolation should, of course, be nearly equal in the two cases.

Example 13. Using Table 3.3, extrapolate for the values of the log of 1.07, 1.08, and 1.09. Estimate the maximum error to be expected in each case solely from a knowledge of the tabulated values.

Letting $x_0 = 1.06$ and $y_0 = 253.059 \times 10^{-4}$ and using Eq. (3.57),

$$y = [253.059 + 41.166u - \tfrac{1}{2}0.394u(u + 1) + \tfrac{1}{6}0.007u(u + 1)(u + 2)]10^{-4} \quad (3.66)$$

Letting $u = 1, 2$, and 3, one obtains for the next three entries

$$
\begin{aligned}
\log 1.07 &= 0.02938\ 38 \\
\log 1.08 &= 0.03342\ 37 \\
\log 1.09 &= 0.03742\ 63
\end{aligned}
$$

The third-degree interpolating polynomial of Eq. (3.66) is associated with the difference triangle lying above the line in Table 3.11 (this table represents a portion of Table 3.3) and therefore gives exactly the values tabulated for y_{-3}, y_{-2}, y_{-1}, and y_0. Its third difference is seen to be constant and equal to 7×10^{-7}. Thus the value of the polynomial at $u = 1, 2, 3$, etc., could have been obtained by writing down 7 in the remaining portion of the $\nabla^3 y$ column and extending the table as described in Sec. 3.2. One needs, therefore, only to know how nearly the values $\bar{y}_1$, $\bar{y}_2$, $\bar{y}_3$, etc., so obtained (which, of course, are the same as the values found above) agree with the actual values of the logarithm.

Let the actual third differences for the logarithm below the line be $7 + \epsilon_1$, $7 + \epsilon_2$, $7 + \epsilon_3$; then one can extend the table as shown in Table 3.11. Since, as one would judge from Table 3.3, the deviation ϵ_1 of the third difference from 7 is probably $-1, 0$, or 1, it is seen that the extrapolated value for log 1.07 should be in error by 1 or less in the seventh decimal place. Likewise, log 1.08 might be in error by as much as 4 in the last place but is very apt to be closer, since ϵ_1 and ϵ_2 may be of opposite sign. Finally log 1.09 may be off as much as 10 in the seventh decimal place; but again it is likely to be closer.

It is clear from Table 3.11 that the errors in the extrapolated entries can be expected to grow rapidly as one extends the table farther and farther from the initial entries.

If one had extrapolated for the last three values in Table 3.3 from a knowledge of the first four, the errors in the third differences would be $\epsilon_1 = -1$, $\epsilon_2 = 0$, and $\epsilon_3 = 1$. This leads to errors of -1, -3, and -5 in the seventh decimal place of the last three tabulated values.

A more cautious estimate is made by taking all the ϵ's equal to the maximum observed error of 1; then the magnitude of the errors in the extrapolated values, as seen by Table 3.11, are 1,4, 10 in the seventh decimal place.

As a check on these estimates note that the correct values to seven decimal places are

$$\log 1.07 = 0.02938\ 38$$
$$\log 1.08 = 0.03342\ 38$$
$$\log 1.09 = 0.03742\ 65$$

Thus the errors in the extrapolated values are, respectively, 0, -1, and -2 in the seventh decimal place. This corresponds to deviations of $\epsilon_1 = 0$, $\epsilon_2 = -1$, and $\epsilon_3 = 1$ in the third differences.

TABLE 3.11. ACCUMULATION OF ERROR IN EXTRAPOLATION

x	$y = \log x$	∇y	$\nabla^2 y$	$\nabla^3 y$
1.03	0.0128 372			
		41 961		
1.04	0.0170 333		-401	
		41 560		
1.05	0.0211 893		-394	7
		41 166		
1.06	0.0253 059		$\nabla^2 \bar{y}_1 + \epsilon_1$	$7 + \epsilon_1$
		$\nabla \bar{y}_1 + \epsilon_1$		
1.07	$\bar{y}_1 + \epsilon_1 10^{-7}$		$\nabla^2 \bar{y}_2 + \epsilon_1 + \epsilon_2$	$7 + \epsilon_2$
		$\nabla \bar{y}_2 + 2\epsilon_1 + \epsilon_2$		
1.08	$\bar{y}_2 + (3\epsilon_1 + \epsilon_2)10^{-7}$		$\nabla^2 \bar{y}_3 + \epsilon_1 + \epsilon_2 + \epsilon_3$	$7 + \epsilon_3$
		$\nabla \bar{y}_3 + 3\epsilon_1 + 2\epsilon_2 + \epsilon_3$		
1.09	$\bar{y}_3 + (6\epsilon_1 + 3\epsilon_2 + \epsilon_3)10^{-7}$			

PROBLEMS

1. Extend Table 3.2 to $x = 12$ using the fact that the third difference is constant.

2. Find any error that may be present in the following table:

x	$f(x)$
0.0	50.326
0.1	50.113
0.2	49.757
0.3	49.263
0.4	48.837
0.5	47.885
0.6	47.019
0.7	46.033
0.8	44.951
0.9	43.780
1.0	42.532

3. Prove that $\Delta u^{[-k]} = -k u^{[-(k+1)]}$.

4. Find the first descending difference of the following functions:
   (a) e^x (b) $\cos x$
   (c) x^n (d) 2^u
   (e) $\sin^{-1} x$ (f) $e^{u^{[2]}}$

5. Show that
   (a) $\Delta f(x)g(x) = f(x)\,\Delta g(x) + g(x + h)\,\Delta f(x)$
   (b) $\Delta^n u^n = n!$ (c) $\Delta^n a^x = (a^h - 1)^n a^x$

6. Find the first ascending difference of the functions in Prob. 4.

7. Derive the formulas analogous to those in Prob. 5 using ascending differences.

8. While it is true that $\Delta 2^u = 2^u$, show that there is no function of u (or x) such that $\nabla F(u) = F(u)$, except the trivial solution $F(u) = 0$.

9. Determine the value of the polynomial $P(u) = 3u^{[5]} + u^{[3]} - 7u^{[2]} + 10u - 8$ at $u = 4$.

10. Write the polynomial $P(u) = u^4 - 3u^3 + u^2 - 5u - 1$ in terms of factorials, as in Prob. 9. (See Example 5 for a rapid method of accomplishing this.)

11. Devise a scheme for obtaining an expansion in powers of u given the factorial form of the polynomial.

12. By making use of the properties of determinants, show that

$$
\begin{vmatrix}
1 & x & x^2 & \cdots & x^n & y \\
1 & x_0 & x_0^2 & \cdots & x_0^n & y_0 \\
1 & x_1 & x_1^2 & \cdots & x_1^n & y_1 \\
\multicolumn{6}{c}{\dotfill} \\
1 & x_n & x_n^2 & \cdots & x_n^n & y_n
\end{vmatrix} = 0
$$

represents a polynomial of the nth degree passing through the $n + 1$ points (x_0, y_0), (x_1, y_1), $\ldots$, (x_n, y_n).

13. Prove that

$$
\sum_{s=0}^{n} (-1)^s s^k \frac{n!}{s!(n-s)!} = \begin{cases} 0 & 0 \le k \le n - 1 \\ n! & k = n \end{cases}
$$

14. Reduce the interval in the table of Example 8 to half, and, using again a sixth-degree polynomial for interpolation, find the value of $f(2.25)$. Compare it with the correct value obtained from Eq. (3.42).

15. Evaluate the remainder term for the interpolation of Example 8.

16. Prove Eq. (3.51) using the operator techniques of Sec. 3.5.

17. Show directly that the mth-degree polynomial of Eq. (3.57) passes through the $m + 1$ points (x_{-m}, y_{-m}), (x_{-m+1}, y_{-m+1}), $\ldots$, (x_{-1}, y_{-1}), and (x_0, y_0).

CHAPTER 4

CENTRAL-DIFFERENCE INTERPOLATION FORMULAS

When one uses the Gregory-Newton formula for forward interpolation [Eq. (3.43)] to determine the value of $f(x)$, one approximates $f(x)$ by an mth-degree polynomial that fits the tabulated values $y = f(x_s)$ at x_0, $x_1, \ldots, x_m$. Since the subscripting of x_s is purely relative, one usually selects the reference value x_0 so that $x_0 < x < x_1$. This means that one is interpolating near the beginning of the interval of interpolation, which in this case is the interval $x_0 < x < x_m$.† As one would suspect, and as will be shown later in this chapter, the interpolating polynomial usually approximates $f(x)$ less accurately near the ends than near the middle of this interval of interpolation. In this chapter, therefore, we turn to those expressions for the interpolating polynomial that lead naturally to interpolation near the middle of the interval. These are the so-called central-difference formulas. We also consider formulas that make use of the average of two neighboring interpolating polynomials.

4.1. Central-difference Notation. It is convenient to express the formulas of this chapter in terms of central differences. The *first central difference* of $f(x)$ is defined as

$$\delta f(x) = f\left(x + \frac{h}{2}\right) - f\left(x - \frac{h}{2}\right) \tag{4.1}$$

If one introduces the convention that $y_r = f(x_r)$, where $x_r = x_0 + rh$ and r is not restricted to integral values, one may express Eq. (4.1) in the form

$$\delta y_r = y_{r+\frac{1}{2}} - y_{r-\frac{1}{2}} \tag{4.2}$$

By treating the differences themselves as functions and assuming the associative law of multiplication for δ we have as defining equations for the higher differences

$$\delta^k f(x) = \delta^{k-1} f\left(x + \frac{h}{2}\right) - \delta^{k-1} f\left(x - \frac{h}{2}\right) \tag{4.3}$$

or alternatively,

$$\delta^k y_r = \delta^{k-1} y_{r+\frac{1}{2}} - \delta^{k-1} y_{r-\frac{1}{2}} \tag{4.4}$$

† Similarly in the Gregory-Newton backward formula the interval of interpolation is $x_{-m} < x < x_0$, but one customarily selects x_0 so that $x_{-1} < x < x_0$.

If $f(x)$ is tabulated at $y_s = f(x_s)$, where s assumes only integral values, and if a difference table is constructed on the basis of these values, it will have the general appearance of Table 4.1. It is clear from this

TABLE 4.1. A CENTRAL-DIFFERENCE TABLE

u	x	y	δy	$\delta^2 y$	$\delta^3 y$	$\delta^4 y$	$\delta^5 y$	$\delta^6 y$
-3	x_{-3}	y_{-3}						
-2	x_{-2}	y_{-2}	$\delta y_{-\frac{5}{2}}$	$\delta^2 y_{-2}$				
-1	x_{-1}	y_{-1}	$\delta y_{-\frac{3}{2}}$	$\delta^2 y_{-1}$	$\delta^3 y_{-\frac{3}{2}}$	$\delta^4 y_{-1}$		
0	x_0	y_0	$\delta y_{-\frac{1}{2}}$	$\delta^2 y_0$	$\delta^3 y_{-\frac{1}{2}}$	$\delta^4 y_0$	$\delta^5 y_{-\frac{1}{2}}$	$\delta^6 y_0$
1	x_1	y_1	$\delta y_{\frac{1}{2}}$	$\delta^2 y_1$	$\delta^3 y_{\frac{1}{2}}$	$\delta^4 y_1$	$\delta^5 y_{\frac{1}{2}}$	
2	x_2	y_2	$\delta y_{\frac{3}{2}}$	$\delta^2 y_2$	$\delta^3 y_{\frac{3}{2}}$			
3	x_3	y_3	$\delta y_{\frac{5}{2}}$					

table that only those odd differences which have subscripts of the form $s + \frac{1}{2}$, where s assumes integral values, occur in a table of central differences. Likewise, only those even differences occur which have integral subscripts.

Table 4.1 differs from the ascending- and descending-difference tables merely in the notation used for the differences. In fact, by comparing this table with Tables 3.1 and 3.8 we see that

$$\delta^k y_r = \Delta^k y_{r-k/2} = \nabla^k y_{r+k/2} \tag{4.5}$$

This latter result also follows from the operator equations

$$\begin{aligned} \Delta &= E - 1 \\ \nabla &= 1 - E^{-1} \\ \delta &= E^{\frac{1}{2}} - E^{-\frac{1}{2}} \end{aligned} \tag{4.6}$$

obtained directly from the definitions, since from these

$$\delta = \Delta E^{-\frac{1}{2}} = \nabla E^{\frac{1}{2}} \tag{4.7}$$

Thus the kth power of these operators applied to y_r gives

$$\delta^k y_r = \Delta^k E^{-k/2} y_r = \nabla^k E^{k/2} y_r$$

which, by making use of the property of the operator E, reduce to Eqs. (4.5).

An interesting property of the three difference operators is the following: let $P_m(x) = \Sigma_{r=0}^m A_r x^r$; then

$$\Delta^m P_m(x) = \nabla^m P_m(x) = \delta^m P_m(x) = h^m D^m P_m(x) = h^m m! A_m \tag{4.8}$$

where D represents the operation of taking the derivative with respect to x. This result follows immediately from Eqs. (4.7) and (3.20) in view of the commutative property of the operators and the fact that any power of E operating on a constant leaves the constant unchanged.

4.2. Polynomials Associated with Null Triangles. It is convenient for the purposes of this chapter to associate with every difference of degree k a polynomial of degree $k + 1$ which has these two properties:

1. Its difference triangle (that with vertex at the given difference) is a *null triangle;* by this is meant that all differences within the triangle are zero—this includes the tabulated values.

2. Its $(k + 1)$th difference is unity.

Since the first condition determines the $k + 1$ roots of the polynomial and the second condition determines the constant multiplier, the polynomial associated with each particular difference is unique.

The factorial polynomial

$$(u - s)^{[k+1]} = (u - s)(u - s - 1) \cdots (u - s - k)$$

vanishes at $u = s, s + 1, \ldots, s + k$, and therefore its difference table contains a null triangle with apex at $\Delta^k y_s$. The $(k + 1)$th difference of the above factorial is, by Eq. (4.8), equal to $(k + 1)!$. Obviously then

$$\frac{1}{(k + 1)!} (u - s)^{[k+1]} \sim \text{null triangle with apex at } \Delta^k y_s \qquad (4.9)$$

and in central-difference notation

$$\frac{1}{(k + 1)!} \left(u + \frac{k}{2} - s \right)^{[k+1]} \sim \text{null triangle with apex at } \delta^k y_s \quad (4.10)$$

4.3. The Forward-interpolation Formula of Gauss. Suppose a function $f(x)$ is tabulated at $y_s = f(x_s)$, where $x_s = x_0 + sh$ and s assumes integral values, and one wishes to construct a polynomial $P_m(x)$ having in common with $f(x)$ the first $m + 1$ differences below:

$$y_0, \delta y_{\frac{1}{2}}, \delta^2 y_0, \delta^3 y_{\frac{1}{2}}, \ldots, \delta^{2k-1} y_{\frac{1}{2}}, \delta^{2k} y_0, \ldots \qquad (4.11)$$

It is readily seen by inspection of the central-difference table (see Table 4.1) that, given these differences, one can complete the entire difference triangle whose vertex is at the last difference employed.

If one introduces the operators†

$$\Delta_k = \begin{cases} \delta^k & \text{when } k \text{ is even} \\ \delta^k E^{\frac{1}{2}} & \text{when } k \text{ is odd} \end{cases} \qquad (4.12)$$

one may write for the $m + 1$ conditions on $P_m(x)$ that

$$\Delta_k P_m(x_0) = \Delta_k y_0 \qquad k = 0, 1, 2, \ldots, m \qquad (4.13)$$

† Note that $\Delta_0 = \delta^0 = 1$.

Assuming that the interpolating polynomial is linear in these differences and introducing the variable $u = (x - x_0)/h$, one may write

$$y = P_m(x) = \sum_{r=0}^{m} a_r(u) \, \Delta_r y_0 \qquad (4.14)$$

The $a_r(u)$ are required to be, as in the Gregory-Newton formulas, independent of the choice of m or the numerical values of the differences $\Delta_k y_0$. For this to be true it is clear that $a_k(u)$ must be a polynomial of degree k.

If one takes the difference corresponding to Δ_k of both sides of Eq. (4.14) and sets $u = 0$, one obtains, by virtue of Eq. (4.13), as conditions on the $a_k(u)$

$$\Delta_k P_m(x_0) = \sum_{r=0}^{m} [\Delta_k a_r(0)] \, \Delta_r y_0 = \Delta_k y_0$$

Since this is to hold for arbitrary values of the differences,

$$\Delta_k a_r(0) = \delta_{kr} = \begin{cases} 1 & \text{when } k = r \\ 0 & \text{when } k \neq r \end{cases} \qquad (4.15)$$

Consider a particular polynomial $a_r(u)$ and the value of its differences in the difference triangle with vertex at $\Delta_{r-1} y_0$.† Since the differences $\Delta_s a_r(0)$, $s = 0, 1, 2, \ldots, r - 1$ are all zero by Eq. (4.15), and since one can complete this difference triangle from just these r differences, one sees that all the differences within the triangle are zero. Thus Eq. (4.15) requires that $a_r(u)$ be the polynomial associated with the null triangle with apex at $\Delta_{r-1} y_0$.

By Eq. (4.12), therefore, $a_{2k-1}(u)$ is associated with $\delta^{2k-2} y_0$ and $a_{2k}(u)$ with $\delta^{2k-1} y_{\frac{1}{2}}$; therefore by Eq. (4.10)

$$a_{2k-1}(u) = \frac{1}{(2k - 1)!} (u + k - 1)^{[2k-1]}$$

and
$$a_{2k}(u) = \frac{1}{(2k)!} (u + k - 1)^{[2k]} \qquad (4.16)$$

The mth-degree interpolating polynomial thus becomes, by Eqs. (4.12), (4.14), and (4.16),

$$y = y_0 + u \, \delta y_{\frac{1}{2}} + \frac{1}{2!} u^{[2]} \, \delta^2 y_0 + \frac{1}{3!} (u + 1)^{[3]} \, \delta^3 y_{\frac{1}{2}} + \frac{1}{4!} (u + 1)^{[4]} \, \delta^4 y_0$$

$$+ \cdots + \frac{1}{(2k - 1)!} (u + k - 1)^{[2k-1]} \, \delta^{2k-1} y_{\frac{1}{2}}$$

$$+ \frac{1}{(2k)!} (u + k - 1)^{[2k]} \, \delta^{2k} y_0 + \cdots \qquad (4.17)$$

† Only the position of $\Delta_{r-1} y_0$, not its numerical value, is involved in this terminology.

where the series is terminated after the mth difference. This is the *forward-interpolation formula of Gauss* and is briefly designated as Gauss (I).

As stated earlier, one can complete the difference triangle with vertex at $\Delta_m y_0$ by using only the differences given in Eq. (4.13). Thus the polynomial $y = P_m(x)$ agrees with $f(x)$ over this triangle. This means, with regard to the tabulated values, that $y_s = P_m(x_s) = f(x_s)$ for $-k \leq s \leq k$ when $m = 2k$, and for $-k \leq s \leq k + 1$ when $m = 2k + 1$. If, as is ordinarily the case, one chooses x_0 in each interpolation in such a way that $x_0 < x < x_1$, then $0 < u < 1$, and one is using the polynomial $P_m(x)$ near the middle of its interval of interpolation above.

4.4. The Backward-interpolation Formula of Gauss. If one chooses to express an interpolating polynomial in terms of the central differences of y_0 and $y_{-\frac{1}{2}}$, rather than in terms of the differences of y_0 and $y_{\frac{1}{2}}$, one needs simply to redefine the Δ_k of the last section. Thus if

$$\Delta_k = \begin{cases} \delta^k & \text{when } k \text{ is even} \\ \delta^k E^{-\frac{1}{2}} & \text{when } k \text{ is odd} \end{cases} \tag{4.18}$$

the mth-degree interpolating polynomial of Eq. (4.14) will have the required differences $\Delta_k y_0$, provided, again, that Eq. (4.15) holds. Moreover, these equations are seen to hold, provided, as before, that $a_r(u)$ is the polynomial associated with the null triangle with vertex at $\Delta_{r-1} y_0$.

By Eq. (4.18), therefore, $a_{2k}(u)$ is associated with $\delta^{2k-1} y_{-\frac{1}{2}}$ and $a_{2k+1}(u)$ with $\delta^{2k} y_0$; hence, by Eq. (4.10),

$$a_{2k} = \frac{1}{(2k)!} (u + k)^{[2k]}$$

$$a_{2k+1} = \frac{1}{(2k + 1)!} (u + k)^{[2k+1]} \tag{4.19}$$

and

The mth-degree interpolating polynomial of Eq. (4.14) then becomes, by Eqs. (4.18) and (4.19),

$$y = y_0 + u\, \delta y_{-\frac{1}{2}} + \frac{1}{2!} (u + 1)^{[2]}\, \delta^2 y_0 + \frac{1}{3!} (u + 1)^{[3]}\, \delta^3 y_{-\frac{1}{2}} + \cdots$$

$$+ \frac{1}{(2k)!} (u + k)^{[2k]}\, \delta^{2k} y_0 + \frac{1}{(2k + 1)!} (u + k)^{[2k+1]}\, \delta^{2k+1} y_{-\frac{1}{2}}$$

$$+ \cdots \tag{4.20}$$

where again the series is terminated after the term involving the mth difference. This is the *backward-interpolation formula of Gauss* and is briefly designated as Gauss (II).

The interpolating polynomial represented by Eq. (4.20) fits $f(x)$ over the difference triangle with vertex at $\delta^m y_0$ for m even and at $\delta^m y_{-\frac{1}{2}}$ for m odd. Thus the interval of interpolation (see Sec. 3.9) is x_{-k} to x_k if $m = 2k$

and $x_{-(k+1)}$ to x_k if $m = 2k + 1$. For m even, one obtains the same polynomial from the backward- as from the forward-interpolation formula of Gauss, namely, that associated with the difference triangle with vertex at $\delta^m y_0$.

4.5. Stirling's Interpolation Formula. If one writes down the average of interpolating polynomials represented by the forward- and the backward-interpolation formulas of Gauss [see Eqs. (4.17) and (4.20)], one obtains *Stirling's interpolation formula*,

$$y = y_0 + u\,\mu\delta y_0 + \frac{1}{2!}\,u^2\,\delta^2 y_0 + \frac{1}{3!}\,u(u^2 - 1)\,\mu\delta^3 y_0$$

$$+ \frac{1}{4!}\,u^2(u^2 - 1)\,\delta^4 y_0 + \cdots + \frac{1}{(2k - 1)!}\,u(u^2 - 1)(u^2 - 4)$$

$$\cdots\,[u^2 - (k - 1)^2]\,\mu\delta^{2k-1} y_0 + \frac{1}{(2k)!}\,u^2(u^2 - 1)(u^2 - 4)$$

$$\cdots\,[u^2 - (k - 1)^2]\,\delta^{2k} y_0 + \cdots \qquad (4.21)$$

where
$$\mu\delta^{2k-1} y_0 = \tfrac{1}{2}(\delta^{2k-1} y_{\frac{1}{2}} + \delta^{2k-1} y_{-\frac{1}{2}}) \qquad (4.22)$$

is termed a *mean difference*.

Equation (4.21) makes use of the fact that

$$(u + k - 1)^{[2k-1]} = u(u^2 - 1)(u^2 - 4)\,\cdots\,[u^2 - (k - 1)^2] \qquad (4.23)$$

and

$$\tfrac{1}{2}[(u + k - 1)^{[2k]} + (u + k)^{[2k]}]$$
$$= \tfrac{1}{2}[(u - k)(u + k - 1)^{[2k-1]}$$
$$+ (u + k)(u + k - 1)^{[2k-1]}]$$
$$= u(u + k - 1)^{[2k-1]}$$
$$= u^2(u^2 - 1)(u^2 - 4)\,\cdots\,[u^2 - (k - 1)^2] \qquad (4.24)$$

The operator μ is called the *mean* and is defined by the operator equation

$$\mu = \tfrac{1}{2}(E^{\frac{1}{2}} + E^{-\frac{1}{2}}) \qquad (4.25)$$

Thus, for example,

$$\mu f(x) = \tfrac{1}{2}\left[f\left(x + \frac{h}{2}\right) + f\left(x - \frac{h}{2}\right)\right] \qquad (4.26)$$

$$\mu y_r = \tfrac{1}{2}(y_{r+\frac{1}{2}} + y_{r-\frac{1}{2}}) \qquad (4.27)$$

and
$$\mu\delta^k y_s = \tfrac{1}{2}(\delta^k y_{s+\frac{1}{2}} + \delta^k y_{s-\frac{1}{2}}) \qquad (4.28)$$

Note that by Eqs. (4.6) and (4.25)

$$\mu^2 = 1 + \frac{\delta^2}{4} \qquad (4.29)$$

When the last term in the right-hand side of Eq. (4.21) contains an even difference $\delta^m y_0$, it represents the same polynomial as obtained by either the forward- or backward-interpolation formulas of Gauss. The interval of interpolation (see Sec. 3.10) in this case is $-m/2 < u < m/2$, the interval corresponding to the base of the difference triangle whose vertex is at $\delta^m y_0$. When the last term contains the mean difference $\mu \delta^m y_0$, where m is odd, the approximating polynomial given by Eq. (4.21) represents the average of the two interpolating polynomials corresponding to the two interpolation formulas of Gauss. These two interpolating polynomials are associated with the two difference triangles, one with vertex at $\delta^m y_{\frac{1}{2}}$ and the other with vertex at $\delta^m y_{-\frac{1}{2}}$. The average of these polynomials agrees with $f(x)$ only at the $m = 2k + 1$ values of x between x_{-k} and x_k, but it is influenced by the values of $f(x)$ at $x_{-(k+1)}$ and x_{k+1}. Thus in either case, m even or m odd, Stirling's formula represents an interpolating polynomial that is based on tabulated values symmetrically placed with respect to x_0.

Example 1. Using Table 4.2, find the value of sin 1.808.

TABLE 4.2

x	$y = \sin x$	Δy	$\Delta^2 y$	$\Delta^3 y$	$\Delta^4 y$
1.70	0.9916 6481				
		−27 7504			
1.72	0.9888 8977		−3 9555		
		−31 7059		128	
1.74	0.9857 1918		−3 9427		14
		−35 6486		142	
1.76	0.9821 5432		−3 9285		16
		−39 5771		158	
1.78	0.9781 9661		−3 9127		17
		−43 4898		175	
1.80	0.9738 4763		−3 8952		14
		−47 3850		189	
1.82	0.9691 0913		−3 8763		15
		−51 2613		204	
1.84	0.9639 8300		−3 8559		18
		−55 1172		222	
1.86	0.9584 7128		−3 8337		14
		−58 9509		236	
1.88	0.9525 7619		−3 8101		
		−62 7610			
1.90	0.9463 0009				

The interval in this table is $h = 0.02$. If one lets $x_0 = 1.80$,

$$y_0 = 0.97384\ 763$$

$$\mu \delta y_0 = \frac{-0.00434\ 898 - 0.00473\ 850}{2} = -0.00454\ 374$$

$$\delta^2 y_0 = -0.00038\ 952 \qquad \mu \delta^3 y_0 = 0.00000\ 182$$

$$\delta^4 y_0 = 0.00000\ 014$$

$$u = \frac{x - x_0}{h} = \frac{1.808 - 1.80}{0.02} = 0.4$$

$$u^2 = 0.16 \qquad u(u^2 - 1) = -0.336 \qquad u^2(u^2 - 1) = 0.1344$$

Therefore, applying Stirling's formula [see Eq. (4.21)],

$$y = 0.97384\ 763 - 0.00181\ 7496 - 0.00003\ 1162 - 0.00000\ 0102$$
$$- 0.00000\ 0001 = 0.97199\ 887$$

Observe that the third difference affects the answer only 10 units in the eighth decimal place and that the fourth difference does not influence this decimal place at all.   This illustrates the rapid convergence of Stirling's formula.

4.6. Bessel's Interpolation Formula.   One may use the backward-interpolation formula of Gauss to interpolate relative to y_1 instead of y_0 by letting $y_s = y'_{s-1}$ and introducing $u' = (x - x_1)/h$.   In terms of the primed quantities the formula appears as

$$y = y'_0 + u'\ \delta y'_{-\frac{1}{2}} + \frac{1}{2!}\ (u' + 1)^{[2]}\ \delta^2 y'_0 + \frac{1}{3!}\ (u' + 1)^{[3]}\ \delta^3 y'_{-\frac{1}{2}} + \cdots$$

in keeping with Eq. (4.20).   Since $\delta^k y'_s = \delta^k y_{s+1}$ and $u' = u - 1$, one may write

$$y = y_1 + (u - 1)\ \delta y_{\frac{1}{2}} + \frac{1}{2!}\ (u)^{[2]}\ \delta^2 y_1 + \frac{1}{3!}\ (u)^{[3]}\ \delta^3 y_{\frac{1}{2}}$$
$$+ \frac{1}{4!}\ (u + 1)^{[4]}\ \delta^4 y_1 + \cdots + \frac{1}{(2k)!}\ (u + k - 1)^{[2k]}\ \delta^{2k} y_1$$
$$+ \frac{1}{(2k + 1)!}\ (u + k - 1)^{[2k+1]}\ \delta^{2k+1} y_{\frac{1}{2}} + \cdots \quad (4.30)$$

If one now averages the y obtained by using this formula with the y obtained from the first formula of Gauss [see Eq. (4.17)], one obtains the interpolation formula

$$y = \mu y_{\frac{1}{2}} + (u - \tfrac{1}{2})\ \delta y_{\frac{1}{2}} + \frac{1}{2!}\ u^{[2]}\ \mu \delta^2 y_{\frac{1}{2}} + \frac{1}{3!}\ (u - \tfrac{1}{2}) u^{[2]}\ \delta^3 y_{\frac{1}{2}}$$
$$+ \frac{1}{4!}\ (u + 1)^{[4]}\ \mu \delta^4 y_{\frac{1}{2}} + \frac{1}{5!}\ (u - \tfrac{1}{2})(u + 1)^{[4]}\ \delta^5 y_{\frac{1}{2}} + \cdots$$
$$+ \frac{1}{(2k)!}\ (u + k - 1)^{[2k]}\ \mu \delta^{2k} y_{\frac{1}{2}}$$
$$+ \frac{1}{(2k + 1)!}\ (u - \tfrac{1}{2})(u + k - 1)^{[2k]}\ \delta^{2k+1} y_{\frac{1}{2}} + \cdots \quad (4.31)$$

where the series is terminated at any desired difference.   This equation follows from the fact that

$$\tfrac{1}{2}[(u + k)^{[2k+1]} + (u + k - 1)^{[2k+1]}]$$
$$= \tfrac{1}{2}[(u + k) + (u - k - 1)](u + k - 1)^{[2k]}$$
$$= (u - \tfrac{1}{2})(u + k - 1)^{[2k]} \quad (4.32)$$

(See Sec. 3.7.)

Equation (4.31) is *Bessel's interpolation formula.*   It can be written more symmetrically by introducing the variable

$$v = u - \tfrac{1}{2} \quad (4.33)$$

Noting that

$$(u + k - 1)^{[2k]} = (v + k - \tfrac{1}{2})^{[2k]}$$

$$= (v^2 - \tfrac{1}{4})(v^2 - \tfrac{9}{4}) \cdots \left[v^2 - \frac{(2k - 1)^2}{4} \right] \quad (4.34)$$

one may, in fact, write Bessel's equation in the form

$$y = \mu y_{\frac{1}{2}} + v \, \delta y_{\frac{1}{2}} + \frac{1}{2!} (v^2 - \tfrac{1}{4}) \mu \delta^2 y_{\frac{1}{2}} + \frac{1}{3!} v(v^2 - \tfrac{1}{4}) \delta^3 y_{\frac{1}{2}}$$

$$+ \frac{1}{4!} (v^2 - \tfrac{1}{4})(v^2 - \tfrac{9}{4}) \mu \delta^4 y_{\frac{1}{2}} + \frac{1}{5!} v(v^2 - \tfrac{1}{4})(v^2 - \tfrac{9}{4}) \delta^5 y_{\frac{1}{2}}$$

$$+ \cdots + \frac{1}{(2k)!} (v^2 - \tfrac{1}{4})(v^2 - \tfrac{9}{4}) \cdots \left[v^2 - \frac{(2k - 1)^2}{4} \right] \mu \delta^{2k} y_{\frac{1}{2}}$$

$$+ \frac{1}{(2k + 1)!} v(v^2 - \tfrac{1}{4})(v^2 - \tfrac{9}{4}) \cdots \left[v^2 - \frac{(2k - 1)^2}{4} \right] \delta^{2k+1} y_{\frac{1}{2}}$$

$$+ \cdots \quad (4.35)$$

where for mth-order interpolation one terminates the series with the mth difference.

Example 2. Using Bessel's formula and Table 4.2, find the value of sin 1.808.
As in Example 1, let $x_0 = 1.80$; then, since $h = 0.02$, one has $u = 0.4$ and $v = -0.1$. Moreover,

$$\mu y_{\frac{1}{2}} = \tfrac{1}{2}(y_0 + y_1) = 0.97147\ 838$$
$$\delta y_{\frac{1}{2}} = -0.00473\ 850$$
$$\mu \delta^2 y_{\frac{1}{2}} = \tfrac{1}{2}(\delta^2 y_0 + \delta^2 y_1) = -0.00038\ 8575$$
$$\delta^3 y_{\frac{1}{2}} = 0.00000\ 189$$
$$\mu \delta^4 y_{\frac{1}{2}} = \tfrac{1}{2}(\delta^4 y_0 + \delta^4 y_1) = 0.00000\ 0145$$

Therefore, by Eq. (4.35),

$$y = 0.97147\ 838 + (-0.1)(-0.00473\ 850) + \tfrac{1}{2}(-0.24)(-0.00038\ 8575)$$
$$+ \tfrac{1}{6}(-0.1)(-0.24)(0.00000\ 189) + \tfrac{1}{24}(-0.24)(-2.24)(0.00000\ 0145)$$

or, multiplying out the individual terms,

$$y = 0.97147\ 838 + 0.00047\ 3850 + 0.00004\ 6629 + 0.00000\ 0008 + 0.00000\ 0003$$
$$= 0.97199\ 887$$

The effect of the last term is negligible, as in Example 1. In addition, the next to the last term affects the answer only one unit in the eighth decimal place.

4.7. The Laplace-Everett Formula. One can eliminate the odd differences in the first interpolation formula of Gauss [see Eq. (4.17)] by the use of the relation

$$\delta^{2k+1} y_{\frac{1}{2}} = \delta^{2k} y_1 - \delta^{2k} y_0 \quad (4.36)$$

When this is done, one obtains the expression

$$y = (1 - u)y_0 + uy_1 + \left[\frac{1}{2!} u^{[2]} - \frac{1}{3!} (u + 1)^{[3]}\right] \delta^2 y_0$$

$$+ \frac{1}{3!} (u + 1)^{[3]} \delta^2 y_1 + \left[\frac{1}{4!} (u + 1)^{[4]} - \frac{1}{5!} (u + 2)^{[5]}\right] \delta^4 y_0$$

$$+ \frac{1}{5!} (u + 2)^{[5]} \delta^4 y_1 + \cdots + \left[\frac{1}{(2k)!} (u + k - 1)^{[2k]}\right.$$

$$\left. - \frac{1}{(2k + 1)!} (u + k)^{[2k+1]}\right] \delta^{2k} y_0$$

$$+ \frac{1}{(2k + 1)!} (u + k)^{[2k+1]} \delta^{2k} y_1 + \cdots$$

If one introduces the variable $w = 1 - u$, then (see Sec. 3.7)

$$\frac{1}{(2k)!} (u + k - 1)^{[2k]} - \frac{1}{(2k + 1)!} (u + k)^{[2k+1]}$$

$$= [(2k + 1) - (u + k)] \frac{1}{(2k + 1)!} (u + k - 1)^{[2k]}$$

$$= - \frac{1}{(2k + 1)!} (u + k - 1)^{[2k+1]}$$

$$= \frac{-1}{(2k + 1)!} (k - w)^{[2k+1]} = \frac{1}{(2k + 1)!} (w + k)^{[2k+1]}$$

$$= \frac{1}{(2k + 1)!} w(w^2 - 1)(w^2 - 4) \cdots (w^2 - k^2)$$

Therefore, one may write

$$y = wy_0 + \frac{1}{3!} w(w^2 - 1) \delta^2 y_0 + \frac{1}{5!} w(w^2 - 1)(w^2 - 4) \delta^4 y_0 + \cdots$$

$$+ \frac{1}{(2k + 1)!} w(w^2 - 1)(w^2 - 4) \cdots (w^2 - k^2) \delta^{2k} y_0 + \cdots$$

$$+ uy_1 + \frac{1}{3!} u(u^2 - 1) \delta^2 y_1 + \frac{1}{5!} u(u^2 - 1)(u^2 - 4) \delta^4 y_1 + \cdots$$

$$+ \frac{1}{(2k + 1)!} u(u^2 - 1)(u^2 - 4) \cdots (u^2 - k^2) \delta^{2k} y_1 + \cdots \quad (4.37)$$

This is the *Laplace-Everett interpolation formula*. When terminated at the $2k$th difference it is algebraically equivalent to the forward interpolation formula of Gauss [see Eq. (4.17)] when the latter is terminated at the $(2k + 1)$th difference.

4.8. The Lozenge Diagram. Figure 4.1 shows a scheme for writing down any of the standard interpolation formulas by associating it with a prescribed path through the diagram, called a *lozenge diagram*. The paths enter at the left and may either lie along the sides of the individual

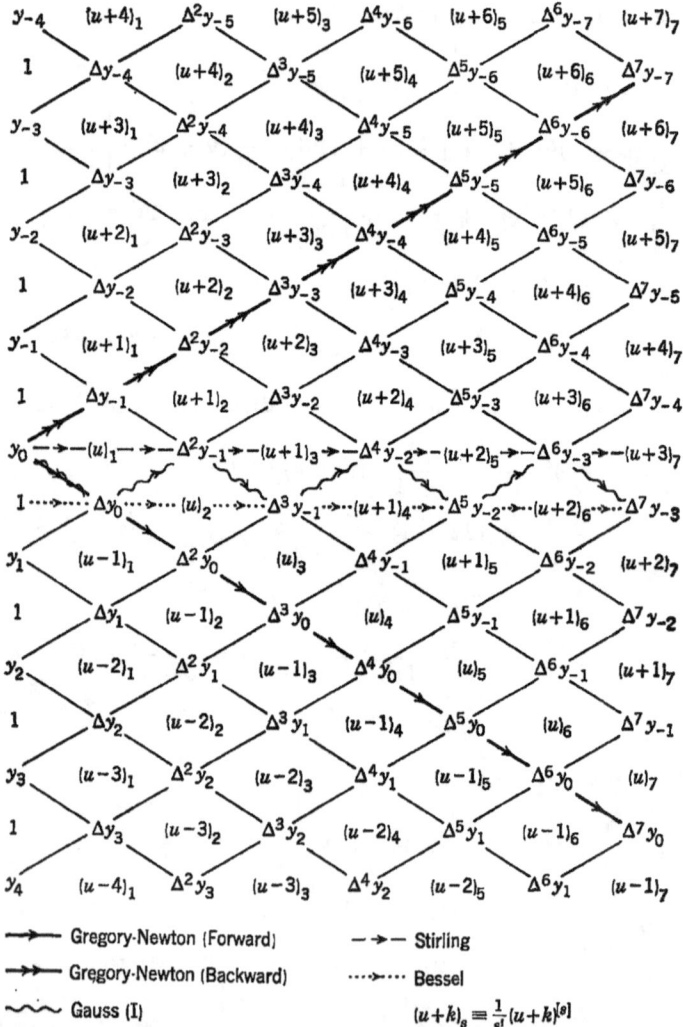

FIG. 4.1. Lozenge diagram showing paths for some of the standard interpolation formulas.

lozenges or pass horizontally through them, as is done in the case of Stirling's and Bessel's formulas. Since many other paths are available besides those shown, the diagram makes it possible for one to write down a large number of valid interpolation formulas, some of which may be useful in special problems.

First note that a new notation for the factorials is used, namely,

$$(u + s)_k = \frac{1}{k!} (u + s)^{[k]} \tag{4.38}$$

To convert a path through the lozenge to an interpolation formula the following rules are formulated:

1. Each time a difference column is crossed from left to right a term is added.

2. If a path enters a difference column (from the left) at a positive slope, the term added is the product of the difference, say $\Delta^k y_{-p}$, at which the column is crossed and the factorial $(u + p - 1)_k$ lying just below that difference.

3. If a path enters a difference column (from the left) at a negative slope, the term added is the product of the difference, say $\Delta^k y_{-p}$, at which the column is crossed and the factorial $(u + p)_k$ lying just above that difference.

4. If a path enters a difference column horizontally (from the left), the term added is the product of the difference, say $\Delta^k y_{-p}$, at which the column is crossed and the average of the two factorials $(u + p)_k$ and $(u + p - 1)_k$ lying, respectively, just above and just below that difference.

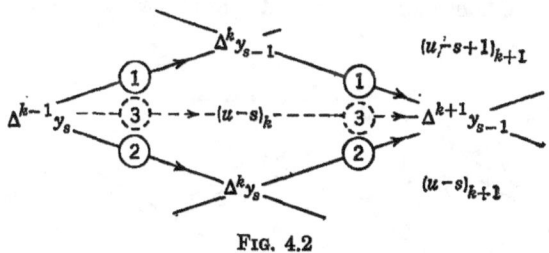

FIG. 4.2

5. If a path crosses a difference column (from left to right) between two differences, say $\Delta^k y_{-(p+1)}$ and $\Delta^k y_{-p}$, then the added term is the product of the average of these two differences and the factorial $(u + p)_k$ at which the column is crossed.

6. Any portion of a path traversed from right to left gives rise to the same terms as would arise from going along this portion from left to right except that the sign of each term is changed.

7. The zero-difference column corresponding to the tabulated values may be treated by the same rules as any other difference columns provided one thinks of the lozenge as being entered just to the left of this column. Thus this column can be crossed by a path making a positive, negative, or zero slope, just as is true of the other columns.

LEMMA 1. The sum of the terms arising from a closed trip around an individual lozenge, or around just the upper (or lower) half of it, is zero.

Proof: Consider the general lozenge in Fig. 4.2. By the rules formulated above, the contribution I_1 in going from $\Delta^{k-1} y_s$ to $\Delta^{k+1} y_{s-1}$ along path 1 is

$$I_1 = (u - s)_k \, \Delta^k y_{s-1} + (u - s + 1)_{k+1} \, \Delta^{k+1} y_{s-1}$$

the contribution I_2 along path 2 is

$$I_2 = (u - s)_k \, \Delta^k y_s + (u - s)_{k+1} \, \Delta^{k+1} y_{s-1}$$

and that along path 3 is

$$I_3 = (u - s)_k \frac{\Delta^k y_{s-1} + \Delta^k y_s}{2} + \frac{(u - s + 1)_{k+1} + (u - s)_{k+1}}{2} \Delta^{k+1} y_{s-1}$$

$$= \tfrac{1}{2}(I_1 + I_2)$$

From these formulas

$$I_1 - I_2 = (u - s)_k (\Delta^k y_{s-1} - \Delta^k y_s)$$
$$+ [(u - s + 1)_{k+1} - (u - s)_{k+1}] \Delta^{k+1} y_{s-1}$$
$$= (u - s)_k (-\Delta^{k+1} y_{s-1}) + (u - s)_k \Delta^{k+1} y_{s-1} = 0$$

Therefore, $I_1 = I_2 = I_3$. A clockwise trip around the lozenge gives $I_1 - I_2$, around the top of the lozenge $I_1 - I_3$, around the bottom of the lozenge $I_3 - I_2$; hence all these closed contours give zero, and the lemma is proved.†

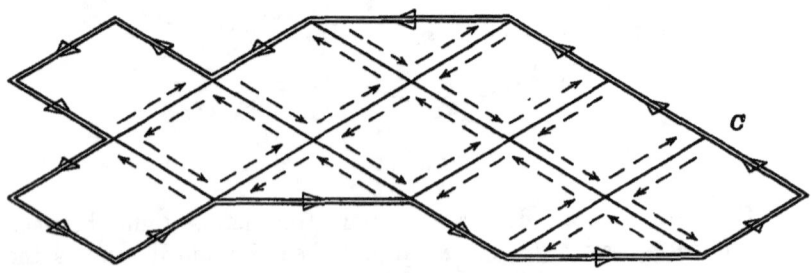

FIG. 4.3

LEMMA 2. The sum of the terms arising from any closed path in the lozenge is zero.

Proof: Consider a typical path as shown in Fig. 4.3. The contribution due to the closed path C is equal to the sum of the contributions made in traveling around the individual lozenges (or in some cases the upper, or lower, portions of a lozenge). This is due to the fact that, if all the lozenges are traversed in the same direction (in this case, counterclockwise), each portion of an interior path is traversed once in each direction, and therefore the only uncanceled contribution is due to the perimeter C. Since the contributions of the individual closed circuits are all zero, the contribution of C, which is equal to their sum, is also zero.

THEOREM I. Any two formulas obtained from paths through the

† Any counterclockwise trip clearly yields just the negative of a clockwise trip and is therefore zero also.

lozenge diagram that terminate on the same difference (or on the mean of the same two differences) are equivalent.†

Proof: First we shall prove that any two paths entering the lozenge at the same entry y_s [or at the same mean entry $\frac{1}{2}(y_s + y_{s-1})$] and ending on a given difference (or mean difference) give rise to equivalent interpolation formulas. Let the sum of the terms obtained by rules 1 to 6 in traversing the distance between y_s and the given difference along any two paths 1 and 2 be S_1 and S_2, respectively; then the two interpolation formulas are $y_s + S_1$ and $y_s + S_2$ [or $\frac{1}{2}(y_s + y_{s-1}) + S_1$ and $\frac{1}{2}(y_s + y_{s-1}) + S_2$] by rule 7 above. By Lemma 2 the sum of terms obtained in traversing the closed circuit formed by the paths 1 and 2 is zero; hence $S_1 - S_2 = 0$, and the formulas are equivalent.

Thus all paths entering the lozenge at a given point yield equivalent formulas, and are said to be *equivalent paths.* If we can show that some path entering the lozenge at y_{s+1} and some path entering at $\frac{1}{2}(y_s + y_{s+1})$ are each equivalent to one entering at y_s (this is done in the next paragraph), then since all paths entering at a given point are equivalent, all paths entering at any one of these three points are equivalent. Moreover, since s is not specified, all adjacent points of entry give rise to equivalent paths. This obviously amounts to the equivalence of all paths that enter at the left and proceed to the same difference.

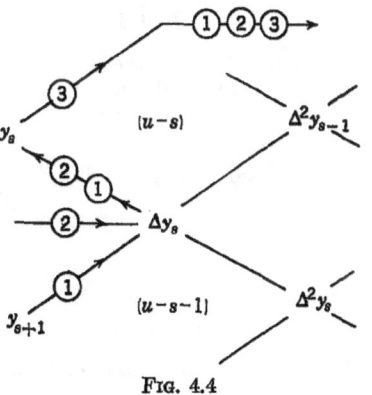

FIG. 4.4

The proof that some path 1 entering at y_{s+1} and some path 2 entering at $\frac{1}{2}(y_s + y_{s+1})$ are each equivalent to a path 3 entering at y_s can be proved by choosing for 1 and 2 (see Fig. 4.4) paths which pass through the point Δy_s, then through y_s, and from there proceed across the lozenge to the given difference. Letting S be the sum of the terms in going from y_s to the given difference, the formulas for the three paths are

Path 1 $= y_{s+1} + (u - s - 1)\,\Delta y_s - (u - s)\,\Delta y_s + S$

Path 2 $= \frac{1}{2}(y_s + y_{s+1}) + \dfrac{(u - s) + (u - s - 1)}{2}\,\Delta y_s - (u - s)\,\Delta y_s + S$

Path 3 $= y_s + S$

Since

$$y_{s+1} - \Delta y_s = \tfrac{1}{2}(y_s + y_{s+1}) - \tfrac{1}{2}\Delta y_s = y_s$$

the three paths are equivalent.

† They are equivalent in the sense that they yield the same interpolated values, i.e., they represent the same interpolating polynomial.

Example 3. Write down the interpolation formulas whose paths are shown on the lozenge diagram.

It is clear that, since the lozenge employs descending differences, all formulas will be in terms of descending differences. If one wishes the central-difference formulas to appear as they have been written in the earlier sections of the chapter, one should use central-difference notation for the differences.

Gregory-Newton (Forward). Applying Rules 1, 3, and 7 to the diagonal path that slopes downward from y_0, one obtains the formula

$$y = y_0 + (u)_1 \Delta y_0 + (u)_2 \Delta^2 y_0 + (u)_3 \Delta^3 y_0 + \cdots + (u)_m \Delta^m y_0 \qquad (4.39)$$

Since $(u)_r = u^{[r]}/r!$, this is the same as Eq. (3.43).

Gregory-Newton (Backward). Employing rules 1, 2, and 7 to the diagonal path that slopes upward from y_0, one obtains the formula

$$y = y_0 + (u)_1 \Delta y_{-1} + (u + 1)_2 \Delta^2 y_{-2} + (u + 2)_3 \Delta^3 y_{-3} + \cdots$$
$$+ (u + m - 1)_m \Delta^m y_{-m} \qquad (4.40)$$

which, by Eq. (4.38), is clearly equivalent to Eq. (3.55).

Gauss (I). Employing rules 1, 2, 3, and 7 to the zigzag path shown, one has

$$y = y_0 + (u)_1 \Delta y_0 + (u)_2 \Delta^2 y_{-1} + (u + 1)_3 \Delta^3 y_{-1} + (u + 1)_4 \Delta^4 y_{-2} + \cdots$$
$$+ (u + k - 1)_{2k-1} \Delta^{2k-1} y_{-k+1} + (u + k - 1)_{2k} \Delta^{2k} y_{-k} + \cdots \qquad (4.41)$$

where, as before, the series is terminated at any desired difference. By employing Eqs. (4.5) and (4.38) this formula is seen to be the same as Eq. (4.17).

Stirling's. Employing rules 1, 4, 5, and 7 to the horizontal path through y_0,

$$y = y_0 + (u)_1 \frac{\Delta y_{-1} + \Delta y_0}{2} + \frac{(u)_2 + (u + 1)_2}{2} \Delta^2 y_{-1}$$

$$+ (u + 1)_3 \frac{\Delta^3 y_{-2} + \Delta^3 y_{-1}}{2} + \cdots$$

$$+ (u + k - 1)_{2k-1} \frac{\Delta^{2k-1} y_{-k} + \Delta^{2k-1} y_{-k+1}}{2}$$

$$+ \frac{(u + k - 1)_{2k} + (u + k)_{2k}}{2} \Delta^{2k} y_{-k} + \cdots$$

Now by Eqs. (4.23), (4.24), and (4.38)

$$(u + k - 1)_{2k-1} = \frac{1}{(2k - 1)!} u(u^2 - 1)(u^2 - 4) \cdots [u^2 - (k - 1)^2]$$

and

$$\frac{(u + k - 1)_{2k} + (u + k)_{2k}}{2} = \frac{1}{2k!} u^2(u^2 - 1)(u^2 - 4) \cdots [u^2 - (k - 1)^2]$$

Therefore, one may write

$$y = y_0 + u \frac{\Delta y_{-1} + \Delta y_0}{2} + \frac{1}{2!} u^2 \Delta^2 y_{-1} + \frac{1}{3!} u(u^2 - 1) \frac{\Delta^3 y_{-2} + \Delta^3 y_{-1}}{2}$$

$$+ \frac{1}{4!} u^2(u^2 - 1) \Delta^4 y_{-2} + \cdots$$

$$+ \frac{1}{(2k - 1)!} u(u^2 - 1)(u^2 - 4) \cdots [u^2 - (k - 1)^2] \frac{\Delta^{2k-1} y_{-k} + \Delta^{2k-1} y_{-k+1}}{2}$$

$$+ \frac{1}{(2k)!} u^2(u^2 - 1)(u^2 - 4) \cdots [u^2 - (k - 1)^2] \Delta^{2k} y_{-k} + \cdots \qquad (4.42)$$

This is the expression of Stirling's formula when descending differences are employed.

By expressing the above differences as central differences one obtains immediately Eq. (4.21).

Bessel's. Employing the rules above to the horizontal path through Δy_0, one has

$$y = \frac{y_0 + y_1}{2} + \frac{(u)_1 + (u-1)_1}{2} \Delta y_0 + (u)_2 \frac{\Delta^2 y_{-1} + \Delta^2 y_0}{2}$$
$$+ \frac{(u+1)_3 + (u)_3}{2} \Delta^3 y_{-1} + \cdots$$
$$+ \frac{(u+k-1)_{2k-1} + (u+k-2)_{2k-1}}{2} \Delta^{2k-1} y_{-k+1}$$
$$+ (u+k-1)_{2k} \frac{\Delta^{2k} y_{-k} + \Delta^{2k} y_{-k+1}}{2} + \cdots$$

and by Eqs. (4.32) and (4.38)

$$y = \frac{y_0 + y_1}{2} + (u - \tfrac{1}{2}) \Delta y_0 + \frac{1}{2!} u^{[2]} \frac{\Delta^2 y_{-1} + \Delta^2 y_0}{2}$$
$$+ \frac{1}{3!} (u - \tfrac{1}{2}) u^{[2]} \Delta^3 y_{-1} + \frac{1}{4!} (u+1)^{[4]} \frac{\Delta^4 y_{-2} + \Delta^4 y_{-1}}{2} + \cdots$$
$$+ \frac{1}{(2k)!} (u+k-1)^{[2k]} \frac{\Delta^{2k} y_{-k} + \Delta^{2k} y_{-k+1}}{2}$$
$$+ \frac{1}{(2k+1)!} (u - \tfrac{1}{2})(u+k-1)^{[2k]} \Delta^{2k+1} y_{-k} + \cdots \quad (4.43)$$

On introducing central differences this equation reduces to Eq. (4.31).

4.9. Remainder Terms. As indicated earlier, any interpolation formula ending on the difference $\delta^m y_r$ represents the mth-degree interpolating polynomial $\mathcal{P}_m(u)$ associated with the difference triangle whose vertex is at this difference. This difference triangle contains all the differences that can be formed from the $m + 1$ tabulated values $y_{r-m/2}$, $y_{r-m/2+1}$, $\ldots$, $y_{r+m/2}$, and the interpolating polynomial $\mathcal{P}_m(u)$ fits the given tabulated function $y = f(x)$ over this triangle. One may therefore write

$$f(x) = \mathcal{P}_m(u) + R_m(u) \quad (4.44)$$

where the remainder term $R_m(u)$ must vanish at $u = r - m/2, r - m/2 + 1, \ldots, r + m/2$. The exact form of $R_m(u)$, however, will be obtained by the following indirect argument.

Using the new notation $(u)_r = u^{[r]}/r!$, the Gregory-Newton formula with remainder term [Eq. (3.48)] becomes

$$y = f(x) = y_0 + (u)_1 \Delta y_0 + (u)_2 \Delta^2 y_0 + \cdots + (u)_m \Delta^m y_0$$
$$+ (u)_{m+1} h^{m+1} f^{(m+1)}(\xi) \quad (4.45)$$

where $x_0 < \xi < x_m$. Since the remainder term depends only on the interpolating polynomial and not on how it is expressed, it must be the same for all interpolation formulas ending on the same difference. Moreover, any one of the tabulated values can be labeled y_0, and hence Eq. (4.45) may be used to determine the general properties of remainder terms.

By reference to Sec. 4.2 we see at once that $(u)_{m+1}$ is exactly the $(m + 1)$th-degree polynomial associated with a null triangle with apex at $\Delta^m y_0$, the last difference in the formula. In other words, $(u)_{m+1}$ has zeros throughout the difference triangle associated with the interpolating polynomial (see Secs. 3.9 and 3.10). This is clearly necessary since the $\mathcal{P}_m(u)$ and y must agree over this triangle. Note also that the range permitted ξ is the interval of interpolation, the range of values of x corresponding to the base of the difference triangle.

Since by Eq. (4.10) the $(m + 1)$th-degree polynomial associated with the null triangle with apex at $\delta^m y_r$ is $(u + m/2 - r)_{m+1}$, the remainder term $R_m(u)$ associated with this difference is, by the above analysis,

$$R_m(u) = \left(u + \frac{m}{2} - r\right)_{m+1} h^{m+1} f^{(m+1)}(\xi) \qquad (4.46)$$

where $x_{r-m/2} < \xi < x_{r+m/2}$.

When one ends an interpolation formula on a mean difference

$$\mu \delta^m y_r = \tfrac{1}{2}(\delta^m y_{r-\frac{1}{2}} + \delta^m y_{r+\frac{1}{2}})$$

one obtains the average of the two interpolating polynomials associated, respectively, with the two differences $\delta^m y_{r-\frac{1}{2}}$ and $\delta^m y_{r+\frac{1}{2}}$, and the remainder is, therefore, just the average of the remainders associated with these two polynomials.

The remainder terms for Stirling's formula and the two interpolation formulas of Gauss are thus given by writing the following relations between the last difference employed in the formula and the corresponding remainder term:†

$$\delta^{2k} y_0 \sim \frac{h^{2k+1}}{(2k + 1)!} (u + k)^{[2k+1]} f^{(2k+1)}(\xi) \qquad x_{-k} < \xi < x_k \quad (4.47)$$

$$\delta^{2k-1} y_{-\frac{1}{2}} \sim \frac{h^{2k}}{(2k)!} (u + k)^{[2k]} f^{(2k)}(\xi) \qquad x_{-k} < \xi < x_{k-1} \quad (4.48)$$

$$\delta^{2k-1} y_{\frac{1}{2}} \sim \frac{h^{2k}}{(2k)!} (u + k - 1)^{[2k]} f^{(2k)}(\xi) \qquad x_{-k+1} < \xi < x_k \quad (4.49)$$

and $\mu \delta^{2k-1} y_0$ corresponds to a remainder term that is the average of the last two.

† For the purposes of Stirling's formula we may again write

$$(u + k)^{[2k+1]} = u(u^2 - 1)(u^2 - 4) \cdots (u^2 - k^2)$$

For Bessel's equation, using $v = u - \frac{1}{2}$, one needs the two following associations of remainder terms with last differences:

$$\delta^{2k-1}y_{\frac{1}{2}} \sim \frac{h^{2k}}{(2k)!}(v^2 - \tfrac{1}{4})(v^2 - \tfrac{9}{4}) \cdots \left[v^2 - \frac{(2k-1)^2}{4} \right] f^{(2k)}(\xi)$$
$$x_{-k+1} < \xi < x_k \quad (4.50)$$

$$\mu\delta^{2k}y_{\frac{1}{2}} \sim \frac{h^{2k+1}}{2(2k+1)!}[(v + k + \tfrac{1}{2})^{[2k+1]}f^{(2k+1)}(\xi_1)$$
$$+ (v + k - \tfrac{1}{2})^{[2k+1]}f^{(2k+1)}(\xi_2)]$$
$$x_{-k} < \xi_1 < x_k \quad \text{and} \quad x_{-k+1} < \xi_2 < x_{k+1} \quad (4.51)$$

Thus, in every case the remainder term for a series terminated on a single difference, say the mth difference, is just the $(m + 1)$th term of the series if the series were extended, with, however, $h^{m+1}f^{(m+1)}(\xi)$ replacing the $(m + 1)$th difference (or mean difference). For a formula terminated on the mean of two differences one must average the error terms for those formulas ending on each of the two differences.

4.10. Accuracy of Interpolation Formulas. A great deal of misinformation is available on the accuracy of interpolation formulas. As was seen in Chap. 3, a set of tabulated values and the difference table formed from them cannot be used to express the error in interpolation. Knowledge of the function tabulated or equivalent information is needed. This follows from the elementary fact that any number of curves can be passed through a discrete set of points. Thus, statements to the effect that "all formulas obtained by a path in the lozenge which terminates on the constant-difference column are exact" are of necessity suspect, as are attempts to express the remainder term *in terms of the $(m + 1)$th differences.* This does not mean that such attempts to judge the accuracy of an interpolation formula are without value. They may be valuable in indicating a reasonable guess as to the approximation involved. However, there is only one safe way to determine whether an interpolation is sufficiently accurate, and that is to determine the magnitude of the remainder term.

Since the remainder term always involves the $(m + 1)$th derivative of the particular function tabulated little or nothing can be said about the remainder term unless one specifies something about the function. This means that one cannot be sure which interpolation formula will give the best results in a particular case. Thus the Gregory-Newton interpolation formula could give better results than any of the central-difference formulas even though usually the central-difference formulas are superior. This is particularly true near a singularity of the $(m + 1)$th derivative.

The reason one would expect the central-difference formulas to give better approximations for the interpolated values when $|u| < 1$ and the first m differences are used is that the factorial coefficients of $h^{m+1}f^{(m+1)}(\xi)$ are much smaller for these formulas than they are for the Gregory-Newton formulas. For instance, $(u + k)_{2k+1}$ of Eq. (4.47) is much smaller than the corresponding $(u)_{2k+1}$ of Eq. (4.45) and the $(u + 2k)_{2k+1}$ of Eq. (3.61).† This means that one will have smaller remainder terms for central-difference formulas than for the Gregory-Newton formulas, provided $f^{(m+1)}(\xi)$ does not vary too greatly in the range allowed ξ. This agrees with the intuitive feeling that a polynomial that fits the tabulated values over nearly equal distances on either side of the interpolated value is superior to a polynomial that fits at many points below but at only one point above the interpolated value (or the other way around).‡

One can carry the above argument further. On the assumption that $f^{(m+1)}(\xi)$ does not vary greatly within the permitted ranges of the ξ's involved, compare the remainder terms for formulas ending on the difference $\delta^{2k}y_0$, on the mean difference $\mu\delta^{2k}y_{\frac{1}{2}}$, and on the difference $\delta^{2k}y_1$. On the assumption of constancy of $f^{(m+1)}(\xi)$ in the three remainders, one would predict that the most accurate formula should have the smallest value of the factorial. Since the associated factorials are, respectively, $(u + k)_{2k+1}$, $\frac{1}{2}[(u + k)_{2k+1} + (u + k - 1)_{2k+1}]$, and $(u + k - 1)_{2k+1}$, which are in the ratio of $u + k$, $u - \frac{1}{2}$, and $u - k - 1$, respectively, for the case $0 < u < 1$ (the interpolated values being between y_0 and y_1) and $k > 0$, the middle term is smallest in absolute value. The mean-difference termination, therefore, should ordinarily be better than either of the other two. Thus for termination on an even-difference column, Bessel's formula is more accurate than any of the others. Similarly it can be shown that if one terminates on an odd-difference column, one is likely to obtain the most accurate result if one employs Stirling's formula.§

The consideration of relative accuracy of the interpolation formulas given above applies primarily to those cases in which one uses a lower-order interpolation formula than called for by the difference table. When all formulas are extended to a nearly constant difference column, they yield approximately equal results, and in general they will be found to give the interpolated values nearly as accurately as the tabulated values.

PROBLEMS

1. Prove Eqs. (4.8) both with and without the use of operator techniques.

† In the latter two equations m was replaced by $2k$ so as to correspond to the first.

‡ Of course this description fits the Gregory-Newton formulas only if $|u| < 1$, as it generally is. However, as stated before, a Gregory-Newton formula having the same mth difference as the Gauss is exactly equivalent to it.

§ Whittaker and Robinson use the same kind of an analysis as above but give different conclusions. See "The Calculus of Observations," 4th ed., p. 49, Blackie & Son, Ltd., Glasgow, 1924.

2. Show that a knowledge of the differences used in the forward interpolation formula of Gauss [see Eq. (4.17)] is sufficient to permit one to complete the entire difference triangle whose vertex is at the last difference used.

3. Derive Gauss (II) from Gauss (I) by applying the latter to a table in which the entries are tabulated in reverse order.

4. Prove Eq. (4.29).

5. Find sin 1.7924 from Table 4.2 using each of the following formulas: (a) Gauss (I), (b) Stirling's, (c) Bessel's.

6. Find sin 1.7811 using the Laplace-Everett interpolation formula.

7. Obtain the interpolation formula Gauss (I) directly from the lozenge of Fig. 4.1.

8. Take any new path through the lozenge diagram you like, and write down the corresponding interpolation formula. Assume that the path terminates on some given difference, and write down the points through which the polynomial represented by this formula passes.

9. From Bessel's formula obtain a formula for inserting values halfway between the entries in a table. By expressing the differences in this formula in terms of the entries obtain a set of formulas requiring a knowledge of 2, 4, 6, and 8 of the entries.

10. Using Bessel's formula and Table 3.4, find the sun's declination on Feb. 6.

11. Using Table 3.3, find log π/3.

CHAPTER **5**

LAGRANGE'S INTERPOLATION FORMULA
AND INVERSE INTERPOLATION

The interpolation formulas derived so far, the formulas of Gregory-Newton, Gauss, Stirling, Bessel, and Everett, have been predicated on the assumption that the function is tabulated for equal intervals of the argument. When this interval is not constant, one can no longer construct the difference tables of Chaps. 3 and 4, and hence one is unable to determine the differences needed for the above formulas. One can, however, as before, pass a polynomial through any number of the entries and use this polynomial for approximating the function between these tabulated values.

5.1. Lagrange's Interpolation Formula for Unequal Intervals. Suppose $y_0, y_1, y_2, \ldots, y_n$ are the values of $y = f(x)$ for the arguments $x_0, x_1, x_2, \ldots, x_n$, where the interval between successive x's is not necessarily constant. To find an nth-degree polynomial passing through the $n + 1$ points $(x_0,y_0), (x_1,y_1), (x_2,y_2), \ldots, (x_n,y_n)$ one may proceed as follows. Let $a_s(x)$ be an nth-degree polynomial that is zero for all the tabulated arguments except x_s, and for this argument it is equal to one, i.e.,

$$a_s(x_k) = \begin{cases} 0 & \text{for } k \neq s \\ 1 & \text{for } k = s \end{cases} \tag{5.1}$$

From this equation it follows that $a_s(x)$ has the n zeros $x_0, x_1, \ldots, x_{s-1}, x_{s+1}, \ldots, x_n$, and hence

$$a_s(x) = A_s(x - x_0)(x - x_1) \cdots (x - x_{s-1})(x - x_{s+1}) \cdots (x - x_n)$$

where A_s is a constant. Moreover, since $a_s(x_s) = 1$, one finds from the above equation that

$$A_s = \frac{1}{(x_s - x_0)(x_s - x_1) \cdots (x_s - x_{s-1})(x_s - x_{s+1}) \cdots (x_s - x_n)}$$

and therefore

$$a_s(x)$$
$$= \frac{(x - x_0)(x - x_1) \cdots (x - x_{s-1})(x - x_{s+1}) \cdots (x - x_n)}{(x_s - x_0)(x_s - x_1) \cdots (x_s - x_{s-1})(x_s - x_{s+1}) \cdots (x_s - x_n)} \tag{5.2}$$

86

In particular,

$$a_0(x) = \frac{(x - x_1)(x - x_2) \cdots (x - x_n)}{(x_0 - x_1)(x_0 - x_2) \cdots (x_0 - x_n)}$$

$$a_1(x) = \frac{(x - x_0)(x - x_2) \cdots (x - x_n)}{(x_1 - x_0)(x_1 - x_2) \cdots (x_1 - x_n)}$$

$$a_2(x) = \frac{(x - x_0)(x - x_1)(x - x_3) \cdots (x - x_n)}{(x_2 - x_0)(x_2 - x_1)(x_2 - x_3) \cdots (x_2 - x_n)} \qquad (5.3)$$

$$\cdots\cdots\cdots\cdots\cdots\cdots\cdots\cdots\cdots\cdots$$

$$a_n(x) = \frac{(x - x_0)(x - x_1) \cdots (x - x_{n-1})}{(x_n - x_0)(x_n - x_1) \cdots (x_n - x_{n-1})}$$

Consider the polynomial

$$P(x) = \sum_{s=0}^{n} a_s(x)y_s = a_0(x)y_0 + a_1(x)y_1 + \cdots + a_n(x)y_n \qquad (5.4)$$

Since by Eq. (5.1)

$$P(x_k) = \sum_{s=0}^{n} a_s(x_k)y_s = a_k(x_k)y_k = y_k \qquad (5.5)$$

the polynomial $P(x)$ passes through all the points (x_k, y_k), where $k = 0$, $1, 2, \ldots, n$. Writing in the explicit expression for $a_s(x)$, one has

$$
\begin{aligned}
y &= P(x) \\
&= \sum_{s=0}^{n} \frac{(x - x_0)(x - x_1) \cdots (x - x_{s-1})(x - x_{s+1}) \cdots (x - x_n)}{(x_s - x_0)(x_s - x_1) \cdots (x_s - x_{s-1})(x_s - x_{s+1}) \cdots (x_s - x_n)} y_s \\
&= \frac{(x - x_1)(x - x_2) \cdots (x - x_n)}{(x_0 - x_1)(x_0 - x_2) \cdots (x_0 - x_n)} y_0 \\
&\quad + \frac{(x - x_0)(x - x_2) \cdots (x - x_n)}{(x_1 - x_0)(x_1 - x_2) \cdots (x_1 - x_n)} y_1 \\
&\quad + \frac{(x - x_0)(x - x_1)(x - x_3) \cdots (x - x_n)}{(x_2 - x_0)(x_2 - x_1)(x_2 - x_3) \cdots (x_2 - x_n)} y_2 + \cdots \\
&\quad + \frac{(x - x_0)(x - x_1) \cdots (x - x_{n-1})}{(x_n - x_0)(x_n - x_1) \cdots (x_n - x_{n-1})} y_n \qquad (5.6)
\end{aligned}
$$

which is *Lagrange's interpolation formula.*†

Example 1. Find log 47 to five decimal places from log 2, log 3, log 5, and log 7.

$$
\begin{aligned}
\log 2 &= 0.30103\ 00 \\
\log 3 &= 0.47712\ 13 \\
\log 5 &= 0.69897\ 00 \\
\log 7 &= 0.84509\ 80
\end{aligned}
$$

† The use of this formula usually involves a considerable amount of numerical calculation. If a desk calculator is not available, it is best to make use of logarithms.

Thus

$$\log 40 = 2 \log 2 + \log 10 = 1.60206\ 00 = y_0$$
$$\log 42 = \log 2 + \log 3 + \log 7 = 1.62324\ 93 = y_1$$
$$\log 45 = 2 \log 3 + \log 5 = 1.65321\ 26 = y_2$$
$$\log 47 = y$$
$$\log 48 = 4 \log 2 + \log 3 = 1.68124\ 13 = y_3$$
$$\log 49 = 2 \log 7 = 1.69019\ 60 = y_4$$
$$\log 50 = \log 5 + \log 10 = 1.69897\ 00 = y_5$$

Letting $x_0 = 40$, $x_1 = 42$, $x_3 = 45$, $x_4 = 48$, $x_5 = 49$, $x_6 = 50$, and $x = 47$ and applying Lagrange's interpolation formula,

$$y = \frac{(5)(2)(-1)(-2)(-3)}{(-2)(-5)(-8)(-9)(-10)} y_0 + \frac{(7)(2)(-1)(-2)(-3)}{(2)(-3)(-6)(-7)(-8)} y_1$$
$$+ \frac{(7)(5)(-1)(-2)(-3)}{(5)(3)(-3)(-4)(-5)} y_2 + \frac{(7)(5)(2)(-2)(-3)}{(8)(6)(3)(-1)(-2)} y_3$$
$$+ \frac{(7)(5)(2)(-1)(-3)}{(9)(7)(4)(1)(-1)} y_4 + \frac{(7)(5)(2)(-1)(-2)}{(10)(8)(5)(2)(1)} y_5$$

or

$$y = \tfrac{1}{120}y_0 - \tfrac{1}{24}y_1 + \tfrac{7}{30}y_2 + \tfrac{35}{24}y_3 - \tfrac{5}{8}y_4 + \tfrac{7}{40}y_5$$

Putting in the values of the y's given above,

$$\log 47 = 1.67209\ 80$$

The correct value is 1.67209 79. That it should be off as much as one unit in the seventh decimal place is not surprising, as the logarithms of 42, 45, and 47 calculated above are off one unit in this place.

5.2. Remainder Term in Lagrange's Formula.

Since the approximating polynomial $P(x)$ [see Eq. (5.6)] given by Lagrange's formula has the same values $y_0, y_1, \ldots, y_n$ as does $y = f(x)$ for the arguments $x_0, x_1, \ldots, x_n$, the remainder term must have zeros at these $n + 1$ points. Therefore $(x - x_0)$, $(x - x_1)$, $\ldots$, $(x - x_n)$ must be factors of the remainder, and one may write

$$f(x) = P(x) + \frac{(x - x_0)(x - x_1)(x - x_2) \cdots (x - x_n)}{(n + 1)!} K(x) \quad (5.7)$$

Let us take x to be fixed in value and consider the function

$$W(t) = f(t) - P(t) - \frac{(t - x_0)(t - x_1)(t - x_2) \cdots (t - x_n)}{(n + 1)!} K(x) \quad (5.8)$$

then $W(t)$ has zeros at $t = x_0, x_1, \ldots, x_n$, and x. Since the $(n + 1)$th derivative of the nth-degree polynomial $P(x)$ is zero,†

$$W^{(n+1)}(t) = f^{(n+1)}(t) - K(x) \quad (5.9)$$

As a consequence of Rolle's theorem the $(n + 1)$th derivative of $W(t)$

† One must here assume that $f(x)$ has a continuous derivative of the $(n + 1)$th order.

has at least one real zero $t = \xi$ in the range $x_0 < \xi < x_n$.† Therefore substituting $t = \xi$ in Eq. (5.9),

$$K(x) = f^{(n+1)}(\xi) - W^{(n+1)}(\xi) = f^{(n+1)}(\xi)$$

Using this expression for $K(x)$ and writing out $P(x)$,

$$
\begin{aligned}
f(x) = {} & \frac{(x - x_1)(x - x_2) \cdots (x - x_n)}{(x_0 - x_1)(x_0 - x_2) \cdots (x_0 - x_n)} f(x_0) \\
& + \frac{(x - x_0)(x - x_2)(x - x_3) \cdots (x - x_n)}{(x_1 - x_0)(x_1 - x_2)(x_1 - x_3) \cdots (x_1 - x_n)} f(x_1) \\
& + \cdots + \frac{(x - x_0)(x - x_1) \cdots (x - x_{n-1})}{(x_n - x_0)(x_n - x_1) \cdots (x_n - x_{n-1})} f(x_n) \\
& + \frac{(x - x_0)(x - x_1) \cdots (x - x_n)}{(n + 1)!} f^{(n+1)}(\xi) \quad (5.10)
\end{aligned}
$$

where $x_0 < \xi < x_n$.‡ This is *Lagrange's interpolation formula with a remainder term*.

5.3. Effect of Errors in the Tabulated Values. As seen in Sec. 3.14, one must add to the above class A error, due to replacing the tabulated function $f(x)$ by a polynomial, the class B error due to round-off error in the tabulated values and the class C error involved in making the computation. Class C error can usually be made negligible by carrying along an extra decimal place in the calculation

If ϵ_s is the error in the tabulated value y_s, the class B error ϵ_B, by Eq. (5.4), is

$$\epsilon_B = \sum_{s=0}^{n} a_s(x)\epsilon_s \quad (5.11)$$

Now if e is the maximum absolute value that the ϵ_s may have, the maximum absolute error e_B is

$$e_B = Fe \quad (5.12)$$

where

$$F = \sum_{s=0}^{n} |a_s(x)| \quad (5.13)$$

is called the *class B error factor*. It is, of course, a function of x.

The $a_s(x)$ may be looked upon as the weights to be given the tabulated values y_s in calculating the interpolated value y. By Eq. (5.11) they are also the weights to be given the errors in the tabulated values.

---

† The proof of this statement is given in Sec. 3.13. It is clear that the function $W(t)$ defined by Eq. (5.8) will also depend on x; thus ξ will, in general, be a function of x, the argument for which the interpolation is being made.

‡ This is the range we refer to as the interval of interpolation. For extrapolation the range is increased to include the point x.

If one applies Lagrange's equation (5.10) to the trivial function $f(x) = 1$ and uses the notation of Eq. (5.3), then, since the remainder term vanishes, one finds that

$$\sum_{s=0}^{n} a_s(x) = 1 \qquad (5.14)$$

Thus the sum of the weights is 1. By this equation and Eq. (5.13), *the error factor F is 1 plus twice the sum of the absolute values of the negative* $a_s(x)$.

Example 2. Determine the accuracy of the interpolation for log 47 in Example 1. Since $n = 5$ in this example, by Eq. (5.10), the remainder term is

$$\frac{(7)(5)(2)(-1)(-2)(-3)}{6!} f^{(6)}(\xi) = \frac{-7}{12} f^{(6)}(\xi)$$

where $40 < \xi < 50$. Now†

$$f^{(6)}(x) = \frac{d^6}{dx^6} \log x = (\log e) \frac{d^6}{dx^6} \ln x = -\frac{5!}{x^6} \log e$$

Hence

$$\frac{-7}{12} f^{(6)}(\xi) < \frac{70 \log e}{40^6} = 0.73 \times 10^{-8}$$

and the interpolation error is less than one unit in the eighth decimal place and is therefore negligible.

From Example 1 (see Sec. 5.1) the class B error factor is

$$F = 1 + 2(\tfrac{1}{24} + \tfrac{5}{6}) = 2.75$$

If, therefore, we make use of the observation that the tabulated values may be off by one unit in seventh decimal place, we see by Eq. (5.12) that the class B error is less than 3 in the seventh decimal place.

The class C error was kept negligible by carrying along an extra decimal place.

Example 3. Determine the accuracy with which a linear interpolation based on the data of Example 1 would give the value of $y = \log 47$.

Lagrange's formula with remainder term for this case, by Eq. (5.10), is

$$y = \frac{x - x_1}{x_0 - x_1} y_0 + \frac{x - x_0}{x_1 - x_0} y_1 + \tfrac{1}{2}(x - x_0)(x - x_1) \left[\frac{d^2}{dx^2} \log x \right]_{x=\xi} \qquad (5.15)$$

where $x_0 = 45$, $x_1 = 48$, $x = 47$, and $45 < \xi < 48$. Therefore, the remainder term is

$$\frac{(2)(-1)}{2} \left[\frac{-1}{x^2} \log e \right]_{x=\xi} < \frac{0.43}{45^2} = 2 \times 10^{-4}$$

and the computed value may be too small by as much as two units in the fourth decimal place.

Carrying out the linear interpolation given by the first two terms of Eq. (5.15),

$$y = \tfrac{1}{3} y_0 + \tfrac{2}{3} y_1 = 1.67190$$

† Unless otherwise noted, in this book log x will stand for $\log_{10} x$, and ln x will stand for the natural logarithm.

This value is in error by almost the full two units in the fourth decimal place permitted by the above analysis. Since $F = 1$, the class B error is negligible.

5.4. Lagrange's Formula for Equal Intervals.† When the arguments $x_0, x_1, \ldots, x_n$ are uniformly spaced at an interval h,

$$
\begin{aligned}
x_s &= x_0 + sh \\
x &= x_0 + uh
\end{aligned}
\tag{5.16}
$$

and Eq. (5.10) takes on the simpler form

$$
\begin{aligned}
f(x) = (-1)^n &\frac{(u-1)(u-2)\cdots(u-n)}{n!} f(x_0) \\
+ (-1)^{n-1} &\frac{u(u-2)(u-3)\cdots(u-n)}{1!(n-1)!} f(x_1) \\
+ (-1)^{n-2} &\frac{u(u-1)(u-3)\cdots(u-n)}{2!(n-2)!} f(x_2) + \cdots \\
+ &\frac{u(u-1)(u-2)\cdots(u-n+1)}{n!} f(x_n) \\
+ &\frac{u(u-1)(u-2)\cdots(u-n)}{(n+1)!} h^{n+1} f^{(n+1)}(\xi) \quad (5.17)
\end{aligned}
$$

where, as before, $x_0 < \xi < x_n$. This can be written

$$
f(x) = \sum_{s=0}^{n} a_s(u) f(x_s) + r(u)
\tag{5.18}
$$

where‡

$$
a_s(u) = \frac{(-1)^{n-s} u^{[n+1]}}{s!(n-s)!(u-s)} \qquad s = 0, 1, 2, \ldots, n
\tag{5.19}
$$

and§

$$
r(u) = h^{n+1} \frac{u^{[n+1]}}{(n+1)!} f^{(n+1)}(\xi) = h^{n+1}(u)_{n+1} f^{(n+1)}(\xi)
\tag{5.20}
$$

for $x_0 < \xi < x_n$.

If one changes the subscript designation and writes $x'_{-m}, x'_{-m+1}, \ldots, x'_{-1}, x'_0, x'_1, \ldots, x'_{n-m}$ in place of $x_0, x_1, x_2, \ldots, x_n$, then Eqs. (5.16) are replaced by the equations

$$
\begin{aligned}
x'_k &= x'_0 + kh = x_{k+m} \\
x &= x'_0 + u'h = x_0 + (u' + m)h
\end{aligned}
\tag{5.21}
$$

† See "Tables of Lagrangian Interpolation Coefficients," p. xv, Columbia University Press, New York, 1944.

‡ $u^{[n+1]} = u(u-1)(u-2)\cdots(u-n)$ (see Sec. 3.7).

§ $(u)_{n+1} = u^{[n+1]}/(n+1)!$.

Thus s is replaced by $k + m$ and u by $u' + m$, and Eqs. (5.18) to (5.20) become

$$f(x') = \sum_{k=-m}^{n-m} A_k(u')f(x'_k) + R(u') \tag{5.22}$$

$$A_k(u') = a_{k+m}(u' + m) = \frac{(-1)^{n-m-k}(u' + m)^{[n+1]}}{(m + k)!(n - m - k)!(u' - k)} \tag{5.23}$$

and $$R(u') = r(u' + m) = h^{n+1}(u' + m)_{n+1}f^{(n+1)}(\xi) \tag{5.24}$$

where $x'_{-m} < \xi < x'_m$.

By a proper choice of m it is now possible to fix the reference point x'_0 at any of the tabulated points $x_0, x_1, \ldots, x_n$ through which the polynomial passes. It is usually best to choose m to be the largest integer in $n/2$. This places the reference point at the middle of the interval if n is even and only slightly off-center for n odd. Since one will customarily use these reference points, the primes can be dropped, and Eqs. (5.22) to (5.24) can be written

$$f(x) = \begin{cases} \sum\limits_{k=-m}^{m} A_k(u)f(x_k) + R(u) & \text{for } n = 2m \\ \sum\limits_{k=-m}^{m+1} A_k(u)f(x_k) + R(u) & \text{for } n = 2m + 1 \end{cases} \tag{5.25}$$

where

$$A_k(u) = \begin{cases} \dfrac{(-1)^{m-k}(u + m)^{[2m+1]}}{(m + k)!(m - k)!(u - k)} & \text{for } n = 2m \\ \dfrac{(-1)^{m-k+1}(u - m)^{[2m+2]}}{(m + k)!(m - k + 1)!(u - k)} & \text{for } n = 2m + 1 \end{cases} \tag{5.26}$$

and

$$R(u) = \begin{cases} h^{2m+1}(u + m)_{2m+1}f^{(2m+1)}(\xi) & \text{for } n = 2m \\ h^{2m+2}(u + m)_{2m+2}f^{(2m+2)}(\xi) & \text{for } n = 2m + 1 \end{cases} \tag{5.27}$$

The $A_k(u)$ are the Lagrangian interpolation coefficients given in the "Tables of Lagrangian Interpolation Coefficients" referred to above.

In particular for $n = 2$, $m = 1$,

$$A_{-1} = \frac{(u + 1)^{[3]}}{0!2!(u + 1)} = \frac{u(u - 1)}{2}$$

$$A_0 = -\frac{(u + 1)^{[3]}}{1!1!(u)} = -(u^2 - 1) \tag{5.28}$$

$$A_1 = \frac{(u + 1)^{[3]}}{2!0!(u - 1)} = \frac{u(u + 1)}{2}$$

For $n = 3$, $m = 1$,

$$A_{-1} = -\frac{(u+1)^{[4]}}{0!3!(u+1)} = \frac{-u(u-1)(u-2)}{3!}$$

$$A_0 = \frac{(u+1)^{[4]}}{1!2!(u)} = \frac{(u^2-1)(u-2)}{2!}$$

$$A_1 = -\frac{(u+1)^{[4]}}{2!1!(u-1)} = -\frac{(u+1)u(u-2)}{2!}$$ \qquad (5.29)

$$A_2 = \frac{(u+1)^{[4]}}{3!0!(u-2)} = \frac{u(u^2-1)}{3!} .$$

Lagrange's interpolation formula for equal intervals of the argument [Eqs. (5.25) to (5.27)] has the advantage over the other interpolation formulas of permitting one to interpolate without first constructing a difference table. It has the disadvantage of requiring one to decide ahead of time the degree of the polynomial needed to achieve the accuracy desired. In the other interpolation formulas the degree of the polynomial increases as each additional term of the series is added, and one can stop when the additional terms give promise of being negligible (for the accuracy required).

The remainder term given in Eq. (5.27) for Lagrange's formula is the same as the remainder term for the forward interpolation formula of Gauss (see Sec. 4.9). This is because they represent the same interpolation polynomial, that associated with the difference triangle with apex at $\delta^n y_0$ if n is even and at $\delta^n y_{\frac{1}{2}}$ if n is odd.

Example 4. Determine the four-point formula for finding the value of a function halfway between two tabulated values.

Applying Eqs. (5.25) to (5.27) with $n = 3$ and $m = 1$,

$$f(x) = \sum_{k=-1}^{2} A_k(u)f(x_k) + h^4(u+1)_4 f^{(4)}(\xi)$$

where $x_{-1} < \xi < x_2$ and the $A_k(u)$ are given by Eqs. (5.29). Since $x = \frac{1}{2}(x_0 + x_1)$, $u = (x - x_0)/h = \frac{1}{2}$, and

$$A_{-1} = \frac{-(\frac{1}{2})(-\frac{1}{2})(-\frac{3}{2})}{6} = -\frac{1}{16}$$

$$A_0 = \frac{(-\frac{3}{4})(-\frac{3}{2})}{2} = \frac{9}{16}$$

$$A_1 = \frac{-(\frac{3}{2})(\frac{1}{2})(-\frac{3}{2})}{2} = \frac{9}{16}$$

$$A_2 = \frac{(\frac{1}{2})(-\frac{3}{4})}{6} = -\frac{1}{16}$$

$$(u+1)_4 = \frac{(\frac{3}{2})(\frac{1}{2})(-\frac{1}{2})(-\frac{3}{2})}{4!} = \frac{3}{128}$$

NUMERICAL ANALYSIS

Therefore,

$$f\left(\frac{x_0 + x_1}{2}\right) = \tfrac{1}{16}[-f(x_{-1}) + 9f(x_0) + 9f(x_1) - f(x_2)] + \tfrac{3}{128}h^4 f^{(4)}(\xi) \quad (5.30)$$

where $x_{-1} < \xi < x_2$.

Example 5. Find the value of $J_0(x)$ for $x = 9.274$ by interpolation in the table of the zero-order Bessel function given below:

x	$y = J_0(x)$
9.0	$-0.09033\ 361 = y_{-2}$
9.1	$-0.11423\ 923 = y_{-1}$
9.2	$-0.13674\ 837 = y_0$
9.3	$-0.15765\ 519 = y_1$
9.4	$-0.17677\ 157 = y_2$
9.5	$-0.19392\ 875 = y_3$

Since six points are given, it is possible to pass a fifth-degree polynomial through these points, i.e., $n = 5$ in the Lagrangian interpolation equations (5.25) to (5.27). By the second equation of Eqs. (5.25) one obtains the approximation

$$J_0(x) = A_{-2}(u)y_{-2} + A_{-1}(u)y_{-1} + A_0(u)y_0 + A_1(u)y_1 + A_2(u)y_2 + A_3(u)y_3$$

where, since $u = 0.74$, the $A(u)$'s can be looked up in a table of six-point Lagrangian interpolation coefficients.† They are

$$A_{-2} = 0.00794\ 423$$
$$A_{-1} = -0.06254\ 943$$
$$A_0 = 0.29415\ 138$$
$$A_1 = 0.83720\ 010$$
$$A_2 = -0.08637\ 779$$
$$A_3 = 0.00963\ 151$$

Therefore, the interpolated value is

$$J_0(9.274) = -0.00071\ 7631 + 0.00714\ 5599 - 0.04022\ 4722 - 0.13198\ 8941$$
$$+ 0.01526\ 9138 - 0.00186\ 7827$$
$$= -0.15238\ 438$$

5.5. Inverse Interpolation. The problem of direct interpolation is to find the value of $y = f(x)$ for an x lying between two tabulated values of the argument, which can be taken to be x_0 and x_1. The problem in inverse interpolation is to find the value of x corresponding to a y lying between two tabulated values of y, which can be designated by y_0 and y_1. Since

$$x = f^{-1}(y) \quad (5.31)$$

inverse interpolation is ordinary interpolation of the inverse function $f^{-1}(y)$; however, the interpolation formulas of Chap. 4 are not applicable, since the interval of the new argument y is not constant. Therefore,

† See "Tables of Lagrangian Interpolation Coefficients" of previous footnote.

of necessity, a direct interpolation of the inverse function demands the use of Lagrange's interpolation formula for unequal intervals of the argument [see Eq. (5.10)]. This use of Lagrange's formula will be illustrated in Example 6.

The second approach to inverse interpolation is to approximate the function $y = f(x)$ by an mth-degree polynomial using any of the standard interpolation formulas. Thus one sets

$$y = \mathcal{P}_m(u) \tag{5.32}$$

and uses any of the methods of Chaps. 1 and 2 to find the value of u corresponding to the given value of y. For example, if five terms of the Gregory-Newton formula [see Eq. (3.43)] give a sufficiently close approximation to $f(x)$,

$$y = y_0 + u\,\Delta y_0 + \frac{u(u-1)}{2!}\,\Delta^2 y_0 + \frac{u(u-1)(u-2)}{3!}\,\Delta^3 y_0$$
$$+ \frac{u(u-1)(u-2)(u-3)}{4!}\,\Delta^4 y_0 \tag{5.33}$$

Since y is given, one needs to solve for a real root of this equation or the equivalent equation,

$$\frac{\Delta^4 y_0}{24}\,u^4 + \left(\frac{\Delta^3 y_0}{6} - \frac{\Delta^4 y_0}{4}\right)u^3 + \left(\frac{\Delta^2 y_0}{2} - \frac{\Delta^3 y_0}{2} + \frac{11\Delta^4 y_0}{24}\right)u^2$$
$$+ \left(\Delta y_0 - \frac{\Delta^2 y_0}{2} + \frac{\Delta^3 y_0}{3} - \frac{\Delta^4 y_0}{4}\right)u + (y_0 - y) = 0 \tag{5.34}$$

If one designates by y_0 and y_1 the tabulated values lying just above and just below the given value y, then the required root u lies between zero and one.

One can solve Eq. (5.33) by successive approximations. Let $u_0, u_1, u_2, \ldots$ be defined by the equations

$$0 = (y_0 - y) + u_0\,\Delta y_0$$
$$0 = (y_0 - y) + u_1\,\Delta y_0 + u_0(u_0 - 1)\tfrac{1}{2}\Delta^2 y_0$$
$$0 = (y_0 - y) + u_2\,\Delta y_0 + u_1(u_1 - 1)\tfrac{1}{2}\Delta^2 y_0 + u_1(u_1 - 1)(u_1 - 2)\tfrac{1}{6}\Delta^3 y_0$$
$$0 = (y_0 - y) + u_3\,\Delta y_0 + u_2(u_2 - 1)\tfrac{1}{2}\Delta^2 y_0$$
$$+ u_2(u_2 - 1)(u_2 - 2)\tfrac{1}{6}\Delta^3 y_0 + u_2(u_2 - 1)(u_2 - 2)(u_2 - 3)\tfrac{1}{24}\Delta^4 y_0 \tag{5.35}$$
$$\cdots \cdots \cdots \cdots \cdots \cdots \cdots \cdots \cdots \cdots$$
$$0 = (y_0 - y) + u_{i+1}\,\Delta y_0 + u_i(u_i - 1)\tfrac{1}{2}\Delta^2 y_0 + u_i(u_i - 1)(u_i - 2)\tfrac{1}{6}\Delta^3 y_0$$
$$+ u_i(u_i - 1)(u_i - 2)(u_i - 3)\tfrac{1}{24}\Delta^4 y_0$$

The first equation can be solved for u_0, the second for u_1, the third for u_2, etc. If the differences $\Delta^2 y_0$, $\Delta^3 y_0$, $\Delta^4 y_0$ decrease rather rapidly, the sequence $u_0, u_1, u_2, \ldots, u_i, \ldots$ converges quickly to the desired root of Eq. (5.33).

A little neater procedure that lends itself to the use of a synthetic calculational scheme is the following. The Gregory-Newton interpolation formula [see Eq. (3.43)] can be written in the form

$$y - y_0 = \sum_{r=1}^{m} \frac{u^{[r]}}{r!} \Delta^r y_0 = u \sum_{r=1}^{m} (u - 1)^{[r-1]} \left(\frac{1}{r!} \Delta^r y_0 \right) \qquad (5.36)$$

Therefore, one may use as the iteration equations

$$u_{i+1} = \frac{y - y_0}{\sum_{r=1}^{m_i} (u_i - 1)^{[r-1]} \left(\frac{1}{r!} \Delta^r y_0 \right)} \qquad i = 0, 1, 2, \ldots \qquad (5.37)$$

where m_i starts at 1 for $i = 0$ and increases to m after the first few iterations. These first few values of m_i can be chosen in any way one likes.

The iterates will generally approach a limiting value, which by Eq. (5.36) must be u, and the desired value of the argument corresponding to y is given approximately by $x = x_0 + uh$.

In addition to the methods of Chaps. 1 and 2, which can be applied to solve Eq. (5.34), there is the *reversion-of-the-series* method. For a discussion of the method Eq. (5.34) can be written in a more general form as

$$y - y_0 = a_1 u + a_2 u^2 + a_3 u^3 + a_4 u^4 + \cdots \qquad (5.38)$$

Let

$$w = \frac{y - y_0}{a_1} \qquad (5.39)$$

and assume that u can be expanded as a power series in w,

$$u = c_1 w + c_2 w^2 + c_3 w^3 + c_4 w^4 + \cdots \qquad (5.40)$$

Substituting this expression for u and $a_1 w$ for $y - y_0$ in Eq. (5.38) and equating the coefficients of like powers of w, one finds that

$$c_1 = 1$$

$$c_2 = -\frac{a_2}{a_1}$$

$$c_3 = -\frac{a_3}{a_1} + 2\left(\frac{a_2}{a_1}\right)^2$$

$$c_4 = -\frac{a_4}{a_1} + 5\frac{a_2 a_3}{a_1^2} - 5\left(\frac{a_2}{a_1}\right)^3 \qquad (5.41)$$

$$c_5 = -\frac{a_5}{a_1} + 6\frac{a_2 a_4}{a_1^2} + 3\left(\frac{a_3}{a_1}\right)^2 - 21\frac{a_2^2 a_3}{a_1^3} + 14\left(\frac{a_2}{a_1}\right)^4$$

$$c_6 = -\frac{a_6}{a_1} + 7\frac{a_2 a_5 + a_3 a_4}{a_1^2} - 28\frac{a_2^2 a_4 + a_2 a_3^2}{a_1^3} + 84\frac{a_2^3 a_3}{a_1^4} - 42\left(\frac{a_2}{a_1}\right)^5 \quad \text{etc.}$$

This method is not a very practical method of inverse interpolation.

5.6. Accuracy of Inverse Interpolation. The accuracy of inverse interpolation depends on the method used but in any case is limited by the following consideration. Suppose $y = f(x)$ is tabulated to N decimal places; then y is known to lie in an interval of length $\epsilon_y = 10^{-N}$ (see Fig. 5.1). By reference to the triangle of this figure it is seen that the range of possible values of x, denoted by ϵ_x, is approximately

$$\epsilon_x = \frac{\epsilon_y}{|f'(x)|} \qquad (5.42)$$

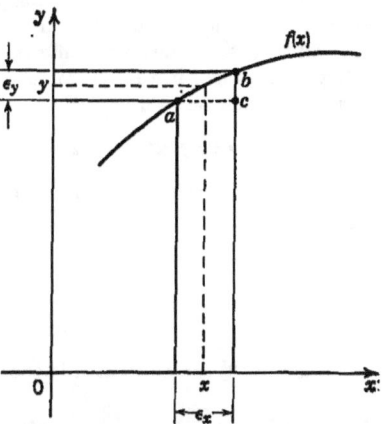

This means that x may be determined to a greater or a smaller number of decimal places than y depending, respectively, on whether

$$|f'(x)| = \frac{df(x)}{dx}.$$

is greater or less than one. Since this accuracy applies to the tabulated values of x, it is the upper limit

Fig. 5.1. Ultimate accuracy of any method of inverse interpolation.

on the accuracy with which inverse interpolation can yield the value of x for a given value of y and a given table. This error takes the place of the class B error of Sec. 3.14.

If one uses Lagrange's interpolation formula, one adds to the above uncertainty in the value of x the class A and class C errors arising from approximating the inverse function by a polynomial. It is important to note that the degree of the polynomial necessary to approximate the direct function is no guide to the degree of the polynomial required in approximating the inverse function.

If one uses the method of successive approximation, an additional error may be introduced in the process of solving the interpolating polynomial for x (or for u). This error, however, can be kept negligible by determining the root to a sufficient number of decimal places. On the other hand, one gains in this method by the fact that accidental and round-off errors tend to die out in subsequent calculations; moreover, the direct function is perhaps more likely to be approximated adequately by a polynomial than the inverse function.

Example 6. Given $y = \sin x$, find x when $y = 0.98000\ 000$ using Table 4.2.

Since the tabulated values of y are known to only eight decimal places, the range of uncertainty in their value is $\epsilon_y = 10^{-8}$. Therefore, by Eq. (5.42) the range of uncertainty in x is

$$\epsilon_x = \frac{10^{-8}}{\cos x} = \frac{10^{-8}}{(1 - 0.98^2)^{\frac{1}{2}}} = 5 \times 10^{-8}$$

Thus one can determine x to a little better than seven significant figures. If the table included values of cos x, it would be better to apply inverse interpolation to the cos x values, for in that case x could be determined to the full eight decimal places.

Method 1. *Method of Successive Approximation.* Write y as a polynomial in u and solve the polynomial for u when $y = 0.98000\ 000$.

By reference to Example 1 of Chap. 4 it is seen that a third-degree polynomial approximates the function $y = \sin x$ sufficiently well for the points tabulated. The contribution due to the fourth difference did not affect the eighth decimal places when interpolating for y in that example. We shall therefore assume that the Gregory-Newton formula [see Eq. (3.48)],

$$y = y_0 + u\,\Delta y_0 + u(u-1)(\tfrac{1}{2}\Delta^2 y_0) + u(u-1)(u-2)(\tfrac{1}{6}\Delta^3 y_0) \quad (5.43)$$

approximates $y = \sin x$ to eight decimal places.†
From Table 4.2,

$$
\begin{aligned}
y_0 &= 0.98215\ 432 \\
\Delta y_0 &= -0.00395\ 771 \\
\tfrac{1}{2}\Delta^2 y_0 &= -0.00019\ 5635 \\
\tfrac{1}{6}\Delta^3 y_0 &= 0.00000\ 0875
\end{aligned}
$$

Equation (5.43) can be written in the form

$$u = -\frac{1}{\Delta y_0}[(y - y_0) - u(u-1)(\tfrac{1}{2}\Delta^2 y_0) - u(u-1)(u-2)(\tfrac{1}{6}\Delta^3 y_0)]$$

which, on substituting the numerical values, becomes

$$u = 0.54433\ 498 - 0.04943\ 136u(u-1) - 0.00007\ 370u(u-1)(u-2) \quad (5.44)$$

The maximum value of $|u(u-1)|$ occurs at the value of u, assuming $0 < u < 1$, which makes

$$\frac{d}{du}u(u-1) = 2u - 1 = 0$$

Therefore
$$|u(u-1)| \leq \tfrac{1}{4} \quad (5.45)$$
This being true,
$$|u(u-1)(u-2)| \leq \tfrac{1}{4}|u-2| < \tfrac{1}{2} \quad (5.46)$$

From these inequalities it is seen that the first term in Eq. (5.44) is by far the dominant term; hence as an approximation for u, we take

$$u_0 = 0.54433$$

† This is confirmed by finding an upper limit for the remainder term

$$R = \frac{u(u-1)(u-2)(u-3)}{4!}h^4\left[\frac{d^4}{dx^4}\sin x\right]_{x=\xi}$$

Since by Eq. (5.45) $|u(u-1)| \leq \tfrac{1}{4}$ for $0 < u < 1$,

$$|u(u-1)(u-2)(u-3)| = |u(u-1)|\,|u-2|\,|u-3| < \tfrac{6}{4}$$

Moreover, since $|\sin \xi| < |$,

$$|R| < \left(\frac{1}{4!}\right)\left(\frac{6}{4}\right)(0.02)^4 = 10^{-8}$$

A better approximation is

$$u_1 = 0.54433 + 0.04943u_0(u_0 - 1) = 0.55659$$

and successively better approximations are

$$u_2 = 0.54433\ 498 - 0.04943\ 136u_1(u_1 - 1) - 0.00007\ 370u_1(u_1 - 1)(u_1 - 2)$$
$$= 0.55656\ 07$$
$$u_3 = 0.54433\ 498 - 0.04943\ 136u_2(u_2 - 1) - 0.00007\ 370u_2(u_2 - 1)(u_2 - 2)$$
$$= 0.55655\ 97$$

Since there is little change between u_2 and u_3, we shall take for the value of u

$$u = u_3 = 0.55655\ 97$$

Now

$$x = x_0 + hu = 1.76 + (0.02)(0.55655\ 97) = 1.76 + 0.01113\ 1194$$

which on rounding to seven decimal places gives

$$x = 1.77113\ 12$$

This is correct to all seven places.

Method 2. Use of Lagrange's General Interpolation Formula on the Inverse Function $x = \arcsin y$. If one uses a third-degree polynomial for this purpose,

$$x = \frac{(y - y_1)(y - y_2)(y - y_3)}{(y_0 - y_1)(y_0 - y_2)(y_0 - y_3)} x_0 + \frac{(y - y_0)(y - y_2)(y - y_3)}{(y_1 - y_0)(y_1 - y_2)(y_1 - y_3)} x_1$$
$$+ \frac{(y - y_0)(y - y_1)(y - y_3)}{(y_2 - y_0)(y_2 - y_1)(y_2 - y_3)} x_2 + \frac{(y - y_0)(y - y_1)(y - y_2)}{(y_3 - y_0)(y_3 - y_1)(y_3 - y_2)} x_3 \quad (5.47)$$

where

$$\begin{array}{ll} x_0 = 1.74 & y_0 = 0.98571\ 918 \\ x_1 = 1.76 & y_1 = 0.98215\ 432 \\ x_2 = 1.78 & y_2 = 0.97819\ 661 \\ x_3 = 1.80 & y_3 = 0.97384\ 763 \end{array}$$

Since $y = 0.98000\ 00$,

$$y - y_0 = -0.00571\ 918$$
$$y - y_1 = -0.00215\ 432$$
$$y - y_2 = 0.00180\ 339$$
$$y - y_3 = 0.00615\ 237$$

and

$$\begin{array}{ll} y_0 - y_1 = 0.00356\ 486 & y_1 - y_2 = 0.00395\ 771 \\ y_0 - y_2 = 0.00752\ 257 & y_1 - y_3 = 0.00830\ 669 \\ y_0 - y_3 = 0.01187\ 155 & y_2 - y_3 = 0.00434\ 898 \end{array}$$

Therefore

$$x = -0.07508\ 033x_0 + 0.54144\ 134x_1 + 0.58544\ 864x_2 - 0.05180\ 967x_3$$

These coefficients add up to $0.99999\ 998 \cong 1$, which serves as a check on our calculations. Putting in the values of the x's,

$$x = -0.13063\ 9774 + 0.95293\ 6758 + 1.04209\ 8579 - 0.09325\ 7406 = 1.77113\ 82$$

This answer is off by seven in the sixth decimal place.

If one goes to a fifth-degree polynomial, instead of a third, the improvement is not great (one then gets $x = 1.77113\ 02$), and the amount of computation involved is almost prohibitive. Thus this method fails to give sufficient accuracy. The explanation for this is that for the intervals involved the arcsin y is not sufficiently well approximated by a polynomial, as can be seen by evaluating the remainder term.

Since the higher derivatives of inverse functions are usually difficult to evaluate, one often does not know ahead of time whether the Lagrange method will give a sufficiently accurate value of x; therefore, it is generally better to employ the method of successive approximation.

5.7. Divided Differences. When the arguments are unequally spaced, the ordinary concept of differences becomes meaningless; however, it is possible to generalize the concept of a difference to take care of arbitrarily spaced arguments. Let the *first divided difference* of $f(x)$ for the points x_0 and x_1 be defined as

$$f(x_0,x_1) = \frac{f(x_0) - f(x_1)}{x_0 - x_1} \tag{5.48}$$

The *second divided difference* of $f(x)$ for the three points x_0, x_1, and x_2 is defined as

$$f(x_0,x_1,x_2) = \frac{f(x_0,x_1) - f(x_1,x_2)}{x_0 - x_2} \tag{5.49}$$

and the *nth divided difference* as

$$f(x_0,x_1,x_2, \ldots ,x_n) = \frac{f(x_0,x_1,x_2, \ldots ,x_{n-1}) - f(x_1,x_2, \ldots ,x_n)}{x_0 - x_n} \tag{5.50}$$

A divided-difference table can be constructed for any set of tabulated values of $f(x)$; it has the appearance of Table 5.1.

TABLE 5.1. DIVIDED-DIFFERENCE TABLE

x_0	$f(x_0)$			
x_1	$f(x_1)$	$f(x_0,x_1)$	$f(x_0,x_1,x_2)$	
x_2	$f(x_2)$	$f(x_1,x_2)$	$f(x_1,x_2,x_3)$	$f(x_0,x_1,x_2,x_3)$
x_3	$f(x_3)$	$f(x_2,x_3)$		

It can be shown that

$$f(x_0,x_1) = \frac{f(x_0)}{x_0 - x_1} + \frac{f(x_1)}{x_1 - x_0}$$

$$f(x_0,x_1,x_2) = \frac{f(x_0)}{(x_0 - x_1)(x_0 - x_2)} + \frac{f(x_1)}{(x_1 - x_0)(x_1 - x_2)}$$
$$+ \frac{f(x_2)}{(x_2 - x_0)(x_2 - x_1)}$$

$$\cdots \cdots \cdots \cdots \cdots \cdots \cdots \cdots$$

$$f(x_0,x_1,x_2, \ldots ,x_n) = \frac{f(x_0)}{(x_0 - x_1)(x_0 - x_2) \cdots (x_0 - x_n)} \tag{5.51}$$
$$+ \frac{f(x_1)}{(x_1 - x_0)(x_1 - x_2) \cdots (x_1 - x_n)}$$
$$+ \cdots + \frac{f(x_{n-1})}{(x_{n-1} - x_0) \cdots (x_{n-1} - x_{n-2})(x_{n-1} - x_n)}$$
$$+ \frac{f(x_n)}{(x_n - x_0) \cdots (x_n - x_{n-1})}$$

and that therefore the divided differences are symmetrical in all their arguments. This means that the arguments can be written in any order; thus

$$f(x_0,x_1,x_2) = f(x_1,x_0,x_2) = f(x_2,x_1,x_0) = \cdots$$

Since by Eq. (5.48)

$$f(x,x_0) = \frac{f(x) - f(x_0)}{x - x_0}$$

when $f(x)$ is a polynomial of degree n, $f(x,x_0)$ is a polynomial of degree $n - 1$. Likewise the second divided difference

$$f(x,x_0,x_1) = \frac{f(x,x_0) - f(x_0,x_1)}{x - x_1}$$

is of degree $n - 2$. Proceeding in this way, it is seen that the nth divided difference of an nth-degree polynomial is a constant and that all higher divided differences are zero.

If two or more of the arguments coincide, the divided difference can still be given a meaning. The meaning assigned is obtained by taking the limit; thus

$$f(x_0,x_0) = \lim_{x_1 \to x_0} \frac{f(x_0) - f(x_1)}{x_0 - x_1} = f'(x_0) \tag{5.52}$$

By a similar treatment†

$$f(\underset{r+1 \text{ arguments}}{x_0,x_0, \ldots ,x_0}) = \frac{1}{r!} \frac{d^r}{dx^r} f(x_0) \tag{5.53}$$

and

$$f(x,x,x_0,x_1, \ldots ,x_n) = \lim_{x' \to x} \frac{f(x',x_0,x_1, \ldots ,x_n) - f(x,x_0,x_1, \ldots ,x_n)}{x' - x}$$

$$= \frac{d}{dx} f(x,x_0,x_1, \ldots ,x_n) \tag{5.54}$$

From the definition of divided differences contained in Eqs. (5.48) to (5.50),

$$\begin{aligned}
f(x) &= f(x_0) + (x - x_0)f(x,x_0) \\
f(x,x_0) &= f(x_0,x_1) + (x - x_1)f(x,x_0,x_1) \\
f(x,x_0,x_1) &= f(x_0,x_1,x_2) + (x - x_2)f(x,x_0,x_1,x_2)
\end{aligned} \tag{5.55}$$

$$\cdots\cdots\cdots\cdots\cdots\cdots\cdots\cdots\cdots$$

$$f(x,x_0,x_1, \ldots ,x_{n-1}) = f(x_0,x_1, \ldots ,x_n) + (x - x_n)f(x,x_0,x_1, \ldots ,x_n)$$

Multiplying the second equation by $x - x_0$, the third by $(x - x_0)(x - x_1)$,

† See J. F. Steffensen, "Interpolation," 2d ed., pp. 20–21, Chelsea Publishing Company, New York, 1950.

the fourth by $(x - x_0)(x - x_1)(x - x_2)$, etc., the last being multiplied by $\Pi_{i=0}^{n-1}(x - x_i)$, one obtains, on adding, the equation

$$f(x) = f(x_0) + (x - x_0)f(x_0,x_1) + (x - x_0)(x - x_1)f(x_0,x_1,x_2)$$
$$+ (x - x_0)(x - x_1)(x - x_2)f(x_0,x_1,x_2,x_3) + \cdots$$
$$+ (x - x_0)(x - x_1) \cdots (x - x_{n-1})f(x_0,x_1, \ldots ,x_n) + R \quad (5.56)$$

where the remainder R is

$$R = f(x,x_0,x_1, \ldots ,x_n) \prod_{i=0}^{n} (x - x_i) \quad (5.57)$$

This formula, due to Newton, is called *Newton's divided-difference interpolation formula.*

Equation (5.56) can be written in the form

$$f(x) = P_n(x) + R \quad (5.58)$$

If we assume that the derivative $f^{(n+1)}(x)$ exists and is continuous, the function $f(x,x_0,x_1, \ldots ,x_n)$ is finite, and therefore, by Eq. (5.57), $R \equiv R(x)$ vanishes at the $n + 1$ points $x_0, x_1, \ldots, x_n$. Thus, by Rolle's theorem, $R'(x)$ vanishes at n points lying between the smallest and largest of these arguments, $R''(x)$ at, at least, $n - 1$ points, etc., and finally $R^{(n)}(x)$ vanishes for some point $x = \xi$. Taking the nth derivative of Eq. (5.56) and substituting $x = \xi$, one has $R^{(n)}(\xi) = 0$, and therefore

$$f^{(n)}(\xi) = n!f(x_0,x_1,x_2, \ldots ,x_n)$$

This may be written

$$f(x_0,x_1,x_2, \ldots x_n) = \frac{1}{n!}f^{(n)}(\xi) \quad (5.59)$$

where ξ lies somewhere in the interval of the set of arguments $x_0, x_1, \ldots,$ x_n. This important formula provides one with an upper and lower limit for the value of a divided difference. By this formula

$$f(x,x_0,x_1, \ldots ,x_n) = \frac{1}{(n + 1)!}f^{(n+1)}(\xi) \quad (5.60)$$

where ξ lies in the interval of the set of arguments $x, x_0, x_1, \ldots, x_n$. Using this result, the remainder given in Eq. (5.57) can be expressed as

$$R = \frac{1}{(n + 1)!}f^{(n+1)}(\xi) \prod_{i=0}^{n} (x - x_i) \quad (5.61)$$

where ξ lies in the interval stated.

This is the same remainder term as given in Eq. (5.10) for Lagrange's formula. That they must agree can be seen from the fact that they each represent the difference between $f(x)$ and a polynomial of the nth degree that has the same value as $f(x)$ at $x = x_0, x_1, \ldots,$ and x_n. Since the polynomial is unique, this difference is unique.

Moreover, formula (5.56) includes as special cases the Gregory-Newton formulas and the formulas of Gauss. For example, if $x_{2i-1} = ih$ and $x_{2i} = -ih$, the formula reduces to the forward-interpolation formula of Gauss. The remainder term will, however, ordinarily involve a $(n + 1)$th divided difference in place of the $(n + 1)$th derivative at the unprescribed point ξ [see, however, Eq. (5.60)].

We may summarize by saying that the remainder term for any interpolation formula representing a polynomial that fits $f(x)$ at $x_0, x_1, \ldots, x_n$ can always be expressed in either of the two forms

$$R = \frac{1}{(n + 1)!} f^{(n+1)}(\xi) \prod_{i=0}^{n} (x - x_i) = f(x, x_0, \ldots, x_n) \prod_{i=0}^{n} (x - x_i)$$

$$(5.62)$$

where ξ lies between the least and the greatest of the numbers $x, x_0, \ldots, x_n$.

If the interval is constant and equal to h, the divided-difference table (see Table 5.1) reduces not to an ordinary difference table but to one like Table 5.2.

TABLE 5.2. DIVIDED-DIFFERENCE TABLE FOR EQUALLY SPACED ARGUMENTS

x_0	$f(x_0)$			
		$\frac{1}{h}\Delta f(x_0)$		
x_1	$f(x_1)$		$\frac{1}{2!h^2}\Delta^2 f(x_0)$	
		$\frac{1}{h}\Delta f(x_1)$		$\frac{1}{3!h^3}\Delta^3 f(x_0)$
x_2	$f(x_2)$		$\frac{1}{2!h^2}\Delta^2 f(x_1)$	
		$\frac{1}{h}\Delta f(x_2)$		
x_3	$f(x_3)$			

An nth-order divided difference has the same physical dimensions as the corresponding derivative and not those of an ordinary difference.

PROBLEMS

1. Using the technique of Example 1 and a four-point Lagrangian interpolation formula, make a table of the logarithms of all integers up to 20.
2. Determine the accuracy of the table made in Prob. 1, evaluating the class A, B, and C errors in your computation (see also Appendix A).

3. Determine directly for what values $x = x_k$ the right-hand sides of Eqs. (5.25) yield the correct value of $f(x)$.

4. Obtain from Eqs. (5.25) the formula for linear interpolation. What is the error term in this case?

5. Find the value of x for which $J_0(x) = \pi/20$ using the table of Example 5 and (a) Lagrange's interpolation formula and (b) the method of successive approximations.

6. Solve Prob. 5 using the reversion-of-the-series method.

7. Determine the ultimate accuracy with which the inverse interpolation in Prob. 5 can be made.

8. Illustrate by means of a graph the meaning of the first and second divided differences.

9. Prove by mathematical induction that a divided difference of any order is a symmetrical function of its arguments as given in Eq. (5.51).

10. Show that Eq. (5.53) follows from the limiting process.

11. Construct a divided-difference table from a table of the fourth power of all prime numbers less than 20.

12. Find the fourth power of 23 by adding on a diagonal line of divided difference to the table in Prob. 11. (Start with the fourth difference column and work to the left.)

13. The area of a regular polygon of n sides may be written in the form $A = C(n)s^2$, where s is the length of one of the sides. From the value of $C(n)$ for $n = 3, 4$, and 6 interpolate and extrapolate for $C(5)$ and $C(7)$, respectively. Compare these to the correct values.

CHAPTER 6

SUMMATION OF SERIES

Summation plays the same role with respect to differencing that integration plays to differentiation. In fact, as will be brought out in this chapter, there is a very close analogy between the set of operators ∇ and $\nabla^{-1} \equiv \Sigma$ and the set $D = d/dx$ and $D^{-1} \equiv \int dx$, respectively. This analogy involves the use of the factorials $u^{(k)}$ in place of the powers x^k.

6.1. Properties of Sums. By definition, for m and n integers and $m \leq n$,†

$$\sum_{m}^{n} A_i = A_m + A_{m+1} + A_{m+2} + \cdots + A_n = \sum_{i=m}^{n} A_i \qquad (6.1)$$

and

$$\sum_{m}^{n} A_i = \sum_{m}^{k-1} A_i + \sum_{k}^{n} A_i \qquad (6.2)$$

We shall generalize the definition of Eq. (6.1) by requiring that Eq. (6.2) hold for all integral values of k, m, and n.

Letting $k = m$ in Eq. (6.2), we find that

$$\sum_{m}^{m-1} A_i = 0 \qquad (6.3)$$

---

† A closer analogy to the limits in integration can be had by using the definition, for $m \leq n$,

$$\sum_{m}^{n} A_i = A_m + A_{m+1} + \cdots + A_{n-1}$$

since then

$$\sum_{m}^{n} A_i = \sum_{m}^{k} A_i + \sum_{k}^{n} A_i \qquad \sum_{n}^{n} A_i = 0 \qquad \text{and} \qquad \sum_{n}^{m} A_i = -\sum_{m}^{n} A_i$$

This procedure is used by many authors of books on finite differences, e.g., Charles Jordan, "Calculus of Finite Differences," 2d ed., p. 117, Chelsea Publishing Company, New York, 1950. The main objection to this procedure is the almost universal use in most fields of mathematics of the convention adopted in Eq. (6.1).

By virtue of this result, if we let $m = n + 1$ in Eq. (6.2),

$$\sum_{n+1}^{k-1} A_i + \sum_{k}^{n} A_i = \sum_{n+1}^{n} A_i = 0$$

or

$$\sum_{k}^{n} A_i = - \sum_{n+1}^{k-1} A_i \qquad (6.4)$$

This formula permits one to assign a meaning to a summation in which the upper limit is less than the lower. For instance, let $n = k - 2$; then

$$\sum_{k}^{k-2} A_i = - \sum_{k-1}^{k-1} A_i = -A_{k-1} \qquad (6.5)$$

Whenever one wishes to emphasize the fact that the A_i are functions of the discrete variable i, one may write $F(i)$ in place of A_i.

6.2. Indefinite Sums. If $F(i)$ and $G(i)$ are functions of the discrete variable i that takes on only integral values and

$$\nabla G(i) = G(i) - G(i - 1) = F(i) \qquad (6.6)$$

then $F(i)$ is the *first ascending difference* of $G(i)$ (see Sec. 3.15). On the other hand, the general solution $G(i)$ of this equation corresponding to a given $F(i)$ is called the *indefinite sum of* $F(i)$ and is written

$$G(i) = \nabla^{-1}F(i) = \Sigma F(i) \qquad (6.7)$$

Suppose $G_1(i)$ is a particular solution of Eq. (6.6); then $G_1(i) + C$, where C is an arbitrary constant, is also a solution. Moreover, $G_1(i) + C$ is the general solution. This follows from the fact that, if

$$G_2(i) = G_1(i) + H(i)$$

is any other solution of Eq. (6.6),

$$\nabla H(i) = \nabla[G_2(i) - G_1(i)] = \nabla G_2(i) - \nabla G_1(i) = 0$$

Thus $H(i) = H(i - 1)$ for all values of i, and hence $H(i)$ is a constant. The indefinite sum, like the indefinite integral, is therefore undefined to the extent of an arbitrary addition constant C.

A useful solution of Eq. (6.6) can be had by changing the variable to j and summing both sides from $j = 1$ to $j = i$, for then one obtains the equation

$$\sum_{j=1}^{i} F(j) = \sum_{j=1}^{i} G(j) - \sum_{j=1}^{i} G(j - 1) = G(i) - G(0)$$

and as a consequence

$$\Sigma F(i) = G(i) = G(0) + \sum_{j=1}^{i} F(j) \qquad (6.8)$$

Here $G(0)$ plays the role of the arbitrary constant. This equation can also be used to obtain a definite sum from the indefinite sum. Thus

$$\sum_{j=1}^{i} F(j) = G(i) - G(0) = \left[\sum F(i)\right]_0^i \qquad (6.9)$$

and in general $\qquad \sum_{i=m}^{n} F(i) = \left[\sum F(i)\right]_{i=m-1}^{i=n} \qquad (6.10)$

Note that, since the arbitrary constant in the indefinite sum drops out in these equations, one may use any particular value of the constant one finds expedient.

Example 1. Find the formula for the sum of the first n odd numbers.

Since $\qquad\qquad\qquad \nabla i^2 = 2i - 1 \qquad\qquad\qquad (6.11)$
the indefinite sum of $2i - 1$ is

$$\Sigma(2i - 1) = i^2 + C \qquad (6.12)$$

Therefore by Eq. (6.9)

$$\sum_{i=1}^{n} (2i - 1) = [i^2 + C]_0^n = n^2 \qquad (6.13)$$

As mentioned above, one may take $C = 0$ and eliminate it immediately from the calculations.

6.3. Summation for a Continuous Argument. Since the indefinite sum is defined as the general solution of the difference equation (6.6), the term may be used in connection with a function of a continuous variable. Using the definition of the first ascending difference $\nabla g(x)$ of a continuous function $g(x)$, one obtains, in place of Eq. (6.6), the difference equation

$$\nabla g(x) = g(x) - g(x - h) = f(x) \qquad (6.14)$$

Therefore the indefinite sum of $f(x)$ is

$$\Sigma f(x) = \nabla^{-1} f(x) = g(x) \qquad (6.15)$$

where $g(x)$ is the general solution of Eq. (6.14).

If $g_1(x)$ is a particular solution of Eq. (6.14) and $g_2(x) = g_1(x) + \omega(x)$ is any other solution,

$$\nabla \omega(x) = \nabla g_2(x) - \nabla g_1(x) = 0$$

or from the definition of the ascending difference $\omega(x) = \omega(x - h)$. This latter requires that $\omega(x)$ be a periodic function† of period h. Thus the indefinite sum of a continuous function is a particular solution $g_1(x)$ of Eq. (6.14) plus an arbitrary periodic function $\omega(x)$.

Just as the difference of a given function involves the interval h as a parameter, so does the sum. For this reason it is usually helpful to first

† This includes the special case of $\omega(x)$ being a constant.

express the function to be summed in terms of the variable $u = (x - x_0)/h$, where x_0 is any convenient value of x. In terms of the argument u, Eqs. (6.14) and (6.15) become

$$\nabla G(u) = G(u) - G(u - 1) = F(u) \tag{6.16}$$

and

$$\Sigma F(u) = \nabla^{-1} F(u) = G(u) \tag{6.17}$$

Here $G(u)$ is the general solution of Eq. (6.16) and is given by $G_1(u) + \Omega(u)$, where $G_1(u)$ is a particular solution and $\Omega(u)$ an arbitrary periodic function of period 1.

The *definite sum* for a function $f(x)$ of a continuous variable may be defined as

$$[\Sigma f(x)]_a^b = g(b) - g(a) \tag{6.18}$$

where $g(x)$ is the indefinite sum of $f(x)$. If $b = a + nh$, where n is an integer, the arbitrary periodic function $\omega(x)$ drops out, and the definite sum of (6.18) has a unique value. This value is $g_1(b) - g_1(a)$, where $g_1(x)$ is any particular solution of Eq. (6.14).

6.4. Ascending Factorials. The ascending factorials $u^{[k]}$ introduced in Chap. 3 to simplify writing the Gregory-Newton formula for backward interpolation are very useful in summation problems. They are defined in terms of the usual factorial by the formula

$$u^{[k]} = (u + k - 1)^{[k]} \tag{6.19}$$

Thus for n a positive integer

$$u^{[n]} = u(u + 1)(u + 2) \cdots (u + n - 1) \tag{6.20}$$

$$u^{[-n]} = \frac{1}{(u - 1)(u - 2) \cdots (u - n)} \tag{6.21}$$

and $u^{[0]} = 1$.

The important characteristic of these factorials is that

$$\nabla u^k = k u^{[k-1]} \tag{6.22}$$

and hence for $k \neq -1$

$$\sum u^{[k]} = \frac{u^{[k+1]}}{k + 1} + \Omega \tag{6.23}$$

where $\Omega = \Omega(u)$ is again an arbitrary periodic function of unit period. If one introduces the discrete variable i in place of u, this equation becomes, again for $k \neq -1$,

$$\sum i^{[k]} = \frac{i^{[k+1]}}{k + 1} + C \tag{6.24}$$

where C is an arbitrary constant.

As a practical consequence of Eq. (6.24), by virtue of Eq. (6.9),

$$\sum_{i=1}^{n} i^{\{m\}} = \left[\frac{i^{\{m+1\}}}{m+1}\right]_{0}^{n} = \frac{1}{m+1}\, n^{\{m+1\}} \quad \text{for } m \geq 0 \quad (6.25)$$

Thus, for example, for $m = 1$ and $m = 2$ one has

$$\tfrac{1}{2}n(n+1) = 1 + 2 + 3 + \cdots + n$$

and

$$\tfrac{1}{3}n(n+1)(n+2) = (1)(2) + (2)(3) + (3)(4) + \cdots + n(n+1)$$

Since any polynomial in i can be expressed in the form

$$P_m(i) = \sum_{k=0}^{m} a_k i^{\{k\}} \qquad (6.26)$$

one may readily obtain the sum of any series whose terms are expressible as a polynomial. Thus by Eq. (6.25)

$$\sum_{i=1}^{n} P_m(i) = \sum_{k=0}^{m} a_k \left(\sum_{i=1}^{n} i^{\{k\}}\right) = \sum_{k=0}^{m} \frac{a_k}{k+1}\, n^{\{k+1\}} \qquad (6.27)$$

Example 2. Find the formula for the sum of the first n cubes.
First one expresses the cube of a number i in factorial form. One lets

$$i^3 = ai^{\{3\}} + bi^{\{2\}} + ci^{\{1\}} + d$$
$$= i\{(i+1)[a(i+2) + b] + c\} + d$$

and observes that d, c, b, a are obtained as remainders in dividing in turn by $i, i + 1$, and $i + 2$. This may be done very conveniently using the following scheme (see related schemes of Chap. 3):

$$
\begin{array}{cccc}
1 & 0 & 0 & 0 = d \\
  & -1 & 1 & \\
\hline
1 & -1 & 1 = c & \\
  & -2 & & \\
\hline
1 = a & -3 = b & &
\end{array}
$$

Thus
$$i^3 = i^{\{3\}} - 3i^{\{2\}} + i$$

Therefore by Eq. (6.27)

$$\sum_{i=1}^{n} i^3 = \tfrac{1}{4}n^{\{4\}} - n^{\{3\}} + \tfrac{1}{2}n^{\{2\}} = \left[\frac{n(n+1)}{2}\right]^2 \qquad (6.28)$$

Example 3. Evaluate the series

$$S_n = \sum_{i=1}^{n} \frac{1}{4i^2 - 1} = \frac{1}{3} + \frac{1}{15} + \frac{1}{35} + \cdots + \frac{1}{4n^2 - 1}$$

Observe that

$$S_n = \tfrac{1}{4} \sum_{i=1}^{n} \frac{1}{(i + \tfrac{1}{2})(i - \tfrac{1}{2})} = \tfrac{1}{4} \sum_{i=1}^{n} (i + \tfrac{3}{2})^{[-2]}$$

Therefore, since by Eq. (6.24)

$$\sum (i + \tfrac{3}{2})^{[-2]} = -(i + \tfrac{3}{2})^{[-1]} + C = \frac{-1}{i + \tfrac{1}{2}} + C$$

one has by Eq. (6.9)

$$S_n = \sum_{i=1}^{n} \frac{1}{4i^2 - 1} = \frac{1}{4} \left[\frac{-1}{i + \tfrac{1}{2}} \right]_0^n = \frac{n}{2n + 1} \qquad (6.29)$$

Note that the sum of the infinite series obtained by letting $n \to \infty$ is $S_\infty = \tfrac{1}{2}$.

6.5. Operators Δ^{-1} and δ^{-1}.

Consider the difference equations

$$\Delta r(x) = \nabla r(x + h) = f(x) \qquad (6.30)$$

and

$$\delta s(x) = \nabla s\left(x + \frac{h}{2}\right) = f(x) \qquad (6.31)$$

By definition, the solutions of these equations, for a given $f(x)$, are $r(x) = \Delta^{-1}f(x)$ and $s(x) = \delta^{-1}f(x)$. On the other hand, by Eqs. (6.14) and (6.15),

$$\Sigma f(x) = r(x + h) = \Delta^{-1}f(x + h) \qquad (6.32)$$

and

$$\sum f(x) = s\left(x + \frac{h}{2}\right) = \delta^{-1}f\left(x + \frac{h}{2}\right) \qquad (6.33)$$

If one deals with a function of $u = (x - x_0)/h$, the above equations become

$$\Sigma F(u) = \Delta^{-1}F(u + 1) = \delta^{-1}F(u + \tfrac{1}{2}) \qquad (6.34)$$

and the same equations hold, of course, when u is replaced by the discrete argument i that assumes only integral values. Thus the indefinite sum can be simply obtained from any of the inverse difference operators; in operator form

$$\Sigma = \nabla^{-1} = \Delta^{-1}E = \delta^{-1}E^{\frac{1}{2}} \qquad (6.35)$$

Since by Eq. (3.23)

$$\Delta u^{[k+1]} = (k + 1)u^{[k]}$$

one has†

$$\Delta^{-1}u^{[k]} = \frac{u^{[k+1]}}{k + 1} + \Omega$$

† $\Omega = \Omega(u)$ is a periodic function of unit period (see Sec. 6.3).

and therefore†

$$\sum u^{[k]} = \Delta^{-1}(u+1)^{[k]} = \frac{(u+1)^{[k+1]}}{k+1} + \Omega \qquad (6.36)$$

Thus one may sum readily any polynomial expressed either in the ascending factorials $u^{[k]}$ or the descending factorials $u^{[k]}$.

Example 4. Show that

$$\sum_{i=1}^{n} \sin i\lambda = \frac{\sin(n\lambda/2) \sin[(n+1)\lambda/2]}{\sin(\lambda/2)} \qquad (6.37)$$

By elementary trigonometry

$$\delta \cos i\lambda = \cos(i + \tfrac{1}{2})\lambda - \cos(i - \tfrac{1}{2})\lambda = -2\sin\frac{\lambda}{2}\sin i\lambda$$

and hence

$$\delta^{-1}\sin i\lambda = -\frac{\cos i\lambda}{2\sin(\lambda/2)} + C$$

Thus by Eq. (6.34), with u replaced by i,

$$\sum_{i=1}^{n}\sin i\lambda = \left[\sum\sin i\lambda\right]_0^n = [\delta^{-1}\sin(i + \tfrac{1}{2})\lambda]_0^n = \frac{-1}{2\sin(\lambda/2)}[\cos(i + \tfrac{1}{2})\lambda]_0^n$$

$$= \frac{\cos(\lambda/2) - \cos(n + \tfrac{1}{2})\lambda}{2\sin(\lambda/2)} = \frac{\sin(n\lambda/2)\sin[(n+1)\lambda/2]}{\sin(\lambda/2)}$$

Example 5. Derive the formula

$$\sum_{i=1}^{\infty}\frac{1}{i^{\{m\}}} = \sum_{i=1}^{\infty}\frac{1}{i(i+1)(i+2)\cdots(i+m-1)} = \frac{1}{(m-1)(m-1)!} \qquad (6.38)$$

for m a positive integer greater than 1.

Since by Eq. (6.36)

$$\sum\frac{1}{i^{\{m\}}} = \sum(i-1)^{[-m]} = -\frac{i^{[-m+1]}}{m-1} + C$$

one has

$$\sum_{i=1}^{\infty}\frac{1}{i^{\{m\}}} = \frac{-1}{m-1}\left[\frac{1}{(i+1)(i+2)\cdots(i+m-1)}\right]_0^{\infty} = \frac{1}{(m-1)(m-1)!}$$

6.6. Summation by Parts. Let

$$\nabla G_1(i) = F_1(i)$$
$$\nabla G_2(i) = F_2(i) \qquad (6.39)$$

Since

$$\nabla[F_1(i)G_2(i)] = F_1(i)G_2(i) - F_1(i-1)G_2(i-1)$$
$$= F_1(i)[G_2(i) - G_2(i-1)] + G_2(i-1)[F_1(i) - F_1(i-1)]$$

† Since $u = (x - x_0)/h$, $Eu = (x + h - x_0)/h = u + 1$.

or, in terms of differences,

$$\nabla[F_1(i)G_2(i)] = F_1(i)\,\nabla G_2(i) + G_2(i-1)\,\nabla F_1(i) \qquad (6.40)$$

by the definition of an indefinite sum (see Sec. 6.2)

$$F_1(i)G_2(i) = \Sigma[F_1(i)\,\nabla G_2(i)] + \Sigma[G_2(i-1)\,\nabla F_1(i)]$$

Likewise from Eqs. (6.39) $G_1(i) = \Sigma F_1(i)$, and $G_2(i) = \Sigma F_2(i)$; therefore this equation may be written

$$\Sigma[F_1(i)F_2(i)] = F_1(i)\Sigma F_2(i) - \Sigma[\nabla F_1(i)\Sigma F_2(i-1)] \qquad (6.41)$$

Instead of letting $G_1(i)$ and $G_2(i)$ be the general solutions of Eqs. (6.39), one may let them be any particular solution. Thus, if one lets $G_2(i) = \Sigma_{k=r}^{i} F_2(k)$, Eq. (6.41) is replaced by

$$\sum F_1(i)F_2(i) = F_1(i)\sum_{k=r}^{i} F_2(k) - \sum\left[\nabla F_1(i)\sum_{k=r}^{i-1} F_2(k)\right]$$

and therefore, by Eq. (6.10),

$$\sum_{i=r}^{n} F_1(i)F_2(i) = \left[F_1(i)\sum_{k=r}^{i} F_2(k)\right]_{i=r-1}^{i=n} - \left[\sum\left(\nabla F_1(i)\sum_{k=r}^{i-1} F_2(k)\right)\right]_{i=r-1}^{i=n}$$

Making use of Eqs. (6.3) and (6.10), this equation reduces at once to the useful equation

$$\sum_{i=r}^{n} F_1(i)F_2(i) = F_1(n)\sum_{k=r}^{n} F_2(k) - \sum_{i=r}^{n}\left[\nabla F_1(i)\sum_{k=r}^{i-1} F_2(k)\right] \qquad (6.42)$$

Since $\nabla F_1(i) = \Delta F_1(i-1)$ and the first term in the brackets vanishes, Eq. (6.42) may be written

$$\sum_{i=r}^{n} F_1(i)F_2(i) = F_1(n)\sum_{k=r}^{n} F_2(k) - \sum_{i=r+1}^{n}\left[\Delta F_1(i-1)\sum_{k=r}^{i-1} F_2(k)\right]$$

which on changing the summation variable simplifies to

$$\sum_{i=r}^{n} F_1(i)F_2(i) = F_1(n)\sum_{k=r}^{n} F_2(k) - \sum_{i=r}^{n-1}\left[\Delta F_1(i)\sum_{k=r}^{i} F_2(k)\right] \qquad (6.43)$$

Example 6. Derive the formula

$$\sum_{i=1}^{n} i2^{i-1} = (n-1)2^n + 1 \qquad (6.44)$$

Since

$$\nabla 2^k = 2^k - 2^{k-1} = 2^{k-1}$$

and, therefore,

$$\Sigma 2^{k-1} = 2^k + C$$

one has

$$\sum_{k=1}^{i} 2^{k-1} = [2^k]_0^i = 2^i - 1 \qquad (6.45)$$

Applying formula (6.42),

$$\sum_{i=1}^{n} i2^{i-1} = n \sum_{k=1}^{n} 2^{k-1} - \sum_{i=1}^{n} \left(\sum_{k=1}^{i-1} 2^{k-1} \right)$$

and by Eq. (6.45)

$$\sum_{i=1}^{n} i2^{i-1} = n(2^n - 1) - \sum_{i=1}^{n} (2^{i-1} - 1)$$

$$= n(2^n - 1) - [(2^n - 1) - n] = (n - 1)2^n + 1$$

6.7. Transformation of Infinite Series. Consider the series

$$S_n = \sum_{i=0}^{n} (-1)^i \frac{1}{i+1} \tag{6.46}$$

and the limiting value of S_n as $n \to \infty$. Calling this limiting value S,

$$S = \lim_{n \to \infty} S_n = \ln 2$$

It is known from the theory of alternating series that S lies between S_n and S_{n+1}. Let $n = 999$; then from Eq. (6.46)

$$S_{999} = S_{1,000} - \tfrac{1}{1000}$$

and S is therefore known to only about three decimal places. It would take a million terms to evaluate S to six decimal places.

In this section we shall consider various transformations of an infinite series that may be used to obtain a more rapidly convergent series.

Euler's Transformation. Consider the power series

$$S = \sum_{i=0}^{\infty} a_i x^i \tag{6.47}$$

under the assumption that the series converges and hence $a_n x^n$ approaches zero as n approaches infinity. Since

$$\nabla x^{i+1} = x^{i+1} - x^i = (x - 1)x^i$$

one has

$$\sum x^i = \frac{x^{i+1}}{x - 1}$$

and by Eq. (6.10)

$$\sum_{k=0}^{i} x^k = \frac{1 - x^{i+1}}{1 - x} \tag{6.48}$$

One may therefore investigate the effect of summing by parts in Eq. (6.47).

To make the procedure more rigorous consider first the sum S_n of the first $n + 1$ terms of (6.47). Thus, applying Eq. (6.43) with $r = 0$, $F_1(i) = a_i$, and $F_2(i) = x^i$,

$$S_n = \sum_{i=0}^{n} a_i x^i = a_n \left(\frac{1 - x^{n+1}}{1 - x} \right) - \sum_{i=0}^{n-1} \left(\Delta a_i \frac{1 - x^{i+1}}{1 - x} \right)$$

$$= \frac{1}{1 - x} \left(a_n - a_n x^{n+1} - \sum_{i=0}^{n-1} \Delta a_i + \sum_{i=0}^{n-1} \Delta a_i \, x^{i+1} \right)$$

Since, however,

$$\sum_{i=0}^{n-1} \Delta a_i = a_n - a_0$$

one has

$$S_n = \frac{a_0 - a_n x^{n+1}}{1 - x} + \frac{x}{1 - x} \sum_{i=0}^{n-1} \Delta a_i \, x^i \tag{6.49}$$

Now, by assumption, $a_n x^n$ approaches zero as n approaches infinity; therefore

$$S = \lim_{n \to \infty} S_n = \frac{a_0}{1 - x} + \frac{x}{1 - x} \sum_{i=0}^{\infty} \Delta a_i \, x^i \tag{6.50}$$

provided, of course, that this series converges.

This transformed series of Eq. (6.50) may converge more rapidly than that of Eq. (6.47); moreover, one may apply this same transformation to the new series. Thus

$$\sum_{i=0}^{\infty} \Delta a_i \, x^i = \frac{\Delta a_0}{1 - x} + \frac{x}{1 - x} \sum_{i=0}^{\infty} \Delta^2 a_i \, x^i$$

and hence

$$S = \frac{1}{1 - x} \left(a_0 + \frac{x}{1 - x} \Delta a_0 \right) + \left(\frac{x}{1 - x} \right)^2 \sum_{i=0}^{\infty} \Delta^2 a_i \, x^i$$

Repeating the integration by parts r times, one obtains the equation

$$S = \frac{1}{1 - x} \sum_{i=0}^{r-1} \Delta^i a_0 \left(\frac{x}{1 - x} \right)^i + \left(\frac{x}{1 - x} \right)^r \sum_{i=0}^{\infty} \Delta^r a_i \, x^i \tag{6.51}$$

If the last term approaches zero as r approaches infinity, one obtains *Euler's transformation*[†]

$$S = \frac{1}{1-x} \sum_{i=0}^{\infty} \Delta^i a_0 \left(\frac{x}{1-x} \right)^i \tag{6.52}$$

of the infinite series in Eq. (6.47).

Note that if a_i is expressible as an mth-degree polynomial in i, one may set $r = m + 1$ in Eq. (6.51) and obtain a closed form for S.

Example 7. Transform the series in Eq. (6.46)

$$\ln 2 = \sum_{i=0}^{\infty} (-1)^i \frac{1}{i+1}$$

into a more rapidly convergent series.

This series is a special case of the power series in Eq. (6.47) since it is obtained from it by letting $x = -1$ and $a_i = 1/(i+1) = i^{[-1]}$. It can be readily shown by mathematical induction that[‡]

$$\Delta^k i^{[-1]} = (-1)^k k! i^{[-(k+1)]} = \frac{(-1)^k k!}{(i+1)(i+2) \cdots (i+k+1)}$$

and therefore

$$\Delta^k a_0 = \Delta^k 0^{[-1]} = \frac{(-1)^k}{k+1}$$

Since $x/(1-x) = -\frac{1}{2}$, by Eq. (6.52) the transformed series is[§]

$$\ln 2 = \frac{1}{2} \sum_{i=0}^{\infty} \frac{1}{(i+1)2^i} = \sum_{i=1}^{\infty} \frac{1}{i 2^i}$$

[†] Letting $y = x/(1-x)$, this series may be written $S = (1+y) \sum_{i=0}^{\infty} \Delta^i a_0 y^i$.

[‡] This equation surely holds for $k = 0$ and $k = 1$; and assuming that it holds for $k-1$, it must hold for k, since

$$\Delta^k i^{[-1]} = \Delta \Delta^{k-1} i^{[-1]} = (-1)^{k-1}(k-1)! \, \Delta i^{[-k]} = (-1)^k k! i^{[-(k+1)]}$$

Therefore by mathematical induction it holds for all k.

[§] The validity of Euler's transformation for this special case may be checked by using the well-known series

$$\ln (1+x) = \sum_{i=1}^{\infty} (-1)^{i-1} \frac{x^i}{i}$$

from which Eq. (6.46) is obtained by letting $x = 1$. If one lets $x = -\frac{1}{2}$, one has

$$\ln \frac{1}{2} = -\ln 2 = -\sum_{i=1}^{\infty} \frac{1}{i 2^i}$$

which verifies the transformation.

Kummer's Transformation. Suppose that one wishes to evaluate an infinite series such as

$$S = \sum_{i=r}^{\infty} a_i \qquad (6.53)$$

which cannot be summed directly. One may be able to find λ_i such that the sum

$$S' = \sum_{i=r}^{\infty} \lambda_i a_i \qquad (6.54)$$

may be obtained by elementary means and such that limit

$$\lim_{i \to \infty} \lambda_i = \lambda \qquad (6.55)$$

exists. In this case one sees directly that

$$S = \frac{1}{\lambda} S' + \sum_{i=r}^{\infty} \left(1 - \frac{\lambda_i}{\lambda}\right) a_i \qquad (6.56)$$

and one obtains thereby a more rapidly convergent series.

Example 8. Transform the series

$$S = \sum_{i=1}^{\infty} \frac{1}{i^2} \qquad (6.57)$$

into a more rapidly convergent series.

First we observe that by Eq. (6.38), for $m = 2$,

$$S' = \sum_{i=1}^{\infty} \frac{1}{i(i+1)} = 1$$

If we set

$$\lambda_i = \frac{i}{i+1}$$

then

$$S' = \sum_{i=1}^{\infty} \lambda_i \frac{1}{i^2} = 1$$

and by Eq. (6.56) (since λ, the limit of the λ_i as $i \to \infty$, is equal to 1)

$$S = \sum_{i=1}^{\infty} \frac{1}{i^2} = 1 + \sum_{i=1}^{\infty} \frac{1}{(i+1)i^2} \qquad (6.58)$$

a more rapidly convergent series.

Since all terms in the series of Eq. (6.58) are positive, any computation based on this series will give a lower bound for S. One may obtain an upper bound as follows.

From the given series

$$S - 1 = \sum_{i=1}^{\infty} \frac{1}{(i+1)^2}$$

and if one sets

$$\lambda_i = \frac{i+1}{i}$$

then

$$S' = \sum_{i=1}^{\infty} \lambda_i \frac{1}{(i+1)^2} = \sum_{i=1}^{\infty} \frac{1}{i(i+1)} = 1$$

Therefore, by Eq. (6.56), since λ is again 1 and $S - 1$ replaces S,

$$S = 2 - \sum_{i=1}^{\infty} \frac{1}{i(i+1)^2} \qquad (6.59)$$

If a set of numbers A_i can be found such that the numbers λ_i defined by the equations

$$\lambda_i a_i = -\Delta(A_i a_i) = A_i a_i - A_{i+1} a_{i+1} \qquad (6.60)$$

approach a definite limit λ as $i \to \infty$, then Eq. (6.54) is summable, and one has

$$S' = A_r a_r - \alpha \qquad (6.61)$$

where

$$\alpha = \lim_{i \to \infty} A_i a_i \qquad (6.62)$$

Having S', one may obtain the required sum S from Eq. (6.56).

Example 9. Derive the formula

$$\sum_{i=1}^{\infty} \frac{1}{i^2} = \sum_{i=1}^{r} \frac{1}{i^2} + r! \sum_{i=1}^{\infty} \frac{1}{i^2(i+1)(i+2) \cdots (i+r)} \qquad (6.63)$$

Equation (6.58) represents the special case $r = 1$ of this formula. One may apply Kummer's transformation to the series in Eq. (6.58) to again increase the speed of convergence. Consider, however, the more general series

$$S_k = \sum_{i=1}^{\infty} \frac{1}{i^2(i+1)(i+2) \cdots (i+k)} \qquad (6.64)$$

where k is a parameter taking on integral values.
If one takes $A_i = i$, then, since

$$\frac{a_{i+1}}{a_i} = \frac{i^2}{(i+1)(i+k+1)}$$

by Eq. (6.60)

$$\lambda_i = i - \frac{i^2}{i+k+1} = \frac{(k+1)i}{i+k+1}$$

and therefore

$$\lambda = \lim_{i \to \infty} \lambda_i = k + 1$$

Now $$A_i a_i = \frac{1}{i(i+1)(i+2) \cdots (i+k)}$$

and hence by Eq. (6.62)

$$A_1 a_1 - \alpha = \frac{1}{(k+1)!}$$

Since $$1 - \frac{\lambda_i}{\lambda} = 1 - \frac{i}{i+k+1} = \frac{k+1}{i+k+1}$$

by Eqs. (6.56) and (6.61)

$$S_k = \frac{1}{(k+1)(k+1)!} + (k+1)S_{k+1} \qquad (6.65)$$

This is a difference equation for S_k of Eq. (6.64).

Multiplying both sides of Eq. (6.65) by $k!$ and summing, we have

$$\sum_{k=0}^{r-1} k! S_k = \sum_{k=0}^{r-1} \frac{1}{(k+1)^2} + \sum_{k=0}^{r-1} (k+1)! S_{k+1}$$

Canceling equal terms on the right- and left-hand sides of this equation, one has

$$S_0 = \sum_{k=0}^{r-1} \frac{1}{(k+1)^2} + r! S_r$$

This is the required equation (6.63).

6.8. Laplace's Summation Formula. We shall next consider summation formulas relating the sum

$$\sum_{i=r}^{n} F(i) = \sum_{i=r}^{n} f(x_0 + ih)$$

to the integral of the function $f(x)$ that assumes the required values at the discrete points $x_i = x_0 + ih$.

As a simple example consider the integral

$$I_i = \int_{x_i}^{x_i+h} f(x) \, dx = h \int_0^1 f(x_i + uh) \, du \qquad (6.66)$$

By the Gregory-Newton formula for forward interpolation [see Eq. (3.48)]

$$f(x_i + uh) = f(x_i) + \sum_{k=1}^{m} \frac{u^{[k]}}{k!} \Delta^k f(x_i) + \frac{u^{[m+1]}}{(m+1)!} h^{m+1} f^{(m+1)}(\xi_i) \qquad (6.67)$$

where $x_i < \xi_i < x_{i+m}$; therefore

$$I_i = h \left[f(x_i) + \sum_{k=0}^{m} L_h \, \Delta^k f(x_i) + R_i \right] \qquad (6.68)$$

where
$$L_k = \frac{1}{k!} \int_0^1 u^{[k]} \, du \tag{6.69}$$

and
$$R_i = \frac{h^{m+1}}{(m+1)!} \int_0^1 f^{(m+1)}(\xi_i) u^{[m+1]} \, du$$

Since $u^{[m+1]}$ does not change sign in the range of integration, one may apply the mean-value theorem to obtain the simple expression

$$R_i = \frac{h^{m+1}}{(m+1)!} f^{(m+1)}(\xi_i') \int_0^1 u^{[m+1]} \, du = h^{m+1} L_{m+1} f^{(m+1)}(\xi_i') \tag{6.70}$$

where again $x_i < \xi_i' < x_{i+m}$.

If one sums both sides of Eq. (6.68), one obtains the equation

$$\sum_{i=0}^{n-1} \int_{x_i}^{x_i+h} f(x) \, dx = h \left[\sum_{i=0}^{n-1} f(x_i) + \sum_{k=1}^{m} L_k \sum_{i=0}^{n-1} \Delta^k f(x_i) + R \right] \tag{6.71}$$

where by Eq. (6.70)

$$R = \sum_{i=0}^{n-1} R_i = h^{m+1} L_{m+1} \sum_{i=0}^{n-1} f^{(m+1)}(\xi_i')$$

If $f^{(m+1)}(x)$ is assumed to be continuous, it is readily shown that there must exist some number ξ such that

$$\frac{1}{n} \sum_{i=0}^{n-1} f^{(m+1)}(\xi_i') = f^{(m+1)}(\xi)$$

where ξ lies between the smallest and largest of the ξ_i'; therefore

$$R = n h^{m+1} L_{m+1} f^{(m+1)}(\xi) \tag{6.72}$$

where $x_0 < \xi < x_{n+m}$.

Since
$$\sum_{i=0}^{n-1} \Delta^k f(x_i) = \Delta^{k-1} f(x_n) - \Delta^{k-1} f(x_0)$$

Eq. (6.71) may be written

$$\sum_{i=0}^{n-1} f(x_i) = \frac{1}{h} \int_{x_0}^{x_n} f(x) \, dx - \sum_{k=1}^{m} L_k [\Delta^{k-1} f(x_n) - \Delta^{k-1} f(x_0)] + R \tag{6.73}$$

where L_k and R are given by Eqs. (6.69) and (6.72), respectively. This is *Laplace's summation formula*. The right-hand side requires a knowledge of $f(x_i)$ for $i = 0, 1, 2, \ldots, m - 1$ and for $i = n, n + 1, \ldots,$ $n + m - 1$. The first few L_k are as follows:

$$L_1 = \frac{1}{2} \qquad L_2 = -\frac{1}{12} \qquad L_3 = \frac{1}{24} \qquad L_4 = -\frac{19}{720}$$

$$L_5 = \frac{3}{160} \qquad L_6 = -\frac{863}{60,480} \qquad L_7 = \frac{275}{24,192}$$

If the remainder term in Eq. (6.67) goes to zero as $m \to \infty$, R in Eq. (6.73) also goes to zero, and one obtains the formula

$$\sum_{i=0}^{n-1} f(x_i) = \frac{1}{h} \int_{x_0}^{x_n} f(x)\, dx - \sum_{k=1}^{\infty} L_k[\Delta^{k-1}f(x_n) - \Delta^{k-1}f(x_0)] \quad (6.74)$$

This formula, therefore, holds for any function that may be expanded in an infinite Gregory-Newton series.

This formula can also be used as a formula of numerical integration, since it expresses the integral in terms of the tabulated values and their differences. Various other numerical-integration formulas will be considered in Chap. 7.

Example 10. Show that

$$\sum_{i=1}^{n} \frac{1}{i} = \gamma + \ln n + \frac{1}{2n} - \sum_{k=2}^{\infty} \frac{a_k}{n(n+1) \cdots (n+k-1)} \quad (6.75)$$

where $a_2 = \frac{1}{12} \qquad a_3 = \frac{1}{12} \qquad a_4 = \frac{19}{120} \qquad a_5 = \frac{9}{20} \cdots$

and in general [see Eq. (6.69)]

$$a_k = \frac{1}{k} \left| \int_0^1 u^{[k]}\, du \right| = (k-1)!|L_k| \quad (6.76)$$

The constant γ is *Euler's constant*, and is given by

$$\gamma = \tfrac{1}{2} + \sum_{k=2}^{\infty} \frac{a_k}{k!} = 0.57721\ 56649 \cdots \quad (6.77)$$

Since the Gregory-Newton series for $1/(u + 1)$ is a convergent series,† one may let $x = u$ and substitute $f(x) = f(u) = 1/(u + 1)$ in Eq. (6.74). The pth difference of $f(i)$ is then, by Eq. (3.23),

$$\Delta^p \frac{1}{i+1} = \Delta^p i^{[-1]} = (-1)^p p!\, i^{[-(p+1)]} = \frac{(-1)^p p!}{(i+1)(i+2) \cdots (i+p+1)}$$

† The Gregory-Newton formula for $1/(u + 1)$ [see Eq. (3.48) for $m \to \infty$] is

$$\frac{1}{u+1} = \sum_{k=0}^{\infty} \frac{u^{[k]}}{k!} \Delta^k (u+1)^{-1} = \sum_{k=0}^{\infty} \frac{(-1)^k}{(k+1)!} u^{[k]}$$

which is convergent by the ratio test, for $u < 2$.

and therefore, by Eq. (6.74),

$$\sum_{i=1}^{n} \frac{1}{i} = \frac{1}{n} + \sum_{i=0}^{n-2} \frac{1}{i+1}$$

$$= \frac{1}{n} + \int_{0}^{n-1} \frac{1}{u+1} \, du - \sum_{k=1}^{\infty} L_k \left[\frac{(-1)^{k-1}(k-1)!}{(i+1)(i+2) \cdots (i+k)} \right]_{i=0}^{n-1} \quad (6.78)$$

Since, for $0 < u < 1$,

$$u^{[k]} = u(u-1) \cdots (u-k+1) = (-1)^{k-1} |u^{[k]}|$$

by Eq. (6.69)

$$(k-1)! L_k = \frac{(-1)^{k-1}}{k} \int_0^1 |u^{[k]}| \, du = (-1)^{k-1} a_k$$

where the u_k are the positive constants defined by Eq. (6.76). Thus, Eq. (6.78) becomes

$$\sum_{i=1}^{n} \frac{1}{i} = \ln n + \frac{1}{n} - \sum_{k=1}^{\infty} \frac{a_k}{n(n+1)(n+2) \cdots (n+k-1)} + \sum_{k=1}^{\infty} \frac{a_k}{k!}$$

This is the same as Eq. (6.75) by virtue of the fact that we have here taken $a_1 = L_1 = \frac{1}{2}$.

6.9. Summation Formulas of Gauss. If one expresses $f(x_i + uh)$ by Stirling's formula and integrates from $u = -\frac{1}{2}$ to $u = \frac{1}{2}$, the coefficients of the odd differences drop out, and one has [see Eq. (4.21)]

$$\int_{x_i - h/2}^{x_i + h/2} f(x) \, dx = h \int_{-\frac{1}{2}}^{\frac{1}{2}} f(x_i + uh) \, du = hf(x_i) + h \sum_{r=1}^{k} K_{2r} \, \delta^{2r} f(x_i)$$
$$+ h^{2k+3} K_{2k+2} f^{(2k+2)}(\xi_i) \quad (6.79)$$

where $x_{i-r} < \xi_i < x_{i+r}$ and

$$K_{2r} = \frac{1}{(2r)!} \int_{-\frac{1}{2}}^{\frac{1}{2}} u^2(u^2 - 1) \cdots [u^2 - (r-1)^2] \, du \quad (6.80)$$

Summing i from 1 to n in Eq. (6.79), one obtains the *first Gaussian summation formula*.

$$\sum_{i=1}^{n} f(x_i) = \frac{1}{h} \int_{x_0 + h/2}^{x_n + h/2} f(x) \, dx$$
$$- \sum_{r=1}^{k} K_{2r} \left[\delta^{2r-1} f\left(x_n + \frac{h}{2}\right) - \delta^{2r-1} f\left(x_0 + \frac{h}{2}\right) \right]$$
$$- h^{2k+2} K_{2k+2} f^{(2k+2)}(\xi) \quad (6.81)$$

where $x_{-k+1} < \xi < x_{n+k}$. In particular,

$$K_2 = \frac{1}{24} \quad K_4 = \frac{-17}{5,760} \quad K_6 = \frac{367}{967,680} \quad K_8 = \frac{-27,859}{464,486,400}$$

The quadrature formula (7.78) obtained in the next chapter by integrating Bessel's formula may be used also as a summation formula; thus, one has

$$\sum_{i=0}^{n} f(x_i) = \frac{1}{h} \int_{x_0}^{x_n} f(x) \, dx + \tfrac{1}{2}[f(x_n) + f(x_0)] + \tfrac{1}{12}[\mu \delta f(x_n) - \mu \delta f(x_0)]$$

$$- \tfrac{11}{720}[\mu \delta^3 f(x_n) - \mu \delta^3 f(x_0)] + \frac{191}{60,480}[\mu \delta^5 f(x_n) - \mu \delta^5 f(x_0)] + \cdots \quad (6.82)$$

This is the *second Gaussian summation formula*.

Note that the Gaussian summation formulas converge more rapidly than Laplace's formula.

6.10. Euler's Summation Formula. The Euler-Maclaurin formula of the next chapter [Eq. (7.92)] may also be employed as a summation formula. For this purpose it may be written

$$\sum_{i=1}^{n} f(x_i) = \frac{1}{h} \int_{x_1}^{x_n} f(x) \, dx + \tfrac{1}{2}[f(x_n) + f(x_1)] + \frac{h}{12}[f^{(1)}(x_n) - f^{(1)}(x_1)]$$

$$- \frac{h^3}{720}[f^{(3)}(x_n) - f^{(3)}(x_1)] + \frac{h^5}{30,240}[f^{(5)}(x_n) - f^{(5)}(x_1)] - \cdots \quad (6.83)$$

where the superscripts indicate the order of the derivative.

In terms of the *Bernoulli numbers*, which are defined by the series expansion†

$$\frac{x}{e^x - 1} = 1 - \frac{x}{2} + B_1 \frac{x^2}{2!} - B_2 \frac{x^4}{4!} + B_3 \frac{x^6}{6!} - \cdots \quad (6.84)$$

Euler's summation formula may be written

$$\sum_{i=1}^{n} f(x_i) = \frac{1}{h} \int_{x_1}^{x_n} f(x) \, dx + \tfrac{1}{2}[f(x_n) + f(x_1)]$$

$$+ \sum_{r=1}^{m} \frac{(-1)^r B_r}{(2r)!} h^{2r-1}[f^{(2r-1)}(x)]_{x_1}^{x_n} + R \quad (6.85)$$

† Particular values of the B_k are

$$B_1 = \tfrac{1}{6} \quad B_2 = \tfrac{1}{30} \quad B_3 = \tfrac{1}{42} \quad B_4 = \tfrac{1}{30} \quad B_5 = \tfrac{5}{66} \quad \text{and} \quad B_6 = \tfrac{691}{2730}$$

For a further discussion of Bernoulli numbers see G. Crystal, "Algebra," 6th ed., vol. II, pp. 228–233, Chelsea Publishing Company, New York, 1952.

where the remainder can be shown to be

$$R = \frac{nB_{m-1}h^{2m-2}}{(2m+2)!} f^{(2m+2)}(\xi) \tag{6.86}$$

Example 11. Obtain Stirling's approximation

$$\ln n! = \tfrac{1}{2}\ln 2\pi + (n + \tfrac{1}{2})\ln n - n + \frac{1}{12n} - \frac{1}{360n^3} + \frac{1}{1,260n^5} - \cdots \tag{6.87}$$

for the factorial.

Since

$$n! = \prod_{i=1}^{n} i$$

one has by Eq. (6.83)

$$\ln n! = \sum_{i=1}^{n} \ln i = \int_{1}^{n} \ln u \, du + \tfrac{1}{2}(\ln n + \ln 1) + \tfrac{1}{12}\left(\frac{1}{n} - 1\right)$$
$$- \frac{2!}{720}\left(\frac{1}{n^3} - 1\right) + \frac{4!}{30,240}\left(\frac{1}{n^5} - 1\right) - \cdots$$

Now

$$\int_{1}^{n} \ln u \, du = [u \ln u - u]_{1}^{n} = n \ln n - n + 1$$

Therefore $\ln n! = C + (n + \tfrac{1}{2})\ln n - n + \dfrac{1}{12n} - \dfrac{1}{360n^3} + \dfrac{1}{1,260n^5} - \cdots$

By substituting $n = 1$ in this equation we obtain an asymptotic series

$$C = 1 - \frac{1}{12} + \frac{1}{360} - \frac{1}{1,260} + \cdots$$

for the unknown constant. It can be shown by other means that $C = \tfrac{1}{2}\ln 2\pi.$†

PROBLEMS

1. Obtain the formula for the sum of the first n integers.
2. Find the sum S_n of the first n terms of the infinite series

$$S = \frac{1}{1\cdot 2} + \frac{1}{2\cdot 3} + \frac{1}{3\cdot 4} + \frac{1}{4\cdot 5} + \cdots$$

Then find S by taking the limit of S_n as n approaches infinity.
3. Show that

$$\sum_{i=0}^{n} \binom{n}{i} = \sum_{i=0}^{n} \frac{n!}{i!(n-i)!} = 2^n$$

4. Obtain the sum

$$\sum_{i=0}^{n} i^k \frac{(-1)^i n!}{i!(n-i)!}$$

for $k = 0, 1, 2, \ldots, n$ by summing by parts.

† See E. Whittaker and G. Robinson, "The Calculus of Observations," 4th ed., p. 139, Blackie & Son, Ltd., Glasgow, 1944.

5. Derive an expression for the sum of the infinite series

$$S(k) = \sum_{i=1}^{\infty} i^k e^{-i}$$

for integral values of k.

6. As a generalization of Example 8, show that

$$\sum_{i=1}^{\infty} \frac{1}{i i^{(k)}} = \frac{1}{k k!} + k \sum_{i=1}^{\infty} \frac{1}{i i^{(k+1)}}$$

and that by repetition of the process

$$\sum_{i=1}^{\infty} \frac{1}{i i^{(k)}} = \frac{1}{(k-1)!} \left[\sum_{i=0}^{m} \frac{1}{(k+i)^2} + (k+m)! \sum_{i=1}^{\infty} \frac{1}{i i^{(k+m+1)}} \right]$$

7. (a) Using the formula in Prob. 6, determine the sum of the infinite series

$$\sum_{i=1}^{\infty} \frac{1}{i^2(i+1)} = \frac{1}{2} + \frac{1}{12} + \frac{1}{36} + \cdots$$

to six decimal places. Take m equal to 4. (b) How many terms would need to be taken if one worked directly with the series?

8. Determine the value of a_6 in Eq. (6.75).

9. Determine the first five Bernoulli numbers, using the defining equation (6.84).

10. Show that the constant term in Stirling's approximation for $\ln n!$ [see Eq. (6.87)] is correctly given by $\frac{1}{2} \ln 2\pi$.

CHAPTER 7

NUMERICAL DIFFERENTIATION AND
NUMERICAL INTEGRATION

The basic idea in numerical differentiation and numerical integration is to approximate the given function $y = f(x)$ over a short range of x by a polynomial $P_m(x)$ and then to differentiate or integrate that polynomial rather than the function $f(x)$. The advantage of this procedure is that, while $f(x)$ may be difficult or practically impossible to integrate (or differentiate), one can always integrate (or differentiate) the polynomial. If the function is tabulated, this approximating polynomial $P_m(x)$ is just the polynomial used in interpolation. Therefore $P_m(x)$, or its equivalent $\mathcal{O}_m(u)$, where $x = x_0 + uh$, is given by any of the interpolation formulas of Chaps. 3 to 5.

7.1. Numerical Differentiation. If the derivative of $y = f(x)$ is desired at the point $x = \bar{x}$, it is always possible to so subscript the x's that $x_0 \leq \bar{x} < x_1$, and therefore $\bar{u}$, the corresponding value of u, lies in the range $0 \leq \bar{u} < 1$. The value of the derivative is found, as stated above, by replacing the function $f(x)$ by a polynomial $P_m(x) = \mathcal{O}_m(u)$; then

$$\frac{df(\bar{x})}{dx} \cong \frac{dP_m(\bar{x})}{dx} = \left[\frac{d\mathcal{O}_m(u)}{du} \frac{du}{dx} \right]_{x=\bar{x}}$$

Since $x = x_0 + uh$,

$$\frac{du}{dx} = \frac{1}{h} \tag{7.1}$$

and hence

$$\frac{df(\bar{x})}{dx} \cong \frac{1}{h} \frac{d\mathcal{O}_m(\bar{u})}{du} \tag{7.2}$$

where $\mathcal{O}_m(u)$ is any of the interpolating polynomials discussed in the previous chapters. These polynomials usually agree with the function $f(x)$ at $m + 1$ adjacent tabulated values in the neighborhood of $\bar{x}$.†

As in the case of interpolation, the polynomials passing through an approximately equal number of points above and below $\bar{x}$ in the table

† In the case of Stirling's or Bessel's formula ending on a mean difference the interpolating polynomial actually passes through only m points. It is the average of two polynomials, each of which passes through $m + 1$ points.

are likely to give better results than those passing through one point above and all the rest below, or vice versa. Since the central-difference formulas lead naturally to the former situation, they are usually to be preferred to the Gregory-Newton formulas. If one uses Lagrange's formula, one obtains, directly, expressions for the derivatives involving the ordinates themselves rather than the differences.

In general the accuracy increases with an increase of the degree m of the polynomial until a certain maximum accuracy is achieved. Beyond this point no appreciable increase in accuracy is obtained by an increase in m because of the effect of round-off errors in the tabulated values of y. This maximum practical value of m is usually equal to the order of the most nearly constant difference column (see Example 1). The relative error in the derivative is usually considerably larger than the errors in the tabulated values of y.

The higher derivatives can be obtained in a similar way,

$$\frac{d^k f(\bar{x})}{dx^k} \cong \frac{1}{h^k} \frac{d^k \mathcal{P}_m(\bar{u})}{du^k} \tag{7.3}$$

but the error in their evaluation increases with their order.

Example 1. Using Table 4.2, find

$$\cos 1.76 = \left[\frac{d}{dx} \sin x \right]_{x=1.76}$$

If one lets $x_0 = 1.76$ and uses the Gregory-Newton formula [see Eq. (3.43)] to approximate $y = \sin x$, one has

$$y \cong \mathcal{P}_m(u) = y_0 + u\,\Delta y_0 + \frac{u(u-1)}{2!}\,\Delta^2 y_0 + \frac{u(u-1)(u-2)}{3!}\,\Delta^3 y_0$$

$$+ \frac{u(u-1)(u-2)(u-3)}{4!}\,\Delta^4 y_0 + \frac{u(u-1)(u-2)(u-3)(u-4)}{5!}\,\Delta^5 y_0 + \cdots$$

$$\left[\frac{dy}{dx}\right]_{x=1.76} \cong \frac{1}{h}\left[\frac{d\mathcal{P}_m(u)}{du}\right]_{u=0} = \frac{1}{h}\left(\Delta y_0 - \tfrac{1}{2}\Delta^2 y_0 + \tfrac{1}{3}\Delta^3 y_0 - \tfrac{1}{4}\Delta^4 y_0 + \tfrac{1}{5}\Delta^5 y_0 + \cdots\right)$$

$$\tag{7.4}$$

$$\left[\frac{dy}{dx}\right]_{x=1.76} = \cos 1.76 \cong \frac{1}{0.02}\,(-0.00395\ 771 + 0.00019\ 5635$$

$$+ 0.00000\ 0583 - 0.00000\ 0035 - 0.00000\ 0002 + \cdots)$$

$$\cos 1.76 = -0.19788\ 55 + 0.00978\ 175 + 0.00002\ 915$$

$$- 0.00000\ 175 - 0.00000\ 010 = -0.18807\ 645$$

Since Δy_0 is accurate, in this example, only to six significant figures, it is clear that dy/dx need be good only to six decimal places. The correct answer is

$$\cos 1.76 = -0.18807\ 680$$

and hence the above calculation is well within the limit of error which might arise from the round-off error in the tabulated values.

The first approximation to the derivative is given by

$$\left(\frac{dy}{dx}\right)_0 \cong \frac{\Delta y_0}{h} = -0.19788\ 55 \tag{7.5}$$

the first term in the series. Note that it is off in the second decimal place.

If one uses Stirling's formula instead of the Gregory-Newton interpolation formula, the result should be better,† since the polynomial then extends approximately the same distance on each side of x_0. By Stirling's formula [see Eq. (4.21)]

$$y \cong y_0 + u\,\frac{\Delta y_{-1} + \Delta y_0}{2} + \tfrac{1}{2}u^2\,\Delta^2 y_{-1} + \tfrac{1}{6}u(u^2 - 1)\,\frac{\Delta^3 y_{-2} + \Delta^3 y_{-1}}{2}$$

$$+ \tfrac{1}{24}u^2(u^2 - 1)\,\Delta^4 y_{-2} + \tfrac{1}{120}u(u^2 - 1)(u^2 - 4)\,\frac{\Delta^5 y_{-3} + \Delta^5 y_{-2}}{2} \tag{7.6}$$

Therefore

$$\left[\frac{dy}{dx}\right]_{x-x_0} \cong \frac{1}{h}\left[\frac{d\varphi_m(u)}{du}\right]_{u=0} = \frac{1}{h}\left(\frac{\Delta y_{-1} + \Delta y_0}{2} - \frac{1}{6}\frac{\Delta^3 y_{-2} + \Delta^3 y_{-1}}{2}\right.$$

$$\left. + \frac{1}{30}\frac{\Delta^5 y_{-3} + \Delta^5 y_{-2}}{2}\right) \tag{7.7}$$

and hence

$$\left[\frac{dy}{dx}\right]_{x-x_0} = \cos 1.76 = -0.18806\ 425 - 0.00001\ 250 = 0.18807\ 675$$

the fifth difference term being too small to affect the result. This answer is correct to seven decimal places.

If Stirling's formula is terminated with the first difference, one obtains as a first approximation for the derivative the equation

$$\left[\frac{dy}{dx}\right]_{x-x_0} \cong \frac{\Delta y_{-1} + \Delta y_0}{2h} = \frac{1}{h}\,\mu\delta y_0 \tag{7.8}$$

Since this yields the value of $-0.18806\ 425$ for the derivative, it differs from the true value in the fifth decimal place. Thus this approximation is very much better than that given by Eq. (7.5). Expressing Eq. (7.8) in ordinate form, we have the important formula

$$\left[\frac{dy}{dx}\right]_{x-x_0} \cong \frac{y_1 - y_{-1}}{2h} \tag{7.9}$$

7.2. Symbolic Derivation of Numerical-differentiation Formulas. In Chap. 3 we showed how the Gregory-Newton interpolation formulas could be obtained from the symbolic equation

$$y = E^u y_0 \tag{7.10}$$

by substituting $1 + \Delta$ and $(1 - \nabla)^{-1}$ for E. To obtain the derivative of y at $x = x_0$ we differentiate this equation to obtain the following result

$$\left[\frac{dy}{dx}\right]_{x-x_0} = \frac{1}{h}\left[\frac{dy}{du}\right]_{u=0} = \frac{1}{h}\,[(\ln E)E^u y_0]_{u=0} = \frac{1}{h}\,(\ln E)y_0 \tag{7.11}$$

† The presence of round-off error in the tabulated values partially obscures this difference.

If one wishes to employ descending differences of y_0, one lets $E = 1 + \Delta$; then

$$\left[\frac{dy}{dx}\right]_{x=x_0} = \frac{1}{h}\left[\ln (1 + \Delta)\right]y_0$$

$$= \frac{1}{h}(\Delta - \tfrac{1}{2}\Delta^2 + \tfrac{1}{3}\Delta^3 - \tfrac{1}{4}\Delta^4 + \cdots)y_0 \quad (7.12)$$

which yields the same expression for the derivative as given in Eq. (7.4). It is convenient to introduce the operator D, defined by the equation

$$Dy_s = \left[\frac{dy}{dx}\right]_{x=x_s} \quad (7.13)$$

Then Eq. (7.11) can be replaced by the operator equation

$$D = \frac{1}{h}\ln E \quad (7.14)$$

If one expresses E in terms of Δ one obtains the following power-series expansion for D in terms of Δ

$$D = \frac{1}{h}\ln (1 + \Delta) = \frac{1}{h}(\Delta - \tfrac{1}{2}\Delta^2 + \tfrac{1}{3}\Delta^3 - \tfrac{1}{4}\Delta^4 + \cdots) \quad (7.15)$$

which is the operator equation corresponding to Eq. (7.12). On the other hand, if one employs ascending differences,

$$D = \frac{1}{h}\ln (1 - \nabla)^{-1} = \frac{1}{h}(\nabla + \tfrac{1}{2}\nabla^2 + \tfrac{1}{3}\nabla^3 + \cdots) \quad (7.16)$$

Applying this operator to y_0, one obtains an expression for the derivative at x_0 in terms of the ascending differences of y_0.

To obtain formulas for the higher derivatives one differentiates Eq. (7.10) the desired number of times. Thus for the kth derivative

$$\left[\frac{d^k y}{dx^k}\right]_{x=x_0} = \frac{1}{h^k}\left[\frac{d^k y}{du^k}\right]_{u=0} = \frac{1}{h^k}\left[\frac{d^k E^u}{dx^k}\right]_{u=0}y_0 = \left(\frac{1}{h}\ln E\right)^k y_0$$

$$= D^k y_0 \quad (7.17)$$

which shows in a purely formal way that the operator corresponding to the kth derivative is just the kth power of D. For instance, by Eq. (7.15),

$$\left[\frac{d^2 y}{dx^2}\right]_{x=x_0} = D^2 y_0$$

$$= \frac{1}{h^2}(\Delta^2 - \Delta^3 + \tfrac{11}{12}\Delta^4 - \tfrac{5}{6}\Delta^5 + \tfrac{137}{180}\Delta^6 - \tfrac{7}{10}\Delta^7 + \cdots)y_0 \quad (7.18)$$

and in terms of ∇, by Eq. (7.16),

$$\left[\frac{d^2y}{dx^2}\right]_{x=x_0} = \frac{1}{h^2}\left(\nabla^2 + \nabla^3 + \tfrac{11}{12}\nabla^4 + \tfrac{5}{6}\nabla^5 + \tfrac{137}{180}\nabla^6\right.$$
$$\left. + \tfrac{7}{10}\nabla^7 + \cdots\right)y_0 \quad (7.19)$$

In practice one terminates all the above series at some suitable order of difference rather than treating them as infinite series. This amounts, of course, to assuming that the function behaves as a polynomial, in keeping with the procedure for interpolation. These polynomial approximations have the important feature of yielding accurate approximations even in those cases where the infinite series fail to converge.†

7.3. Lozenge Diagrams for Obtaining Differentiation Formulas. In Chap. 4 we used a difference table with interspersed factorials in u, the so-called lozenge diagram, to obtain any desired interpolation formula. If one follows the rules there formulated, one obtains a permissible interpolation formula for each path through the lozenge. These rules serve to associate the proper factorial in u with each difference along the path. Moreover, the polynomial represented by the interpolation formula depends only on the difference (or mean difference) on which such a path ends.

Since one may obtain an approximation formula for the derivative at $x = x_p$ by differentiating any chosen interpolation formula with respect to u, dividing by h, and substituting the value of $u = u_p$ corresponding to x_p, it is clear that such a formula is obtainable from the interpolation formula by substituting for the factorials in u their derivatives evaluated at $u = u_p$. It is obvious, therefore, that one may construct a lozenge diagram for the formation of permissible derivative formulas by the simple expedient of replacing each factorial in the lozenge diagram of Chap. 4 by its derivative at, say, $u = 0$.

Such a lozenge diagram for the first derivative at the point having ordinate Y_0 is given in Fig. 7.1. The use of the capital Y's is to call to mind that these ordinate designations are purely relative. Thus Y_0 at which the derivative is given may be any of the regularly designated ordinates y_k. Note that all formulas obtained from this lozenge need to be multiplied by $1/h$ to take account of the relation between the u and x.

It is not necessary to differentiate all the factorials in order to obtain this new lozenge, since one observes that the numbers representing these derivatives form a difference table rotated 180° with respect to the difference table of the Y's, i.e., the difference table proceeds upward and to the left. Moreover, the first column of numbers, all of which are 1, acts as a

† See N. E. Nörlund, "Vorlesungen über Differenzenrechnung," p. 203, Springer-Verlag OHG, Berlin, 1924.

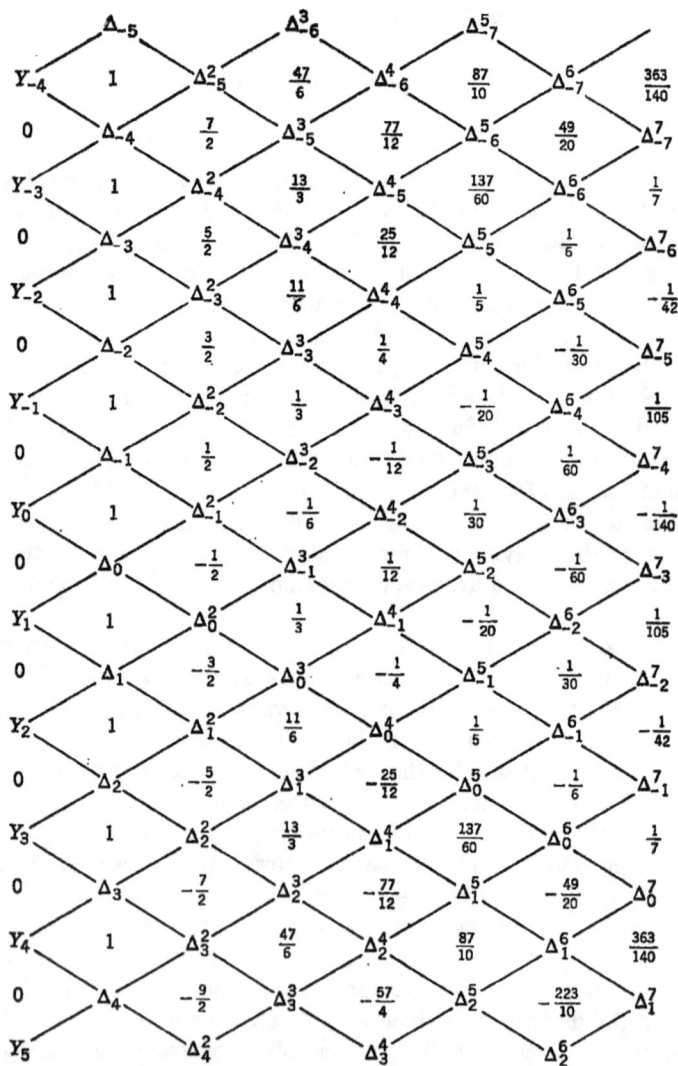

FIG. 7.1. Lozenge for the first derivative at Y_0. All formulas obtained from this diagram are to be multiplied by $1/h$, and Δ^m_s stands for $\Delta^m Y_s$.

constant-difference column. Thus, one needs only the constants contained in either Eq. (7.15) or (7.16) to be able to complete the lozenge.

From this discussion it is clear that the lozenge of Fig. 7.1 may be looked upon as a composite of two difference tables rotated 180° with respect to one another and interlaced. Going back to the original lozenge of Chap. 4, we see that the same is true there. In fact one can readily show that a lozenge formed of any two arbitrary difference tables so related to one another has the characteristic property that an expression obtained by going from some point A to some point B is independent of the path. If, in addition, the farthest-left difference column of the reversed difference table is constant, the expression obtained by entering the lozenge from the left and proceeding to a given point is independent

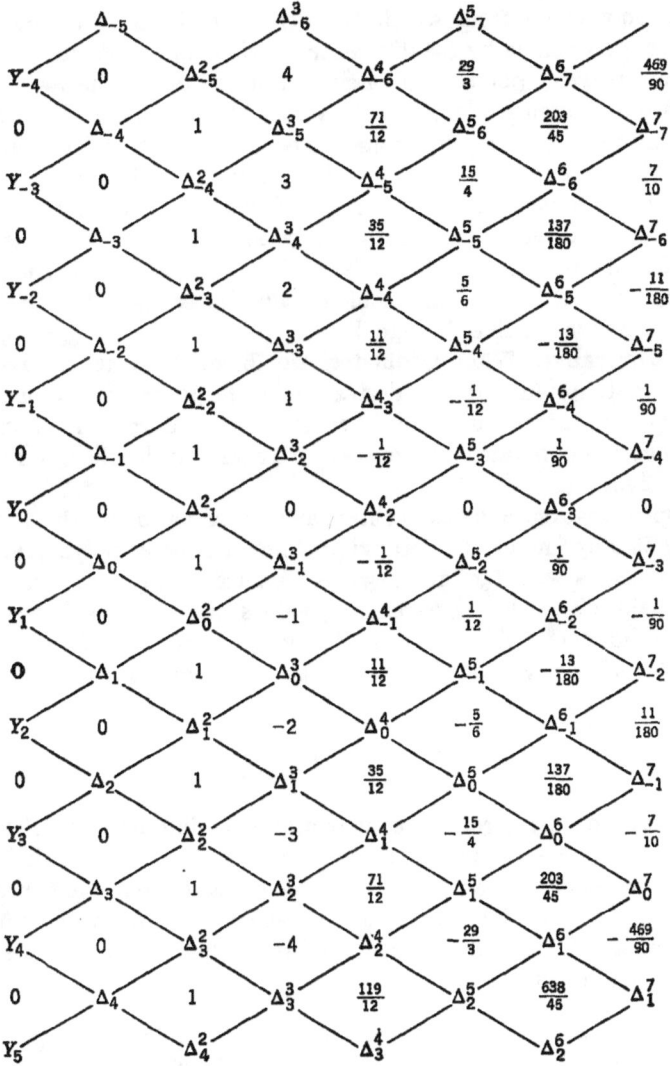

Fig. 7.2. Lozenge diagram for the second derivative at the point Y_0. All formulas obtained should be multiplied by $1/h^2$, and Δ^m_s stands for $\Delta^m Y_s$.

both of the path and the point of entry. This latter property is possessed by all the lozenges considered in this book.

The lozenge shown in Fig. 7.2 is obtained by differentiating the factorial in u twice and setting $u = 0$. Again, one can build up the entire lozenge from these constants contained in either Eq. (7.18) or (7.19). Note that all formulas obtained from this lozenge must be multiplied by the factor $1/h^2$.

The polynomial from which the derivative formula is obtained is uniquely determined by the difference on which one ends, and is, in fact, that interpolating polynomial passing through those ordinates contained within the difference triangle† with vertex at that difference. This fact permits one to make semiquantitative estimates of the relative accuracy of the different formulas. Thus, of those formulas involving the same number of ordinates, one would expect best results from those in which the polynomial passes through a nearly equal number of points before and after the point at which the derivative is desired. This means that, in using Figs. 7.1 and 7.2, one should usually choose a path that terminates near the horizontal line through Y_0.

7.4. Derivatives When Ordinates Are Unequally Spaced. To obtain approximate formulas for the derivative for the general case of unequally spaced ordinates one must use either Lagrange's interpolation formula or the properties of divided differences. We here consider the latter of these alternatives.

Suppose that one chooses to obtain an approximation for the derivative of $f(x)$ at x by finding the derivative of an mth-degree polynomial $P_m(x)$ that coincides with $f(x)$ at the $m + 1$ values $x_0, x_1, \ldots, x_m$. From these values and the corresponding values of y one may construct a divided-difference table similar to Table 5.1. From Eq. (5.60) one sees that the kth derivative of $f(x)$ at x is expressible in terms of divided differences as

$$\frac{d^k f(x)}{dx^k} = k\,!f(\underset{(k+1 \text{ arguments})}{x,x, \ldots ,x}) \tag{7.20}$$

If one is able, therefore, to complete a divided-difference table in which x is repeated $k + 1$ times, one obtains therefrom all the derivatives of the polynomial up to the kth. One must make use of the fact that an mth-degree polynomial has a constant value of the mth divided difference to complete the divided-difference table needed. The example below illustrates the method.

† If in any table one takes any number of consecutive ordinates and forms all the differences obtainable from these ordinates alone, the triangle of differences so formed is termed a *difference triangle*. Clearly such a triangle is uniquely specified by giving the highest difference contained in it; this occurs at the right-hand vertex (see also Secs. 3.9 and 3.10).

Example 2. Given the following pairs of values of x and $y = f(x)$

x	1	2	4	8	10
y	0	1	5	21	27

determine numerically the first and second derivatives of $f(x)$ at $x = 4$.

Since the order of the terms is immaterial in a divided-difference table, one may write the $x = 4$ entry last and construct a divided-difference table from these five pairs of values. The result is that shown above the line in Table 7.1. One may now approximate $f(x)$ by a fourth-degree polynomial $P(x)$ passing through these five pairs of values. The derivatives of $f(x)$ are then given approximately by the derivatives of $P(x)$, which are given in terms of the divided differences by Eq. (7.20); in particular,

$$f'(x) \cong P'(x) = P(x,x)$$
$$f''(x) \cong P''(x) = 2P(x,x,x) \qquad (7.21)$$

TABLE 7.1. DETERMINATION OF DERIVATIVES DIRECTLY FROM A
DIVIDED-DIFFERENCE TABLE

If one repeats the argument 4, as in Table 7.1, one should obtain $P(4,4)$ and $P(4,4,4)$ at the locations shown and hence, by Eq. (7.21), the first and second derivatives of $f(x)$. One can, however, fill in the numerical values of the divided differences of $P(x)$ below the line only by making use of fact that $P(x)$ is a fourth-degree polynomial and has therefore a constant fourth divided difference, i.e.,

$$P(2,8,10,4,4) = P(8,10,4,4,4) = -\tfrac{1}{144}$$

This permits one to work from the right to complete the table; thus by the fundamental property of divided differences [see Eq. (5.55)]

$$P(8,10,4,4) = -\tfrac{1}{16} + (4 - 2)P(2,8,10,4,4) = -\tfrac{11}{144}$$
$$P(10,4,4,4) = P(8,10,4,4) + (4 - 8)P(8,10,4,4,4) = -\tfrac{7}{144}$$

Having in this manner obtained the third divided differences from the fourth differences, one proceeds to obtain the second differences from the third, and then the first differences from the second in a completely similar way; thus

$$P(10,4,4) = -\tfrac{1}{6} + (4 - 8)(-\tfrac{11}{144}) = \tfrac{5}{36}$$
$$P(4,4,4) = \tfrac{5}{36} + (4 - 10)(-\tfrac{7}{144}) = \tfrac{31}{72}$$

and
$$P(4,4) = \tfrac{11}{3} + (4 - 10)\tfrac{5}{36} = \tfrac{17}{6}$$

Thus by Eqs. (7.21)

$$f'(4) \cong \tfrac{17}{6} = 2.833$$
$$f''(4) \cong \tfrac{31}{36} = 0.861$$

7.5. Remainder Term. The error introduced by differentiating the interpolation polynomial instead of the function tabulated can be expressed by means of a remainder term. We shall determine the nature of this remainder for the general case of unequally spaced ordinates.

Newton's interpolation formula employing divided differences [see Eqs. (5.56) and (5.57)] is given by

$$
\begin{aligned}
f(x) &= P_m(x) + R(x) \\
&= f(x_0) + (x - x_0)f(x_0,x_1) + (x - x_0)(x - x_1)f(x_0,x_1,x_2) \\
&\quad + (x - x_0)(x - x_1)(x - x_2)f(x_0,x_1,x_2,x_3) + \cdots \\
&\quad + (x - x_0)(x - x_1) \cdots (x - x_{m-1})f(x_0,x_1,x_2, \ldots ,x_m) \\
&\hspace{7cm} + R(x) \quad (7.22)
\end{aligned}
$$

where
$$R(x) = f(x,x_0,x_1,x_2, \ldots ,x_m) \prod_{i=0}^{m} (x - x_i) \qquad (7.23)$$

Taking the derivative of both sides of Eq. (7.22), one finds that

$$f'(x) = P'_m(x) + R'(x)$$

where
$$R'(x) = \frac{d}{dx}\left[f(x,x_0,x_1, \ldots ,x_m) \prod_{i=0}^{m} (x - x_i)\right]$$

Since $P'_m(x)$ is the approximate numerical derivative of $f(x)$ obtained by some one of the formulas and methods described previously, $R'(x)$ is the remainder term when obtaining the derivative by that formula.

By making use of Eq. (5.54) one may write that

$$
\begin{aligned}
R'(x) &= f(x,x,x_0,x_1, \ldots ,x_m)\prod_{i=0}^{m} (x - x_i) \\
&\quad + f(x,x_0,x_1, \ldots ,x_m)\prod_{i=0}^{m} (x - x_i) \sum_{j=0}^{m} \frac{1}{x - x_j} \quad (7.24)
\end{aligned}
$$

Since one usually obtains the derivative at one of the tabulated points $x = x_k$ (k having any value from 0 to m), the remainder term usually simplifies to

$$R'(x_k) = f(x_k,x_0,x_1, \ldots ,x_m) \prod_{i \neq k} (x_k - x_i) \qquad (7.25)$$

Since, by Eq. (5.60),

$$f(x_k,x_0,x_1, \ldots ,x_m) = \frac{1}{(m + 1)!} f^{(m+1)}(\xi)$$

where ξ lies within the range of x defined by the largest and smallest of the numbers $x_0, x_1, \ldots, x_m$, and x_k, one has

$$R'(x_k) = \frac{1}{(m+1)!} f^{(m+1)}(\xi) \prod_{i \neq k} (x_k - x_i) \qquad (7.26)$$

If the x's are equally spaced, i.e., $x_{j+1} - x_j = h$, then

$$R'(x_k) = (-1)^k \frac{k!(m-k)!}{(m+1)!} h^m f^{(m+1)}(\xi) \qquad (7.27)$$

Where one employs the lozenge of Fig. 7.1 to obtain the expression for the derivative, this remainder term is just the next term that would be obtained if the series were extended but with $h^{m+1} f^{(m+1)}(\xi)$ replacing the $(m+1)$th difference—all terms, of course, being multiplied by the factor $1/h$.

7.6. Notation. It is convenient to designate numerical differentiation formulas in the following way: $D_m^k(n)$ *designates the formula for the kth derivative obtained by differentiating an mth-degree polynomial passing through $m+1$ points of the table and evaluated at a point nh units to the right of the extreme left-hand point.* $\bar{D}_m^k(n)$ *designates this same formula expressed in terms of differences.* The superscript 1 may be omitted in designating the formulas for the first derivative.

If one designates the $m+1$ ordinates by $y_0, y_1, y_2, \ldots, y_m$, the formula $D_m^k(n)$ is of the form

$$D_m^k(n) = \frac{C}{h^k} (A_0 y_0 + A_1 y_1 + \cdots + A_m y_m)$$

where $A_0, A_1, \ldots, A_m$ are integers and C some fraction. It is therefore possible to give full information as to the right-hand side by writing that

$$D_m^k(n) = \frac{C}{h^k} \{ A_0 \quad A_1 \quad A_2 \quad \cdots \quad A_{n-1} \quad \mathbf{A}_n \quad A_{n+1} \quad \cdots \quad A_m \} \qquad (7.28)$$

The set of weights given within the braces is hereafter termed a *weight pattern* or simply a *pattern*. Note that one uses boldface for that weight A_n corresponding to the point $(x = x_n)$ at which the derivative is taken.

As an illustration of this notation,

$$D_4^2(2) = \frac{1}{12h^2} \{ -1 \quad 16 \quad -30 \quad 16 \quad -1 \}$$

represents the five-point formula for the second derivative at the center point, and

$$D_3(0) = \frac{1}{6h} \{ -11 \quad 18 \quad -9 \quad 2 \}$$

represents the four-point formula for the first derivative at the extreme left-hand point.

In Sec. 7.21 we shall consider how one may achieve a direct conversion from one such formula to another.

7.7. Numerical Integration. Consider the problem of evaluating

$$I(x) = \int_a^x f(t) \, dt \qquad (7.29)$$

when the indefinite integral

$$F(x) = \int f(x) \, dx$$

is not expressible in closed form or is so complicated that its evaluation is very laborious. In either case one usually turns to numerical methods.

Three principal types of problems should be mentioned:

1. $I(x)$ is desired for only a single value $x = b$,

$$I(b) = \int_a^b f(t) \, dt$$

In this case a single number $I(b)$ is sought.

2. $I(x)$ is desired for all x's in a given range. For a numerical solution of this problem a table of $I(x)$ is needed with sufficiently small intervals of x to permit interpolation.

3. A third type of problem arises in the numerical solution of a differential equation where one needs to evaluate approximately an integral

$$I(x) = \int_a^x f(t, y(t)) \, dt$$

where only partial information is initially available on $y(t)$.

As one might suppose, the nature of the numerical-integration formula that should be used depends on what type of problem is being solved.

One method of solving the first problem is to expand $f(x)$ in an infinite series, the individual terms of which can be integrated. Thus, if

$$f(x) = \sum_{n=0}^{\infty} A_n \phi_n(x)$$

and one can determine the constants

$$C_n = \int_a^b \phi_n(x) \, dx$$

then

$$I = \int_a^b f(x) \, dx = \sum_{n=0}^{\infty} A_n C_n$$

Such a method is practical provided this series converges rapidly. Then only a few terms need to be calculated to determine I to the desired accuracy.†

A more general procedure, however, is to break the interval a to b up into small subintervals and to replace the function $f(x)$ over each of these subintervals by a polynomial. These polynomials are easily integrated, and the sum used to approximate I. By taking the subintervals small enough the desired degree of accuracy can usually be obtained.

7.8. Quadrature Formulas. The polynomial used to approximate $f(x)$ over a small subinterval of the total range a to b may be any of the polynomials obtained from the interpolation formulas of Gregory-Newton, Gauss, Stirling, Bessel, or Lagrange. Thus one needs to have $f(x)$ tabulated for various values $x = x_s$ in the range. These values of x are usually chosen to be equally spaced. When such an interpolating polynomial is integrated, one obtains a set of weights to be applied to the tabulated values $f_s = f(x_s)$ or to the differences formed from these values. The expression for the integral so obtained is called a *quadrature formula*.

If the tabulation is made at equal intervals, the accuracy of the quadrature formula depends only on the size of the interval, the range of integration, the degree of the polynomial, and the points through which it passes. The particular interpolation formula used affects only the form in which the quadrature formula appears. For instance, if a particular interpolating polynomial is obtained first by using Stirling's formula and second by using Lagrange's formula, the quadrature formula derived from the first of these will reduce to the other if one expresses the differences in the first in terms of the ordinates.

Suppose one uses Lagrange's equation to express $f(x)$ in terms of the known ordinates $f_s = f(x_s)$; then, as seen in Chap. 5,

$$f(x) = \sum_{s=0}^{m} a_s(u)f_s + r(u) \qquad (7.30)$$

where
$$u = \frac{x - x_0}{h}$$

† This method can be used to determine the complete elliptic integral

$$F\left(k, \frac{\pi}{2}\right) = \int_0^{\pi/2} \frac{dx}{\sqrt{1 - k^2 \sin^2 x}}$$

for small values of k. See D. Gibb, "Interpolation and Numerical Integration," p. 64, G. Bell and Sons, London, 1915.

and, by Eqs. (5.19) and (5.20),

$$a_s(u) = \frac{(-1)^{m-s} u^{[m+1]}}{s!(m-s)!(u-s)} \tag{7.31}$$

$$r(u) = h^{m+1} \frac{u^{[m+1]}}{(m+1)!} f^{(m+1)}(\xi) \tag{7.32}$$

Here ξ lies between the smallest and largest of the quantities x_0, x_m, and $x = x_0 + uh$. Therefore, one may obtain a quadrature formula expressed in ordinate form by integrating Eq. (7.30) between any two limits x' and x''. Thus

$$I = \int_{x'}^{x''} f(x)\, dx = h \int_{u'}^{u''} f(x_0 + uh)\, du = h \sum_{s=0}^{m} w_s f_s + R \tag{7.33}$$

where the w_s are constants given by

$$w_s = \int_{u'}^{u''} a_s(u)\, du \tag{7.34}$$

and R is a remainder term given by

$$R = h \int_{u'}^{u''} r(u)\, du \tag{7.35}$$

Since, for a given table of f_s, the quadrature formula (7.33) depends only on the points through which the polynomial passes and the limits of integration x' and x'', it is convenient to designate quadrature formulas by a Q symbol as follows:

$$Q_{mp}(q) = h \sum_{s=0}^{m} w_s f_s \tag{7.36}$$

is the quadrature formula obtained from an mth-degree polynomial that fits $f(x)$ at $x = x_0, x_1, \ldots, x_m$ by integrating from $x' = x_q$ to $x'' = x_{q+p}$. Thus the first subscript on Q designates the degree of the polynomial used; the second subscript designates the number of panels† over which the integration is made; and the number q inside the parentheses gives the number of panels that occur after x_0, the first point of fit, before the integration starts.

For example,

$$Q_{31}(1) = \frac{h}{24}(-f_0 + 13f_1 + 13f_2 - f_3)$$

is the quadrature formula obtained from a third-degree polynomial that agrees with $f(x)$ at the four points x_0, x_1, x_2, and x_3 by integrating between x_1 and x_2, i.e., over the middle panel. Just as in the case of formulas

† A panel is the area under the plot of $f(x)$ between two consecutive ordinates $f(x_i)$ and $f(x_{i+1})$.

for derivatives (see Sec. 7.6) one may express this quadrature formula in terms of a weight pattern; thus

$$Q_{31}(1) = \frac{h}{24} \{-1 \quad \mathbf{13} \quad \mathbf{13} \quad -1\}$$

The boldface numbers indicate the range of integration.

7.9. Symmetrical Formulas. A quadrature formula of the form $Q_{mp}((m - p)/2)$, e.g., the $Q_{31}(1)$ above, has the interval of integration symmetrically placed with respect to the $m + 1$ ordinates. This class of formulas includes most of the well-known quadrature formulas, and since they are fully specified by the subscripts m and p, it is convenient to let $Q_{mp} = Q_{mp}((m - p)/2)$.

Since the weight patterns reflect the symmetry above, it is convenient

TABLE 7.2. SYMMETRICAL QUADRATURE FORMULAS INVOLVING AN ODD NUMBER OF ORDINATES (m EVEN)

$$Q_{mp} = \frac{N}{D} h \sum_{i=-m/2}^{m/2} W_i f_i + \frac{N'}{D'} h^{m+3} f^{(m+2)}(\xi) \dagger$$

Formula‡	N	D	W_0	$W_{\pm 1}$	$W_{\pm 2}$	$W_{\pm 3}$	N'	D'
Q_{01}	1	1	1	......	....	...	1	24
Q_{02}	1	1	2	......	....	...	1	3
Q_{21}	1	24	22	1	....	...	−17	567
Q_{22}	1	3	4	1	....	...	−1	90
Q_{24}	4	3	−1	2	....	...	14	45
Q_{41}	1	5,760	5,178	308	−17	...	367	967,680
Q_{42}	1	90	114	34	−1	...	1	756
Q_{44}	2	45	12	32	7	...	−8	945
Q_{46}	3	10	26	−14	11	...	41	140
Q_{61}	1	967,680	862,564	57,249	−5,058	367	27,859	464,486,400
Q_{62}	1	3,780	4,688	1,503	−72	5	−23	113,400
Q_{64}	2	945	332	612	171	−4	13	14,175
Q_{66}	1	140	272	27	216	41	−9	1,400
Q_{68}	8	945	−2,459	2,196	−954	460	3,956	14,175

† Here ξ lies in the *range of the formula*. By this is meant the range corresponding to the weighted ordinates f_i plus that part of the range of the integration which extends beyond these limits.

‡ Q_{mp} represents the quadrature formula obtained from an mth-degree polynomial that fits the function $f(x)$ at the $m + 1$ points indicated by integrating over the p central panels lying between $f_{-p/2}$ and $f_{p/2}$ or, when $p = 1$, over the center two half panels.

to resubscript the f's to bring out this symmetry. We write in place of Eq. (7.36)

$$Q_{mp} = \frac{N}{D} h \sum_{i=-m/2}^{m/2} W_i f_i \tag{7.37}$$

One can then show that $W_{-i} = W_i$. Tables 7.2 and 7.3 list the weights for these formulas, for $m \le 7$, together with coefficients of their remainder terms.

TABLE 7.3. SYMMETRICAL QUADRATURE FORMULAS INVOLVING AN EVEN NUMBER OF ORDINATES (m ODD)

$$Q_{mp} = \frac{N}{D} h \sum_{i=-m/2}^{m/2} W_i f_i + \frac{N'}{D'} h^{m+2} f^{(m+1)}(\xi) \dagger$$

For-mula‡	N	D	$W_{\pm\frac{1}{2}}$	$W_{\pm\frac{3}{2}}$	$W_{\pm\frac{5}{2}}$	$W_{\pm\frac{7}{2}}$	N'	D'
Q_{11}	1	2	1	....	....	.....	-1	12
Q_{13}	3	2	1	....	....	.....	3	4
Q_{31}	1	24	13	-1	....	.....	11	720
Q_{33}	3	8	3	1	....	.....	-3	80
Q_{35}	5	24	1	11	....	.....	95	144
Q_{51}	1	1,440	802	-93	11	.....	-191	60,480
Q_{53}	3	160	58	23	-1	.....	13	2,240
Q_{55}	5	288	50	75	19	.....	-275	12,096
Q_{57}	7	1,440	562	-453	611	.....	5,257	8,640
Q_{71}	1	120,960	68,323	$-9,531$	1,879	-191	2,497	3,628,800
Q_{73}	1	4,480	4,807	2,049	-149	13	-39	44,800
Q_{75}	5	24,192	4,475	5,805	1,871	-55	425	145,152
Q_{77}	7	17,280	2,989	1,323	3,577	751	$-8,183$	518,400
Q_{79}	9	4,480	$-1,711$	4,967	$-2,803$	1,787	25,713	44,800

† Here ξ lies in the *range of the formula*. By this is meant the range corresponding to the weighted ordinates f_i plus that part of the range of the integration which extends beyond these limits.

‡ Q_{mp} represents the quadrature formula obtained from an mth-degree polynomial that fits the function $f(x)$ at the $m+1$ points indicated by the weighted ordinates f_i by integrating over the p central panels lying between $f_{-p/2}$ and $f_{p/2}$.

7.10. Remainder Terms. From Eqs. (7.32) and (7.35) we see that the remainder R is given by

$$R = \frac{h^{m+2}}{(m+1)!} \int_{u'}^{u''} u^{(m+1)} f^{(m+1)}(\xi) \, du \tag{7.38}$$

where ξ is a function of u. The dependence of ξ on u is indicated by the equivalence between $f^{(m+1)}(\xi)$ and the appropriate divided difference (see Chap. 5).

An upper limit for $|R|$ can always be obtained by the following means:

$$|R| = \frac{h^{m+2}}{(m+1)!}\left| \int_{u'}^{u''} u^{[m+1]}f^{(m+1)}(\xi)\, du \right|$$

$$\leq \frac{h^{m+2}}{(m+1)!} \int_{u'}^{u''} |u^{[m+1]}|\, |f^{(m+1)}(\xi)|\, du$$

Therefore $\quad |R| \leq \dfrac{h^{m+2}}{(m+1)!} |f^{(m+1)}(\xi)|_{\max} \displaystyle\int_{u'}^{u''} |u^{[m+1]}|\, du$

where $|f^{(m+1)}(\xi)|_{\max}$ is the maximum value of $|f^{(m+1)}(\xi)|$ for the range of ξ corresponding to $u' < u < u''$. Since for any particular value of u one knows that ξ lies between the smallest and the largest of the abscissas x_0, x_m, and $x = x_0 + uh$, the range of ξ is determined by the smallest and the largest of the abscissas x_0, x_m, x', and x''. This range we shall henceforth refer to as the *range of the formula*. It represents the range of values of x that play a role in the quadrature formula.

If it happens that $u^{[m+1]}$ does not change sign between u' and u'', one may obtain a more satisfactory expression for R by applying the mean-value theorem to Eq. (7.38). Since $a_s(u)$ and $u^{[m+1]}$ are continuous and $f(x) = f(x_0 + uh)$ is assumed to be continuous, by Eqs. (7.30) and (7.32) $f^{(m+1)}(\xi)$ is continuous in u, and the mean-value theorem applies. Thus

$$R = \frac{h^{m+2}}{(m+1)!} f^{(m+1)}(\xi') \int_{u'}^{u''} u^{[m+1]}\, du \tag{7.39}$$

where ξ' is some value of x in the range of the formula. This expression for R is of the form

$$R = Ch^{m+2}f^{(m+1)}(\xi') \tag{7.40}$$

C being a known numerical constant. One does not know ξ', but upper and lower limits can be put on $f^{(m+1)}(\xi')$ since ξ' is restricted to the range of the formula.

The reduction of the remainder term to the general form of (7.40), in which R is proportional to some derivative of $f(x)$ at some unspecified point $x = \xi'$, is usually possible. When this can be done, the quadrature formula is said to be *simplex*.†

The term simplex may be applied not only to quadrature formulas but also to interpolation formulas, summation formulas, and formulas for numerical differentiation. For this purpose we need a more general

† See P. J. Daniell, Remainders in Interpolation and Quadrature Formulae, *Math. Gaz.*, **24**: 238-244 (1940).

definition than that above. A formula is said to be *simplex* and of degree n, if:

1. The remainder vanishes for all polynomials of degree less than, or equal to, n but not for those of degree $n + 1$;

2. The vanishing of the remainder for any function possessing a continuous $(n + 1)$th derivative implies the vanishing of $f^{(n+1)}(x)$ at some point $x = \xi$ in the range of the formula.

It is helpful for some purposes to use, in place of Eq. (7.38), the following definition for the remainder, obtainable directly from Eq. (7.33),

$$Rf(x) = \int_{x'}^{x''} f(x)\, dx - h \sum_{s=0}^{m} w_s f(x_s) \tag{7.41}$$

We write $Rf(x)$ in place of R, the R now being treated as an operator[†] on $f(x)$. This operator is seen to be linear, i.e.,

$$R[af_1(x) + bf_2(x)] = aRf_1(x) + bRf_2(x) \tag{7.42}$$

Thus, for example, the vanishing of Rx^k when $k < n + 1$ implies the vanishing of $RP_k(x)$, where $P_k(x)$ is any polynomial of degree less than $n + 1$.

THEOREM I. *If a formula is simplex, the remainder term is given by*

$$Rf(x) = \frac{f^{(n+1)}(\xi)}{(n + 1)!}\, Rx^{n+1} \tag{7.43}$$

where n is the degree of the formula and ξ lies in the range of the formula.

Proof: To prove this theorem we note that, since the formula is simplex and of degree n,

$$Rx^k = \begin{cases} 0 & \text{for } k \le n \\ E \ne 0 & \text{for } k = n + 1 \end{cases} \tag{7.44}$$

where E is some constant. Now consider the function

$$g(x) = f(x) - \lambda x^{n+1}$$

The remainder for this function, since R is a linear operator, is

$$Rg(x) = Rf(x) - \lambda E$$

Since $E \ne 0$, we can always make $Rg(x) = 0$ by letting

$$\lambda = \frac{Rf(x)}{E}$$

then

$$Rf(x) = \lambda E \tag{7.45}$$

If $f(x)$ possesses a $(n + 1)$th derivative,

$$g^{(n+1)}(x) = f^{(n+1)}(x) - \lambda(n + 1)!$$

[†] The concept of an operator was introduced in Sec. 3.5.

and since $Rg(x) = 0$ and the formula is simplex, this $(n + 1)$th derivative of $g(x)$ must vanish at some point ξ in the range of the formula. Therefore,

$$g^{(n+1)}(\xi) = f^{(n+1)}(\xi) - \lambda(n + 1)! = 0$$

Substituting the λ from this equation in Eq. (7.45), we have the equivalent of Eq. (7.43), and the theorem is proved.

Example 3. Find the remainder term for the formula Q_{24} on the assumption that the formula is simplex.

The formula Q_{24} may be obtained from Table 7.2. The remainder is thus given by

$$Rf(x) = \int_{-2h}^{2h} f(x)\,dx - \frac{4h}{3}[2f(-h) - f(0) + 2f(h)]$$

Therefore Rx^k is zero for k odd and for $k = 0$ and $k = 2$, while for $k = 4$

$$Rx^4 = 2\frac{(2h)^5}{5} - \frac{4h}{3}4h^4 = \tfrac{1\,1\,2}{1\,5}h^5$$

The degree n of the formula is therefore 3, being exact for all polynomials of degree 3 or less, and by Eq. (7.43) the remainder term is

$$Rf(x) = \tfrac{1\,4}{4\,5}h^5 f^{(4)}(\xi)$$

The range of the formula is here set by the limits of integration since the given ordinates are within that range; therefore $-2h < \xi < 2h$.

7.11. Proving Formulas to Be Simplex. At present, there does not appear to be any general method of establishing the fact that most of our quadrature formulas are simplex. There are, however, methods applicable to certain classes of formulas.

For instance, any formula obtained by integrating over a range in which $u^{[m+1]}$ does not change sign has a remainder term given by Eq. (7.39). Since $u^{[m+1]}$ is constant in sign, the integral in this equation is not zero, and the vanishing of R implies the vanishing of $f^{(m+1)}(\xi)$. Therefore such formulas are simplex and of order m, where m is the degree of the polynomial used to approximate $f(x)$.

Consider next the symmetric formulas Q_{mn} for m and n even integers. These include all the formulas listed in. Table 7.2, except those of the form Q_{m1}. If $P_m(x)$ is a polynomial that agrees with $f(x)$ at the points $x_i = x_0 + ih, i = 0, 1, \ldots, m$, we obtain these formulas by integrating

$$f(x) = P_m(x) + f(x,x_0,x_1, \ldots ,x_m)\prod_{i=0}^{m}(x - x_i) \qquad (7.46)$$

between the limits x_{r-s} and x_{r+s}, where $m = 2r$ and $n = 2s$. The remainder term used here is justified in Eq. (5.62).

The remainder term for the quadrature formula Q_{mn} is therefore

$$Rf(x) = \int_{x_{r-s}}^{x_{r+s}} f(x,x_0,x_1, \ldots ,x_m) \prod_{i=0}^{m} (x - x_i)\, dx \qquad (7.47)$$

If we let

$$G(x) = \int_{x_{r-s}}^{x} \prod_{i=0}^{m} (x - x_i)\, dx \qquad (7.48)$$

then $G(x) = 0$ at x_{r-s} and x_{r+s} but at no points in between.†
Integrating by parts in Eq. (7.47), we have

$$Rf(x) = - \int_{x_{r-s}}^{x_{r+s}} G(x) f(x,x,x_0,x_1, \ldots ,x_m)\, dx$$

Here use has been made of the fact that repeating the argument x in the divided difference is equivalent to differentiation [see Eq. (5.54)]. Since $G(x)$ does not change sign over the range of integration and $f(x,x,x_0,x_1, \ldots ,x_m)$ is continuous, the mean-value theorem is applicable, and hence

$$Rf(x) = -f(\xi,\xi,x_0,x_1, \ldots ,x_m) \int_{x_{r-s}}^{x_{r+s}} G(x)\, dx \qquad (7.49)$$

Replacing the divided difference by the use of Eq. (5.59), we have

$$Rf(x) = - \frac{1}{(m+2)!} f^{(m+2)}(\xi') \int_{x_{r-s}}^{x_{r+s}} G(x)\, dx \qquad (7.50)$$

Thus these formulas are simplex and of degree $m + 1$, since the vanishing of the remainder demands the vanishing of the $(m + 2)$th derivative somewhere in the range of the formula. They are exact for polynomials of degree $m + 1$.

By a somewhat similar reasoning Steffensen shows that the formulas Q_{mn} when m and n are odd (see Table 7.3) are simplex, but in this case they are of order m only. Thus the formulas in Table 7.2 are usually preferable to those of Table 7.3.

It is convenient to divide formulas into *plus* formulas and *minus* formulas, thereby designating the sign in front of the remainder term. Thus in Table 7.2 Q_{22} is a minus formula and Q_{24} a plus. Note that this designation does not specify whether the remainder is actually positive or negative, since the sign of the derivative of $f(x)$ is not known until $f(x)$ is given.

Using this terminology, one can state the following theorem:

† For the proof that $G(x)$ does not vanish between these points, see J. F. Steffensen, "Interpolation," 2d ed., p. 155, Chelsea Publishing Company, New York, 1950. $G(x)$ vanishes at x_{r+s} because the integrand in Eq. (7.48) is an odd function of $x - x_r$.

THEOREM II. *The sum of any number of simplex formulas of the same degree n all of which are plus or all of which are minus results in a simplex formula of this same degree.* Moreover, if one adds N such formulas and the remainder term for the ith formula is $C_i f^{(n+1)}(\xi_i)$, where $x_i' < \xi_i < x_i''$ and C_i is a constant involving the interval h, then the remainder term for the resulting formula is

$$Rf(x) = \sum_{i=1}^{N} C_i f^{(n+1)}(\xi_i) = f^{(n+1)}(\xi) \sum_{i=1}^{N} C_i \qquad (7.51)$$

Here $x' < \xi < x''$, where x' is the least of the x_i' and x'' the largest of the x_i''.

Proof: To prove the above theorem, note first that the remainder of the new formula will be the sum of the remainder terms of the individual formulas, as given in the first equation of (7.51). Clearly then this new formula is of degree n, since

$$Rx^k = \begin{cases} 0 & \text{for } k \leq n \\ (n+1)! \sum\limits_{i=1}^{N} C_i \neq 0 & \text{for } k = n+1 \end{cases} \qquad (7.52)$$

To prove that the formula is simplex, one therefore needs to show that the vanishing of the remainder for any function $f(x)$ having a continuous $(n+1)$th derivative implies the vanishing of its $(n+1)$th derivative somewhere in the range of the formula.

In order for $Rf(x)$ to equal zero, since the C_i are assumed to be all of the same sign, at least two of the derivatives, say $f^{(n+1)}(\xi_j)$ and $f^{(n+1)}(\xi_k)$, in the first equation of (7.51) must be of opposite sign. Moreover, since the $(n+1)$th derivative is assumed continuous, it must therefore vanish at some point ξ between ξ_j and ξ_k. The new composite formula is therefore simplex, and the range of ξ is fixed by the extremes of the ranges permitted ξ_i in the individual formulas.

Knowing that the new formula is simplex, one obtains directly from Eqs. (7.43) and (7.52) the second expression for $Rf(x)$ given in Eq. (7.51), and thus the theorem is proved.

7.12. Newton-Cotes Formulas. The symmetric quadrature formulas $Q_{11}, Q_{22}, \ldots, Q_{mm}$, etc., in which the indices are equal, are known as the Newton-Cotes formulas. They are characterized by the fact that the integration extends from the first to last ordinate employed in the formula —but not beyond.

The first three of these, Q_{11}, Q_{22}, and Q_{33}, have special names. Let $x_i = x_0 + ih$ and $f_i = f(x_i)$; then these are:

Trapezoidal Rule (Q_{11})

$$\int_{x_0}^{x_1} f(x)\, dx = \frac{h}{2}\,(f_0 + f_1) - \tfrac{1}{12}h^3 f^{(2)}(\xi) \tag{7.53}$$

where $x_0 < \xi < x_1$.

The application of this rule to N panels between $x_0 = a$ and $x_N = b$ gives rise to the composite formula

$$\int_{x_0}^{x_1} f(x)\, dx = h(\tfrac{1}{2}f_0 + f_1 + f_2 + \cdots + f_{N-1} + \tfrac{1}{2}f_N)$$
$$- \tfrac{1}{12}(b-a)h^2 f^{(2)}(\xi) \tag{7.54}$$

where $a < \xi < b$. The remainder term is obtained by direct application of Theorem II [see Eq. (7.51)].

Simpson's One-third Rule (Q_{22})

$$\int_{x_0}^{x_1} f(x)\, dx = \frac{h}{3}\,(f_0 + 4f_1 + f_2) - \tfrac{1}{90}h^5 f^{(4)}(\xi) \tag{7.55}$$

where $x_0 < \xi < x_2$.

The application of this rule to the $N = 2k$ panels between $x_0 = a$ and $x_N = b$ gives rise to the composite formula

$$\int_a^b f(x)\, dx = \frac{h}{3}\,(f_0 + 4f_1 + 2f_2 + 4f_3 + 2f_4 + \cdots + 4f_{N-1} + f_N)$$
$$- \tfrac{1}{180}(b-a)h^4 f^{(4)}(\xi) \tag{7.56}$$

where $a < \xi < b$. This formula is used more widely than any other quadrature formula. It is surprisingly accurate for such a simple formula. Needless to say, one calculates the sum in the parentheses as follows:

$$(f_0 + f_N) + 4(f_1 + f_3 + \cdots + f_{N-1}) + 2(f_2 + f_4 + \cdots + f_{N-2})$$

Simpson's Three-eighths Rule (Q_{33})

$$\int_{x_0}^{x_3} f(x)\, dx = \frac{3h}{8}\,(f_0 + 3f_1 + 3f_2 + f_3) - \tfrac{3}{80}h^5 f^{(4)}(\xi) \tag{7.57}$$

where $x_0 < \xi < x_3$.

Again the application of this formula to the $N = 3k$ panels between $x_0 = a$ and $x_N = b$ leads to the composite formula

$$\int_a^b f(x)\, dx$$
$$= \frac{3h}{8}\,(f_0 + 3f_1 + 3f_2 + 2f_3 + 3f_4 + 3f_5 + 2f_6 + \cdots + 3f_{N-1} + f_N)$$
$$- \tfrac{1}{80}(b-a)h^4 f^{(4)}(\xi) \tag{7.58}$$

where $a < \xi < b$.

By comparing the remainder term of this formula with that of Eq. (7.56) one sees that the error to be expected from Simpson's one-third rule is less than half that to be expected from the three-eighths rule.† This fact has led Scarborough to propose that the three-eighths rule, as the more complicated of the two, be relegated to oblivion. This proposal overlooks the fact that for an odd number of panels it is convenient to use both rules. However, even in this case of odd N, one needs to apply the three-eighths rule to only three of the panels, and hence one may restrict oneself to Eq. (7.57).

For the purpose of later reference, it is convenient to write out the formulas Q_{44} and Q_{66}. From Table 7.2, Q_{44} is given by the formula

$$\int_{x_0}^{x_4} f(x)\, dx = \frac{2h}{45}\,(7f_0 + 32f_1 + 12f_2 + 32f_3 + 7f_4) - \tfrac{8}{945}h^7 f^{(6)}(\xi) \qquad (7.59)$$

where $x_0 < \xi < x_4$, and Q_{66} by the formula

$$\int_{x_0}^{x_6} f(x)\, dx = \frac{h}{140}\,(41f_0 + 216f_1 + 27f_2 + 272f_3 + 27f_4 + 216f_5 + 41f_6)$$
$$- \tfrac{9}{1400}h^9 f^{(8)}(\xi) \qquad (7.60)$$

where $x_0 < \xi < x_6$.

7.13. Modifications of the Newton-Cotes Formula Q_{66}. If one adds and subtracts

$$\frac{h}{140}\,\Delta^6 f_0 = \frac{h}{140}\,(f_0 - 6f_1 + 15f_2 - 20f_3 + 15f_4 - 6f_5 + f_6)$$

one obtains for the right-hand side of Eq. (7.60)

$$\frac{h}{140}\,(42f_0 + 210f_1 + 42f_2 + 252f_3 + 42f_4 + 210f_5 + 42f_6)$$
$$- \left[\tfrac{9}{1400}h^9 f^{(8)}(\xi) + \frac{h}{140}\,\Delta^6 f_0 \right]$$

Factoring out 42 from the terms in the parentheses,

$$\int_{x_0}^{x_6} f(x)\, dx = \frac{3h}{10}\,(f_0 + 5f_1 + f_2 + 6f_3 + f_4 + 5f_5 + f_6)$$
$$- \left[\tfrac{9}{1400}h^9 f^{(8)}(\xi) + \frac{h}{140}\,\Delta^6 f_0 \right] \qquad (7.61)$$

where $x_0 < \xi < x_6$.

This formula, known as *Weddle's rule*, is less accurate than Q_{66}, being of degree five while Q_{66} is of degree seven. That is, it is strictly true only for polynomials of degree five or less. It owes its popularity to the simple

† This is only roughly so, since the point $x = \xi$ at which the fourth derivative is to be taken will not ordinarily be the same in the two cases.

coefficients needed. The main objection to its use is that in order to apply it to an integration over a range a to b, the range must contain some multiple of six panels. The dominant term in the remainder is $(h/140)\,\Delta^6 f_0$, which can alternatively be written $(h^7/140)f^{(6)}(\xi')$, where $x_0 < \xi' < x_6$.

It is obvious that if one subtracts the right-sized terms in $\Delta^6 f_0$ from Eq. (7.60), one can eliminate a pair of the coefficients. Since the coefficients of f_2 and f_4 are the smallest pair of coefficients, one may eliminate these by subtracting $(9h/700)\,\Delta^6 f_0$. The formula that results is

$$\int_{x_0}^{x_6} f(x)\,dx = h(0.28f_0 + 1.62f_1 + 2.20f_3 + 1.62f_5 + 0.28f_6)$$
$$+ \frac{9h^7}{1,400}\,[2f^{(6)}(\xi') - h^2 f^{(8)}(\xi)] \quad (7.62)$$

where in the remainder term $h^6 f^{(6)}(\xi')$ has been substituted for $\Delta^6 f_0$, and where $x_0 < \xi < x_6$, $x_0 < \xi' < x_6$. This formula is known as *Hardy's formula*.

It is obvious that a great many modifications could be made in the Newton-Cotes formulas and in other formulas by methods similar to those above; however, one must realize that one sacrifices accuracy and usually complicates the remainder.

7.14. Symmetrical Formulas of the Open Type. In most step-by-step methods for the numerical solution of ordinary differential equations one needs to be able to integrate ahead, i.e., beyond the last value for which the function is known. This may be done either by first extrapolating for the next values and then applying a suitable quadrature formula or by combining these two steps. The latter may be done by using an *open-type* quadrature formula, one in which the integration extends beyond the ordinates employed in the formula. Symmetric formulas of the open type are of the form $Q_{m,m+2}$. Thus, for example, using Table 7.2

$$Q_{02} \sim \int_{x_0}^{x_2} f(x)\,dx = 2hf_1 + \tfrac{1}{3}h^3 f^{(2)}(\xi) \quad (7.63)$$

where $x_0 < \xi < x_2$, and

$$Q_{24} \sim \int_{x_0}^{x_4} f(x)\,dx = \frac{4h}{3}\,(2f_1 - f_2 + 2f_3) + \tfrac{14}{45}h^5 f^{(4)}(\xi) \quad (7.64)$$

where $x_0 < \xi < x_4$.

By changing the interval in Eq. (7.63) to $h/2$, one obtains the *mid-point* formula Q_{01}

$$Q_{01} \sim \int_{x_0}^{x_1} f(x)\,dx = hf\left(x_0 + \frac{h}{2}\right) + \tfrac{1}{24}h^3 f^{(2)}(\xi) \quad (7.65)$$

where $x_0 < \xi < x_1$.

Formulas that are not *open* are referred to as *closed*. Thus Simpson's two formulas are closed quadrature formulas.

7.15. Integration of Newton's Interpolation Formulas. If one starts with Newton's forward-interpolation formula [see Eq. (3.48)],

$$y = y_0 + u\,\Delta y_0 + \frac{u^{[2]}}{2!}\,\Delta^2 y_0 + \cdots + \frac{u^{[m]}}{m!}\,\Delta^m y_0$$

$$+ \frac{u^{[m+1]}}{(m+1)!}\,h^{m+1}f^{(m+1)}(\xi)$$

where $y = f(x)$, $u = (x - x_0)/h$, $u^{[k]} = u(u-1)\cdots[u - (k-1)]$, and $x_0 < \xi < x_m$, and integrates between $u = 0$ and $u = 1$, one has

$$\frac{1}{h}\int_{x_0}^{x_0+h} f(x)\,dx = \int_0^1 y\,du = y_0 + a_1\,\Delta y_1 + a_2\,\Delta^2 y_0 + \cdots$$

$$+ a_m\,\Delta^m y_0 + a_{m+1}h^{m+1}f^{(m+1)}(\xi) \quad (7.66)$$

where $x_0 < \xi < x_0 + mh$ and

$$a_k = \frac{1}{k!}\int_0^1 u^{[k]}\,du \quad (7.67)$$

The remainder term in Eq. (7.66) is readily obtained from the theorem of the mean, since $u^{[m+1]}$ does not change sign in the range of integration.

Note that the remainder term is just the next term that would appear in the series with, however, $h^{m+1}f^{(m+1)}(\xi)$ replacing the difference $\Delta^{m+1}y_0$.

The mth-degree polynomial that is integrated to yield the quadrature formula of Eq. (7.66) is determined by the $m + 1$ ordinates $y_0, y_1, \cdots,$ y_m, and the integration is therefore over the first panel. In conformity with the notation of Sec. 7.6, this formula should be labeled $\tilde{Q}_{m1}(0)$. The tilde on the Q is used to point out that the formula is expressed in terms of differences. If one expresses the differences in terms of the ordinates, one obtains the quadrature formula $Q_{m1}(0)$.

The Gregory-Newton formula for backward interpolation [see Eq. (3.56)] can be written

$$y = \sum_{k=0}^{m}\frac{1}{k!}(u + k - 1)^{[k]}\,\Delta^k y_{-k} + \frac{1}{(m+1)!}(u + m)^{[m+1]}h^{m+1}f^{(m+1)}(\xi)$$

where $x_{-m} = (x_0 - mh) < \xi < x_0$. The mth-degree polynomial represented by the summation agrees with $y = f(x)$ at $u = -m, -(m-1),$ $\cdots, -1, 0$. This follows from the fact that the remainder term vanishes at these points. Therefore, if one integrates from $u = 0$ to $u = 1$, one obtains the quadrature formula $\tilde{Q}_{m1}(m)$

$$\frac{1}{h}\int_{x_0}^{x_0+h} f(x)\,dx = \sum_{k=0}^{m} b_k\,\Delta^k y_{-k} + b_{m+1}h^{m+1}f^{(m+1)}(\xi) \quad (7.68)$$

where $x_0 - mh < \xi < x_0 + h$ and

$$b_k = \frac{1}{k!} \int_0^1 (u + k - 1)^{[k]} \, du \qquad (7.69)$$

This is an *open-type* formula that is very useful in the numerical integration of differential equations.

If one integrates between $u = 0$ and $u = -1$, one obtains the *closed-type* formula $\bar{Q}_{m1}(m - 1)$

$$\frac{1}{h} \int_{x_0-h}^{x_0} f(x) \, dx = \sum_{k=0}^{m} c_k \, \Delta^k y_{-k} + c_{m+1} h^{m+1} f^{(m+1)}(\xi) \qquad (7.70)$$

where in this case $x_0 - mh < \xi < x_0$ and

$$c_k = \frac{1}{k!} \int_{-1}^0 (u + k - 1)^{[k]} \, du \qquad (7.71)$$

It is readily seen that $b_k > 0$ and $c_k\dagger < 0$; therefore Eq. (7.68) is a *plus* and Eq. (7.70) a *minus* quadrature formula. Moreover, by a change of variable,

$$b_k = \frac{1}{k!} \int_{-1}^0 (u + k)^{[k]} \, du = \frac{1}{k!} \int_{-1}^0 |u + k| \, |u + k - 1| \, \cdots \, |u + 1| \, du$$

and

$$|c_k| = \frac{1}{k!} \int_{-1}^0 |u + k - 1| \, |u + k - 2| \, \cdots \, |u| \, du$$

Thus $|c_k| < |b_k|$, for $k \geq 1$, and therefore the closed formula is usually more accurate. It is used in the integration of ordinary differential equations to check and improve the integration initially obtained from the open formula.

7.16. Symbolic Derivation of Quadrature Formulas. To illustrate the use of symbolic methods in deriving quadrature formulas, let us derive the quadrature formula $\bar{Q}_{m1}(0)$ of Sec. 7.15 [see Eqs. (7.66) and (7.67)]. In terms of the operator E defined by the equation $E^k y_0 = y_k$, k any integer, one may express intermediate values of y by (see Sec. 3.5)

$$y = E^u y_0$$

If both sides are integrated between the limits $u = 0$ and $u = 1$, one has for $y = f(x)$

$$\frac{1}{h} \int_{x_0}^{x_0+h} f(x) \, dx = \int_0^1 y \, du = \int_0^1 E^u \, du \, y_0 = \frac{E - 1}{\ln E} y_0 \qquad (7.72)$$

†Except for $k = 0$.

or, in terms of $\Delta = E - 1$,

$$\frac{1}{h} \int_{x_0}^{x_0+h} f(x)\, dx = \left[\frac{\Delta}{\ln (1 + \Delta)} \right] y_0$$
$$= (1 + \tfrac{1}{2}\Delta - \tfrac{1}{12}\Delta^2 + \tfrac{1}{24}\Delta^3 - \tfrac{19}{720}\Delta^4 + \cdots)y_0 \quad (7.73)$$

If one assumes that the function $f(x)$ behaves as a polynomial of degree m, the series terminates with the mth difference and, except for the absence of a remainder, is the same quadrature formula $\bar{Q}_{m1}(0)$ as given by Eq. (7.66).

7.17. Lozenge Diagrams for Numerical Integration. If one obtains an interpolating polynomial from the lozenge diagram of Fig. 4.1 by the rules there formulated for associating the proper factorial in u with the proper difference, one may get a one-panel quadrature formula by integrating each of the factorials in the formula between the limits of $u = 0$ and $u = 1$. Alternatively one can perform such an integration on all the factorials in the lozenge ahead of time and thereby obtain the lozenge diagram of Fig. 7.3.

Much of the discussion on the construction and use of lozenge diagrams for numerical differentiation applies to this diagram, *mutatis mutandis*. For example, one need not perform all the integrations, since the numbers so obtained must form a difference table that is rotated 180° with respect to the difference table formed from the Y's. This means that one needs initially only the constants given in Eq. (7.73). As before, the use of the capital Y's is to point out that their numbering is purely relative. Thus $Y_0 = y_k$, where y_k is any ordinate one chooses.

The difference on which one terminates the path through the lozenge determines the polynomial integrated. For example, suppose one desires the quadrature formula $\bar{Q}_{21}(0)$; then one must end on the difference $\Delta_0^2 \equiv \Delta^2 Y_0$. Any path could be used, but let us choose the path along the diagonal from Y_0. Then by the rules formulated for the use of Fig. 4.1, but with the interspersed numbers replacing the factorials in u, one has immediately that Y_0, Δ_0, Δ_0^2 are to be multiplied respectively by the numbers 1, $\tfrac{1}{2}$, $-\tfrac{1}{12}$. Remembering that all results are to be multiplied by h to take care of the change of variable from x to u, one has for $\bar{Q}_{21}(0)$

$$\int_{x_0}^{x_0+h} f(x)\, dx = h(Y_0 + \tfrac{1}{2}\Delta Y_0 - \tfrac{1}{12}\Delta^2 Y_0) + R$$

Since the integration in all cases in which one uses this lozenge is over just one panel, one may always use the mean-value theorem. This makes it easy to find the remainder term. *The remainder term is obtained by proceeding one more step to the right along any direction in the lozenge diagram and then replacing the $(m + 1)$th difference or mean difference so obtained by $h^{m+1}f^{(m+1)}(\xi)$.* Here ξ is some value of x in the range of the

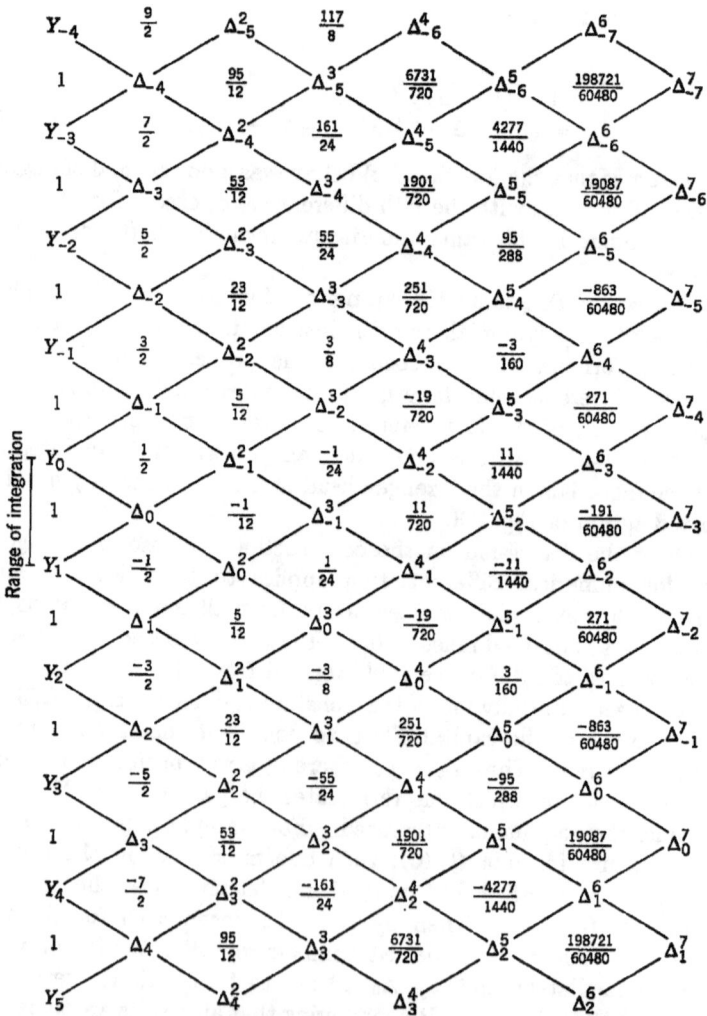

FIG. 7.3. Lozenge diagram for one-panel integration. All formulas obtained from this diagram must be multiplied by h, and Δ_s^m stands for $\Delta^m Y_s$.

formula, by which is meant the region of fit of the polynomial used to derive the formula plus the range of integration lying outside this region.

Applying this rule to the above formula and substituting $f_0 = f(x_0)$ for Y_0, one has for the formula $\bar{Q}_{21}(0)$

$$\int_{x_0}^{x_0+h} f(x)\, dx = h(f_0 + \tfrac{1}{2}\Delta f_0 - \tfrac{1}{12}\Delta^2 f_0) + \tfrac{1}{24}h^4 f^{(3)}(\xi) \qquad (7.74)$$

where $x_0 < \xi < x_0 + 2h$. If one substitutes for the differences their expressions in terms of the ordinates $f_i = f(x_0 + ih)$, one has $Q_{21}(0)$

$$\int_{x_0}^{x_0+h} f(x)\, dx = \frac{h}{12}(5f_0 + 8f_1 - f_2) + \tfrac{1}{24}h^4 f^{(3)}(\xi) \qquad (7.75)$$

This is Simpson's rule for the partial area. Although the form of (7.74) clearly depends on the path in the lozenge chosen, the quadrature formula $Q_{21}(0)$ given by Eq. (7.75) is independent of the path.

If one enters at Y_0 and also terminates there, one has the formula $Q_{01}(0)$

$$\int_{x_0}^{x_0+h} f(x)\, dx = hf_0 + \tfrac{1}{2}h^2 f'(\xi) \qquad (7.76)$$

where $x_0 < \xi < x_0 + h$. This is the so-called *rectangular rule*. It is seen to be an *open-type* quadrature formula.

In Fig. 7.4 is given a lozenge diagram for integrating over two panels. It may be obtained in a way similar to that employed for Fig. 7.3, or it may be obtained from that lozenge by shifting all the numbers one position higher in the diagram and adding them to the original numbers. By this technique one adds to each formula for the integral between Y_0 and Y_1 that for the integral between Y_{-1} and Y_0.

Note that, since one is integrating over two panels, the theorem of the mean cannot be applied directly to obtain the remainder term. There is, therefore, the possibility that some of the formulas obtained from this diagram will not be simplex (see Sec. 7.8). However, if the formula is known to be simplex, its remainder is obtained by proceeding to the right one or two difference columns in the diagram until a nonzero term is added to the formula. If this is the Nth-difference column, one replaces the Nth difference by $h^N f^{(N)}(\xi)$, where ξ is, as always, some value of x in the range of the formula.

7.18. Integration of Bessel's and Stirling's Formulas. If one takes Bessel's interpolation formula given by Eq. (4.35) and integrates v between the ordinates y_0 and y_1, i.e., from $v = -\tfrac{1}{2}$ to $v = \tfrac{1}{2}$, one obtains the symmetric formula $\bar{Q}_{m1}$

$$\int_{x_0}^{x_1} f(x)\, dx = h\left(\frac{f_0 + f_1}{2} - \frac{1}{12}\frac{\delta^2 f_0 + \delta^2 f_1}{2} + \frac{11}{720}\frac{\delta^4 f_0 + \delta^4 f_1}{2} \right.$$
$$\left. - \frac{191}{60{,}480}\frac{\delta^6 f_0 + \delta^6 f_1}{2} + \cdots\right) \qquad (7.77)$$

where $f_i = f(x_i)$ is used in place of y_i. This formula may be obtained directly from the lozenge of Fig. 7.3 by choosing a horizontal path midway between Y_0 and Y_1. One replaces Y_i by f_i and the descending differences by their equivalent central differences (see Chap. 4).

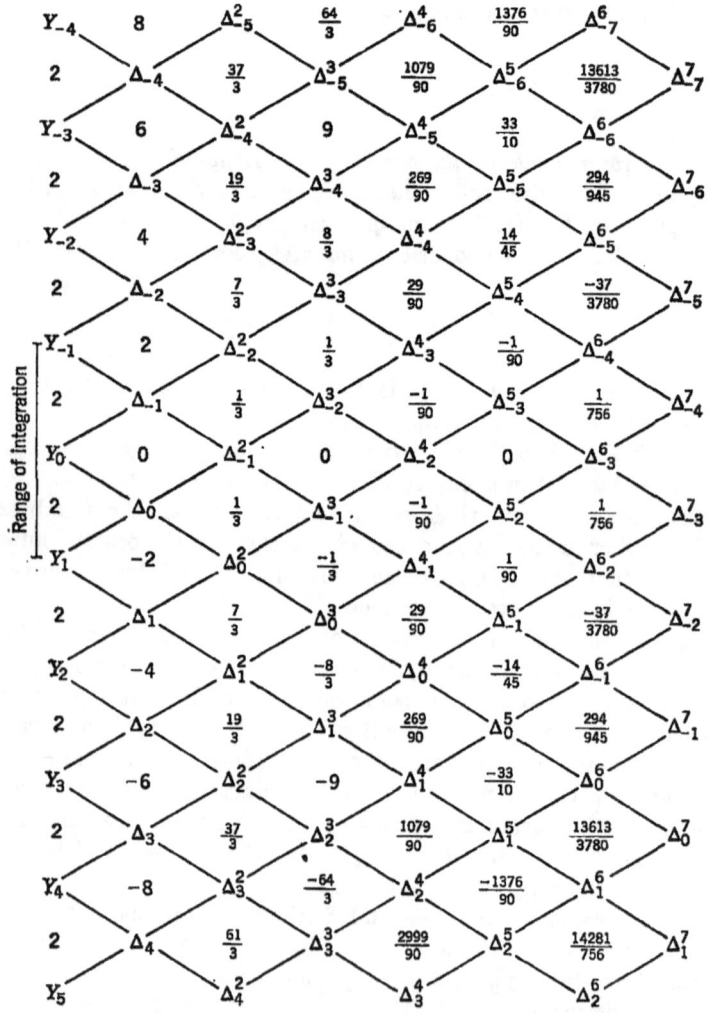

FIG. 7.4. Lozenge diagram for two-panel integration. All formulas obtained from this diagram must be multiplied by h, and Δ_s^m stands for $\Delta^m Y_s$

If one adds up n of these formulas, one has

$$\int_{x_0}^{x_n} f(x)\,dx = \sum_{i=0}^{n-1} \int_{x_i}^{x_{i+1}} f(x)\,dx$$

$$= h \sum_{i=0}^{n-1} \left(1 - \tfrac{1}{12}\delta^2 + \tfrac{11}{720}\delta^4 - \frac{191}{60,480}\delta^6 + \cdots \right)\frac{f_i + f_{i+1}}{2}$$

Now if one makes use of the fact that

$$\sum_{i=0}^{n-1} \delta^{2k} f_i = \delta^{2k-1} f_{n-\frac{1}{2}} - \delta^{2k-1} f_{-\frac{1}{2}}$$

and a similar equation for $\delta^{2k} f_{i+1}$, one obtains the equation

$$\int_{x_0}^{x_n} f(x)\, dx = h\left[(\tfrac{1}{2}f_0 + f_1 + f_2 + \cdots + f_{n-1} + \tfrac{1}{2}f_n) \right.$$
$$- \frac{1}{12}\left(\frac{\delta f_{n+\frac{1}{2}} + \delta f_{n-\frac{1}{2}}}{2} - \frac{\delta f_{\frac{1}{2}} + \delta f_{-\frac{1}{2}}}{2} \right)$$
$$\left. + \frac{11}{720}\left(\frac{\delta^3 f_{n+\frac{1}{2}} + \delta^3 f_{n-\frac{1}{2}}}{2} - \frac{\delta^3 f_{\frac{1}{2}} + \delta^3 f_{-\frac{1}{2}}}{2} \right) - \cdots \right]$$

In terms of the mean difference defined by

$$\mu \delta^k f_r = \frac{\delta^k f_{r+\frac{1}{2}} + \delta^k f_{r-\frac{1}{2}}}{2}$$

(see Chap. 4) the quadrature formula above becomes

$$\int_{x_0}^{x_n} f(x)\, dx = h\left[(\tfrac{1}{2}f_0 + f_1 + f_2 + \cdots + f_{n-1} + \tfrac{1}{2}f_n) - \tfrac{1}{12}(\mu \delta f_n - \mu \delta f_0) \right.$$
$$\left. + \tfrac{11}{720}(\mu \delta^3 f_n - \mu \delta^3 f_0) - \frac{191}{60,480}(\mu \delta^5 f_n - \mu \delta^5 f_0) + \cdots \right] \quad (7.78)$$

This is a very useful quadrature formula. It is the same as the second summation formula of Gauss [see Eq. (6.82)].

The form of this quadrature formula is interesting in that it expresses the integral in terms of the trapezoidal rule [see Eq. (7.54)] with correction terms involving the behavior of the function near the ends of the range of integration. Note that these correction terms require knowledge of $f(x)$ outside the range of integration.

Since the quadrature formula of Eq. (7.77) terminated at any term involves integration over only one panel, it is readily shown to be simplex, and, therefore, by Theorem II [see Eq. (7.51)] the composite formula of Eq. (7.78) is simplex. If Eq. (7.78) is terminated at the term involving the $2k - 1$ differences, it is clearly accurate for polynomials of degree $2k + 1$ and is therefore of degree $2k + 1$; thus, by Eq. (7.43), the remainder is

$$\frac{f^{(2k+2)}(\xi)}{(2k+2)!} \text{ times the remainder for } x^{2k+2}$$

The remainder for x^{2k+2} can be found by taking one more term in the series. It is therefore given by

$$C_{2k+1}[\mu \delta^{2k+1}(x_0 + nh)^{2k+2} - \mu \delta^{2k+1} x_0^{2k+2}]$$

where C_{2k+1} is the numerical coefficient for that term. Since, by Eq. (3.19),

$$\delta^n x^n = \Delta^n x^n = n!h^n$$

and $\delta x = h$, one has that

$$\delta(\delta^{n-1} x^n - n!h^{n-1} x) = 0$$

One knows, therefore, that the quantity in parentheses is a constant or that

$$\delta^{n-1} x^n = n!h^{n-1} x + \text{const} \tag{7.79}$$

Moreover, since $\mu(Ax + B) = Ax + B$, the remainder term for x^{2k+2} becomes

$$C_{2k+1}[(2k + 2)!h^{2k+1}nh]$$

Therefore, the remainder term for Eq. (7.78) is

$$C_{2k+1}nh^{2k+2}f^{(2k+2)}(\xi) \tag{7.80}$$

Note that the coefficient C_{2k+1} is just that for the next term in the series and that the derivative $f^{(2k+2)}(\xi)$ is one order higher than the difference that would occur in that term.

When this derivative is too difficult to determine, one usually calculates the next term in the series and uses it as a remainder term. This is generally more practical but may in some cases lead to incorrect results.

If one takes Stirling's interpolation formula as given in Eq. (4.21) and integrates from $u = -1$ to $u = 1$, one obtains the symmetric quadrature formula $\bar{Q}_{m2}$,

$$\int_{x_0-h}^{x_0+h} f(x)\, dx = h(2f_0 + \tfrac{1}{3}\delta^2 f_0 - \tfrac{1}{90}\delta^4 f_0 + \tfrac{1}{756}\delta^6 f_0 - \cdots) \tag{7.81}$$

This formula may be obtained directly from the lozenge diagram of Fig. 7.4 by choosing a horizontal path through Y_0.

7.19. Integrals Whose Limits Fall between Tabulated Values. Suppose that the limits of an integral to be evaluated do not correspond to the set of equally spaced abscissas for which $f(x)$ is tabulated. For such cases one needs a method of finding the integral over a fraction of a panel. Even where the tabulation of $f(x)$ is yet to be made, one may find it inconvenient, because of the irrational nature of one of the limits, to break up the range into an integral number of equal intervals.

Consider for simplicity the case where only the upper limit falls between tabulated values. Thus, let the integral desired be

$$I(x_n + ph) = \int_{x_0}^{x_n+ph} f(x)\, dx$$

where $x_i = x_0 + ih$ $(i = 1, 2, 3, \ldots)$ are the equally spaced values at which $f(x)$ is tabulated and p is some fraction less than one.

One obvious method of finding $I(x_n + ph)$ is to find $I(x_i)$ for several x_i in the neighborhood of x_n and then interpolate by ordinary formulas for $I(x_n + ph)$. This method is very appealing if one plans to make a table of $I(x)$ anyway.

An improvement on the above procedure is to use an interpolation formula that involves not only the value of $I(x)$ but its derivative $I'(x) = f(x)$.

A second method is to write down an interpolation formula for $f(x)$ in the neighborhood x_n and integrate this formula from $x = x_n$ to

$$x = x_n + ph$$

This amounts to deriving a special quadrature formula for the fractional panel.

7.20. The Method of Extrapolation.   One is often faced with the following problem. One has a very satisfactory quadrature formula, say Q_{31},

$$\int_{x_i}^{x_{i+1}} f(x)\, dx \cong \frac{h}{24} \left(-f_{i-1} + 13f_i + 13f_{i+1} - f_{i+2}\right) \tag{7.82}$$

over the range of integration x_0 to x_n with the exception of the end panels. For these panels, if one is to use this formula, one needs values of $f(x)$ outside the range. Thus for the last panel one needs the value of f_{n+1}.

Of course, one could use the quadrature formula $Q_{31}(2)$,

$$\int_{x_{n-1}}^{x_n} f(x)\, dx \cong \frac{h}{24} \left(f_{n-3} - 5f_{n-2} + 19f_{n-1} + 9f_n\right) \tag{7.83}$$

but this involves looking up, or deriving, a new formula. Moreover, if one is employing an automatic computing machine, this involves a change in the routine.

A more interesting procedure is to use the same formula [Eq. (7.82)] and obtain the needed value of $f(x)$ by extrapolation. Thus for the last panel one writes

$$\int_{x_{n-1}}^{x_n} f(x)\, dx \cong \frac{h}{24} \left(-f_{n-2} + 13f_{n-1} + 13f_n - f_{n+1}\right) \tag{7.84}$$

and finds f_{n+1} by extrapolating from the four values $f_{n-3}, f_{n-2}, f_{n-1},$ and f_n. Since these four values determine a third-degree polynomial, the fourth difference is zero, and hence the extrapolated f_{n+1} is obtained from the formula

$$f_{n-3} - 4f_{n-2} + 6f_{n-1} - 4f_n + f_{n+1} = 0$$

This procedure is just as accurate as the use of Eq. (7.83) even when the extrapolated value of f_{n+1} is known to be completely out of agreement with the actual value of $f(x_{n+1})$. The methods are, in fact, completely equivalent, since if f_{n+1} from the last equation is substituted in Eq. (7.84), one obtains the formula of Eq. (7.83). If one considers carefully, one sees that in each case, basically, one is finding the integral of a third-degree polynomial that fits $f(x)$ at the four values f_{n-3}, f_{n-2}, f_{n-1}, and f_n.

If one is employing a difference table of the f_i and is perhaps using a quadrature formula expressed in terms of the first m differences, one merely extends the difference table by assuming the mth difference constant (see Chap. 3).

Clearly this method of extrapolation can be employed on formulas of numerical differentiation as well.

7.21. Conversion of Formulas. It should be clear that the above reasoning may be used to obtain other formulas of the same order from a given formula. Thus suppose one wants an open formula giving the integral between x_n and x_{n+1} in terms of f_{n-3}, f_{n-2}, f_{n-1}, and f_n. One may proceed as follows. One first finds this integral by the obvious modification of Eq. (7.83); this involves the knowledge of f_{n+1}. One then eliminates f_{n+1} by assuming the fourth difference is zero.

The steps are best carried out by first writing Eq. (7.83) in pattern form

$$Q_{31}(2) \sim \frac{h}{24} \{1 \quad -5 \quad 19 \quad 9\}$$
$$\underset{\text{(range of integration)}}{}$$

This formula, being of degree three, is exact for a polynomial of degree three, and therefore any formula obtained from it by adding any multiple of a fourth difference is of the same degree. Therefore, one may obtain the desired open formula $Q_{31}(3)$ as follows:

$$Q_{31}(2) \sim \frac{h}{24} \{ \quad 0 \quad \quad 1 \quad -5 \quad \quad 19 \quad \quad 9\}$$
$$\frac{h}{24} \{-9 \quad \quad 36 \quad -54 \quad \quad 36 \quad -9\}$$
$$\overline{Q_{31}(3) \sim \frac{h}{24} \{-9 \quad \quad 37 \quad -59 \quad \quad 55 \quad \quad 0\}}$$
$$\underset{\text{(range of integration)}}{}$$

One may also choose to lower the degree of the formula. Thus one can reduce $Q_{31}(2)$ to a formula requiring only three ordinates by adding a third difference instead of a fourth.

$$Q_{31}(2) \sim \frac{h}{24} \{ \quad 1 \quad -5 \quad \quad 19 \quad \quad 9\}$$
$$\frac{h}{24} \{-1 \quad \quad 3 \quad -3 \quad \quad 1\}$$
$$\overline{\frac{h}{24} \{ \quad 0 \quad -2 \quad \quad 16 \quad \quad 10\}}$$

This is Simpson's rule for the partial area

$$Q_{21}(1) \sim \frac{h}{12} \{-1 \quad \mathbf{8} \quad \mathbf{5}\}$$

that is to be paired with Eq. (7.75).

Having obtained $Q_{31}(2)$ and $Q_{31}(0)$† from $Q_{31}(1)$, we can, of course, add these three to find Q_{33}, which is Simpson's three-eighths rule.

7.22. Gregory's Formula. If one starts with the symbolic expression

$$y = f(x_0 + uh) = E^u f_0 \tag{7.85}$$

where E is defined by the property that

$$E y_i = E f(x_0 + ih) = f(x_0 + (i+1)h) = y_{i+1}$$

and integrates both sides between $x = x_0$ and $x = x_0 + nh$, one obtains the formula

$$\int_{x_0}^{x_0+nh} f(x)\, dx = h \int_0^n y\, du = h \int_0^n E^u\, du\, y_0 = h \frac{(E^n - 1)y_0}{\ln E} \tag{7.86}$$

This equation can be written

$$\int_{x_0}^{x_0+nh} f(x)\, dx = h \left[\frac{1}{\ln E} (y_n - y_0) \right]$$

$$= h \left[\frac{-1}{\ln (1 - \nabla)} y_n - \frac{1}{\ln (1 + \Delta)} y_0 \right] \tag{7.87}$$

Since [see Eq. (7.73)]

$$\frac{1}{\ln (1 + \Delta)} = \frac{1}{\Delta} (1 + \tfrac{1}{2}\Delta - \tfrac{1}{12}\Delta^2 + \tfrac{1}{24}\Delta^3 - \tfrac{19}{720}\Delta^4 + \tfrac{3}{160}\Delta^5 - \cdots)$$

$$= \frac{1}{\Delta} + \frac{1}{2} + \sum_{i=2}^{\infty} C_i \Delta^{i-1}$$

the integral becomes

$$\int_{x_0}^{x_0+nh} f(x)\, dx = h \left\{ \left[\frac{1}{\nabla} - \frac{1}{2} - \sum_{i=2}^{\infty} C_i(-\nabla)^{i-1} \right] y_n \right.$$

$$\left. - \left(\frac{1}{\Delta} + \frac{1}{2} + \sum_{i=2}^{\infty} C_i \Delta^{i-1} \right) y_0 \right\}$$

Moreover, since $\nabla = E^{-1}\Delta$,

$$\frac{1}{\nabla} y_n - \frac{1}{\Delta} y_0 = \frac{E^{n+1} - 1}{E - 1} y_0 = y_0 + y_1 + \cdots + y_n$$

† The pattern for $Q_{31}(0)$ is just that for $Q_{31}(2)$ inverted.

the quadrature formula becomes

$$\int_{x_0}^{x_0+nh} f(x)\,dx = h\left\{(\tfrac{1}{2}y_0 + y_1 + y_2 + \cdots + y_{n-1} + \tfrac{1}{2}y_n)\right.$$
$$\left. - \sum_{i=2}^{\infty} C_i[(-1)^{i-1}\nabla^{i-1}y_n + \Delta^{i-1}y_0]\right\}$$
$$= h[(\tfrac{1}{2}y_0 + y_1 + \cdots + y_{n-1} + \tfrac{1}{2}y_n)$$
$$- \tfrac{1}{12}(\nabla y_n - \Delta y_0) - \tfrac{1}{24}(\nabla^2 y_n + \Delta^2 y_0)$$
$$- \tfrac{19}{720}(\nabla^3 y_n - \Delta^3 y_0) - \tfrac{3}{160}(\nabla^4 y_n - \Delta^4 y_0) - \cdots] \quad (7.88)$$

This is *Gregory's formula* for numerical integration. The series is customarily terminated at, or before, the terms involving nth differences. The formula thus involves the knowledge of none of the ordinates $y_i = f(x_0 + ih)$ outside the range of integration.

7.23. The Euler-Maclaurin Formula. The relation between the derivative operator D and the operator E is given in Eq. (7.14), namely,

$$D = \frac{1}{h}\ln E \quad \text{or} \quad E = e^{hD} \quad (7.89)$$

This relationship summarizes the information in Taylor's series, since, if Taylor's series converges,

$$Ef(x_0) = f(x_0 + h)$$
$$= \left(1 + hD + \frac{h^2}{2}D^2 + \frac{h^3}{3!}D^3 + \cdots\right)f(x_0) = e^{hD}f(x_0)$$

Now purely symbolically

$$\frac{E+1}{E-1} = \frac{e^{hD}+1}{e^{hD}-1} = \frac{e^{\frac{1}{2}hD} + e^{-\frac{1}{2}hD}}{e^{\frac{1}{2}hD} - e^{-\frac{1}{2}hD}} = \coth \tfrac{1}{2}hD$$

Therefore by expanding $\coth \tfrac{1}{2}hD$ in a power series, one has

$$\frac{E+1}{E-1} = \frac{2}{hD}\left[1 + \sum_{k=1}^{\infty}(-1)^{k-1}B_{2k-1}\frac{(hD)^{2k}}{(2k)!}\right]$$

where B_n is the *nth Bernoulli number.*† This equation can be rewritten

$$E - 1 = \frac{h}{2}(E+1)D + (E-1)\sum_{k=1}^{\infty}(-1)^k B_{2k-1}\frac{(hD)^{2k}}{(2k)!} \quad (7.90)$$

If this operator equation is applied to the function

$$F(x) = \int f(x)\,dx$$

† See B. O. Peirce, "A Short Table of Integrals," 3d ed., p. 93, Ginn & Company, Boston, 1929.

one obtains the formula

$$\int_{x_0}^{x_1} f(x) \, dx = \frac{h}{2} [f(x_0) + f(x_1)]$$

$$+ \sum_{k=1}^{\infty} (-1)^k B_{2k-1} \frac{h^{2k}}{(2k)!} [f^{(2k-1)}(x_1) - f^{(2k-1)}(x_0)]$$

$$= \frac{h}{2} [f(x_0) + f(x_1)] - \frac{h^2}{12} [f^{(1)}(x_1) - f^{(1)}(x_0)]$$

$$+ \frac{h^4}{720} [f^{(3)}(x_1) - f^{(3)}(x_0)] - \frac{h^6}{30,240} [f^{(5)}(x_1) - f^{(5)}(x_0)] + \cdots \quad (7.91)$$

If one applies this formula to each of the n panels between x_0 and $x_n = x_0 + nh$, one obtains the composite formula

$$\int_{x_0}^{x_n} f(x) \, dx = h(\tfrac{1}{2}f_0 + f_1 + f_2 + \cdots + f_{n-1} + \tfrac{1}{2}f_n)$$

$$- \frac{h^2}{12} [f^{(1)}(x_n) - f^{(1)}(x_0)]$$

$$+ \frac{h^4}{720} [f^{(3)}(x_n) - f^{(3)}(x_0)]$$

$$- \frac{h^6}{30,240} [f^{(5)}(x_n) - f^{(5)}(x_0)] + \cdots \quad (7.92)$$

where $f_k = f(x_k)$. This is the *Euler-Maclaurin* formula for numerical integration. Note that it is expressed as a set of corrections to the trapezoidal rule. These corrections depend on the value of the odd derivatives at the end points of the range of integration.

7.24. Gauss' Formula of Numerical Integration. Instead of requiring that the integral be expressed in terms of ordinates at equally spaced values of the abscissa, let the integral be expressed in the form

$$\int_a^b f(x) \, dx = \frac{b - a}{2} \sum_{i=0}^{n} w_i f(x_i) \quad (7.93)$$

where the x_i are to be determined. Since there are $2n + 2$ of the constants w_i and x_i, it is reasonable to expect that, since a polynomial of degree $2n + 1$ has $2n + 2$ constants, they could be so chosen that the formula is of degree $2n + 1$. That is, the formula would be exact for any polynomial of degree $2n + 1$ or less.

It is convenient to make the transformation

$$t = \frac{2x - (a + b)}{b - a} \quad (7.94)$$

Then Eq. (7.93) becomes

$$\int_{-1}^{1} F(t)\, dt = \sum_{i=0}^{n} w_i F(t_i) \tag{7.95}$$

where

$$F(t) = f\left\{\frac{(b-a)t + (b+a)}{2}\right\} \tag{7.96}$$

and t_i is the value of t corresponding to x_i.

Let $G_n(t)$ be the polynomial of degree n passing through the $n+1$ ordinates $F(t_i)$. This polynomial is unique and by Lagrange's equation [Eq. (5.4)] is of the form

$$G_n(t) = \sum_{i=0}^{n} a_i(t) F(t_i)$$

Integrating both sides, one has

$$\int_{-1}^{1} G_n(t)\, dt = \sum_{i=0}^{n} \int_{-1}^{1} a_i(t)\, dt\, F(t_i)$$

Therefore if the formula (7.95) is to be exact for polynomials of degree n, one must require that [see Eq. (5.3)]

$$w_i = \int_{-1}^{1} a_i(t)\, dt = \frac{1}{\displaystyle\prod_{j \neq i}(t_i - t_j)} \int_{-1}^{1} \prod_{j \neq i}(t - t_j)\, dt \tag{7.97}$$

In other words, once the t_i are fixed, the w_i are uniquely determined by the requirement that the formula hold for all polynomials of degree n.

Suppose one chooses for $F(t)$ an arbitrary polynomial of degree $2n + 1$ and requires that the formula be exact for such a polynomial. Since $G_n(t_i) = F(t_i)$, $F(t)$ must be of the form

$$F(t) = G_n(t) + \phi_n(t) \prod_{i=0}^{n}(t - t_i) \tag{7.98}$$

where $\phi_n(t)$ is an arbitrary polynomial of degree n.

Integrating both sides of Eq. (7.98), one has

$$\int_{-1}^{1} F(t)\, dt = \sum_{i=0}^{n} w_i F(t_i) + \int_{-1}^{1} \phi_n(t) \prod_{i=0}^{n}(t - t_i)\, dt$$

Thus, if the formula (7.95) is to be exact, the roots t_i of the $(n+1)$th-degree polynomial

$$\Pi_{n+1}(t) = \prod_{i=0}^{n}(t - t_i)$$

must be so chosen that

$$\int_{-1}^{1} \phi_n(t)\Pi_{n+1}(t)\, dt = 0 \qquad (7.99)$$

Note that this must hold for an arbitrary polynomial $\phi_n(t)$ of degree n.

The necessary and sufficient condition for $\Pi_{n+1}(t)$ to satisfy Eq. (7.99) is for

$$\Pi_{n+1}(t) = AP_{n+1}(t)$$

where A is an arbitrary constant and $P_{n+1}(t)$ is the *Legendre polynomial* of degree $n + 1$. Thus the t_i in (7.95) must be the roots of $P_{n+1}(t)$.

One may prove this theorem as follows. One assumes the well-known expansion theorem that a function may be expanded in the interval -1 to 1 in terms of Legendre polynomials.[†] Thus

$$\phi_n(t) = \sum_{k=0}^{n} a_k P_k(t)$$

where the a_k are arbitrary constants, and

$$\Pi_{n+1}(t) = \sum_{j=0}^{n+1} b_j P_j(t)$$

where the b_j are to be determined so as to satisfy Eq. (7.99). Since the a_k are arbitrary, Eq. (7.99) requires that[‡]

$$\sum_{j=0}^{n+1} b_j \int_{-1}^{1} P_k(t)P_j(t)\, dt = \frac{2}{2k+1}b_k = 0$$

for $k = 0, 1, 2, \ldots, n$. Therefore $\Pi_{n+1}(t) = b_{n+1}P_{n+1}(t)$, and the theorem is proved.

By the *formula of Rodrigues*[§]

$$P_m(t) = \frac{1}{2^m m!}\frac{d^m}{dx^m}(t^2 - 1)^m$$

hence the even-order Legendre polynomials have only even powers of x and the odd-order have only odd powers of x. Thus the roots t_i of $P_{n+1}(t)$ are symmetrically placed about the origin ($t = 0$).

[†] Charles Jordan, "Calculus of Finite Differences," 2d ed., p. 435, Chelsea Publishing Company, New York, 1950.

[‡] See H. Margenau and G. M. Murphy, "The Mathematics of Physics and Chemistry," p. 100, D. Van Nostrand Company, Inc., Princeton, N.J., 1943.

[§] *Ibid*, p. 98.

The first few Legendre polynomials are

$$P_0(x) = 1$$
$$P_1(x) = x$$
$$P_2(x) = \tfrac{1}{2}(3x^2 - 1)$$
$$P_3(x) = \tfrac{1}{2}(5x^3 - 3x)$$
$$P_4(x) = \tfrac{1}{8}(35x^4 - 30x^2 + 3)$$
$$P_5(x) = \tfrac{1}{8}(63x^5 - 70x^3 + 15x)$$

The t_i in Eq. (7.95) are the roots of $P_{n+1}(x)$, and the w_i are given by Eq. (7.97).

A list of the t_i and w_i for the first few values of n are given below:

$n = 0$: $t_0 = 0$ $w_0 = 2$

$n = 1$: $t_0 = -\dfrac{1}{\sqrt{3}}$ $t_1 = \dfrac{1}{\sqrt{3}}$ $w_0 = 1$ $w_1 = 1$

$n = 2$:

$$t_0 = -\sqrt{\tfrac{3}{5}} \quad t_1 = 0 \quad t_2 = \sqrt{\tfrac{3}{5}} \quad w_0 = \tfrac{5}{9} \quad w_1 = \tfrac{8}{9} \quad w_2 = \tfrac{5}{9}$$

$n = 3$:

$$t_0 = -\left(\tfrac{3}{7} + \tfrac{2}{7}\sqrt{\tfrac{6}{5}}\right)^{\tfrac{1}{2}} \quad t_1 = -\left(\tfrac{3}{7} - \tfrac{2}{7}\sqrt{\tfrac{6}{5}}\right)^{\tfrac{1}{2}} \quad t_2 = \left(\tfrac{3}{7} - \tfrac{2}{7}\sqrt{\tfrac{6}{5}}\right)^{\tfrac{1}{2}}$$

$$t_3 = \left(\tfrac{3}{7} + \tfrac{2}{7}\sqrt{\tfrac{6}{5}}\right)^{\tfrac{1}{2}} \quad w_0 = w_3 = \dfrac{18 - \sqrt{30}}{36} \quad w_1 = w_2 = \dfrac{18 + \sqrt{30}}{36}$$

It is customary to use the transformation

$$x = \tfrac{1}{2}(t + 1)$$

to transform Eq. (7.95) into the formula

$$\int_0^1 f(x)\, dx = \sum_{i=0}^{n} W_i f(x_i) \tag{7.100}$$

A list of the x_i and W_i for this formula is given below:

$n = 0$:	$x_0 = 0.5$	$W_0 = 1$
$n = 1$:	$x_0 = 0.21132\ 487$	$W_0 = \tfrac{1}{2}$
	$x_1 = 0.78867\ 513$	$W_1 = \tfrac{1}{2}$
$n = 2$:	$x_0 = 0.11270\ 167$	$W_0 = \tfrac{5}{18}$
	$x_1 = 0.50000\ 000$	$W_1 = \tfrac{4}{9}$
	$x_2 = 0.88729\ 833$	$W_2 = \tfrac{5}{18}$
$n = 3$:	$x_0 = 0.06943\ 184$	$W_0 = 0.17392\ 74$
	$x_1 = 0.33000\ 948$	$W_1 = 0.32607\ 26$
	$x_2 = 0.66999\ 052$	$W_2 = 0.32607\ 26$
	$x_3 = 0.93056\ 816$	$W_3 = 0.17392\ 74$

$n = 4$:

$$x_0 = 0.04691\ 008 \qquad W_0 = 0.11846\ 34$$
$$x_1 = 0.23076\ 534 \qquad W_1 = 0.23931\ 43$$
$$x_2 = 0.50000\ 000 \qquad W_2 = 0.28444\ 44$$
$$x_3 = 0.76923\ 466 \qquad W_3 = 0.23931\ 43$$
$$x_4 = 0.95308\ 992 \qquad W_4 = 0.11846\ 34$$

Example 4. Obtain an approximate formula for ln x by integrating

$$\ln x = \int_1^x \frac{dt}{t}$$

using *Gauss' three-point* formula above.

By the transformation

$$u = \frac{t-1}{x-1} \qquad \text{or} \qquad t = (x-1)u + 1$$

this integral becomes

$$\ln x = (x-1) \int_0^1 \frac{du}{xu + (1-u)}$$

Therefore, by Gauss' three-point formula

$$\ln x \cong \frac{x-1}{18} \left(\frac{5}{0.11270\ 167x + 0.88729\ 833} + \frac{8}{0.5x + 0.5} \right.$$
$$\left. + \frac{5}{0.88729\ 833x + 0.11270\ 167} \right)$$

This formula may be written

$$\ln x = \frac{z-2}{18} \left[\frac{16}{z} + \frac{5z}{y(z-y)} \right]$$

where
$$z = x + 1$$
and
$$y = 0.11270\ 167x + 0.88729\ 833$$

For $x = 2$, we have $z = 3$, $y = 1.11270\ 167$, and therefore

$$\ln 2 = \frac{1}{18} \left[\frac{16}{3} + \frac{15}{(1.11270\ 167)(1.88729\ 833)} \right] = 0.69312\ 169$$

This result is in error by only three units in the fifth decimal place.

PROBLEMS

1. Determine the relative error in applying Eq. (7.9) to the exponential function e^{-ax}. Assume h small.
2. Check the coefficients in Eq. (7.19).
3. Find the first and second derivatives at $y = 5$ of the inverse function $f^{-1}(y)$ in Example 2 giving x in terms of y. Check your result using the analytical relationships between the derivatives of $f(x)$ and the inverse function.
4. Obtain the formula $D_4^2(2)$ for the second derivative on page 135 directly from a fourth-degree polynomial passing through five consecutive values of $y = f(x)$.
5. Using the method of Sec. 7.7, determine approximately the value of the integral
$$\int_0^\pi \cos(\cos x)\, dx.$$

6. Obtain the quadrature formula $Q_{31}(1)$ directly by taking a third-degree interpolating polynomial passing through four consecutive points of $y = f(x)$ and integrating over the center panel.

7. Prove directly that Simpson's one-third rule is simplex.

8. (a) Find the value of the integral $\int_0^\lambda J_0(x)\, dx$, where λ is the first zero of the Bessel function $J_0(x)$. Use Simpson's rule and an interval of 0.1 for the main portion of the integration. (b) Find an upper limit for the truncation error in part (a).

9. Integrate e^x between 0 and 1 in steps of 0.05 using (a) trapezoidal rule, (b) Simpson's one-third rule, (c) quadrature formula Q_{44}. Compare the results to the analytic solution.

10. Derive the formula

$$\int_{x-h}^{x+h} f(x)e^{-kx}\, dx = h[a(s)y(x-h) + b(s)y(x) + c(s)y(x+h)]$$

where $y(x) = f(x)e^{-kx}$ and the weighting coefficients $a(s)$, $b(s)$, and $c(s)$ are the following functions of $s = kh$:

$$a(s) = \frac{1}{2s^3}[(2s^2 - 3s + 2) - (s+2)e^{-2s}]$$

$$b(s) = \frac{2}{s^3}[(s-1)e^s + (s+1)e^{-s}]$$

$$c(s) = \frac{-1}{2s^3}[(s-2)e^{2s} + (2s^2 + 3s + 2)] = a(-s)$$

HINT: Replace $f(x)$ by an interpolating polynomial of second degree.

11. Derive the formula

$$\int_{x-h}^{x+h} f(x) \cos kx\, dx = h\{[a(s)y(x-h) + b(s)y(x) + a(s)y(x+h)]$$
$$ - c(s)[f(x+h)\sin k(x+h) - f(x-h)\sin k(x-h)]\}$$

where $s = kh$, $y(x) = f(x) \cos kx$, and the weighting functions are

$$a(s) = \frac{1 + \cos^2 s}{s^2} - \frac{\sin 2s}{s^3}$$

$$b(s) = 4\left(\frac{\sin s}{s^3} - \frac{\cos s}{s^2}\right)$$

$$c(s) = \frac{2 \sin^2 s}{s^3} - \frac{\sin 2s}{2s^2} - \frac{1}{s}$$

12. Derive the following quadrature formula

$$\int_{-a}^{b} f(x)\, dx = \frac{a+b}{6ab}[b(2a - b)f(-a) + (a+b)^2 f(0) + a(2b - a)f(b)]$$

applicable when the intervals in a table are unequal.

CHAPTER 8

NUMERICAL SOLUTION OF ORDINARY DIFFERENTIAL EQUATIONS: METHODS OF STARTING THE SOLUTION

8.1. Introduction. We shall start our discussion with a single differential equation of the first order, which can be written in the form

$$\frac{dy}{dx} = f(x,y) \tag{8.1}$$

with the initial condition $y = y_0$ when $x = x_0$. Later it will be seen that any method that is applicable to obtaining a numerical solution of Eq. (8.1) can be quite readily generalized to take care of a set of m simultaneous equations of the first order,

$$\frac{dy_i}{dx} = f_i(x,y_1,y_2, \ldots ,y_m) \qquad i = 1, 2, \ldots , m \tag{8.2}$$

The initial conditions in this case are taken to be

$$y_i = a_i \qquad \text{for } x = x_0 \tag{8.3}$$

It will further be shown that differential equations of higher order than the first can be put in the form of a set of simultaneous equations of the first order, such as given in Eqs. (8.2), by the introduction of new variables. Thus, the solution of Eq. (8.1) is basic in our theory of the numerical solution of differential equations.

In any analytical method one seeks to obtain functional relations between the dependent variables y_i and the independent variable x that will satisfy the differential Eqs. (8.2) and the boundary conditions (8.3). These may involve closed expressions or perhaps infinite series. In contrast, in a numerical solution one may (1) express the functional relationship by tabulating the y_i for a set of closely spaced values of x that cover the required range, (2) express the y_i over various portions of the range by suitable approximating functions, or (3) represent each y_i by a graph having x as the abscissa.†

† Graphical methods, as such, will not be treated in this book. For a discussion of graphical methods see H. Levy and E. A. Baggott, "Numerical Studies in Differential Equations," vol. I, chaps. I and II, C. A. Watts & Co., Ltd., London, 1934.

Where an analytical solution, which we may represent by the set of equations

$$y_i = \phi_i(x) \qquad i = 1, 2, \ldots, n \qquad (8.4)$$

can be found, and where the determination of the functions $\phi_i(x)$ for each value of x is readily made, such an analytical solution is superior to a numerical solution. This superiority lies in the fact that for the analytical expression a new solution, corresponding to different boundary conditions, involves merely a change in the values of the parameters. This is not possible in most numerical work, since one usually must express all the parameters and the boundary values in numerical form before one is able to proceed with the calculations. Any change in the initial conditions or in any of the auxiliary parameters that may occur in the problem necessitates a complete recalculation.

In spite of the above advantage of an analytical solution we find, more often than not, that no such solution is known for the equations we have on hand and that we must resort to a numerical method of solution. Even where an analytical solution exists, it often happens that the functions $y_i = \phi_i(x)$ are difficult to evaluate and may require more computation than the direct numerical solution of the differential equations. This is more likely to be true if we wish the y_i for a great many values of the argument x.

Only the methods that are applicable to finding the first few values of y will be treated here. More efficient methods suitable for continuing the solution once a set of starting values are available will be discussed in Chap. 9.

The methods available for starting a solution fall into two distinct classes. The first class includes Taylor's method and the method of Picard. In these methods y in' Eq. (8.1) is approximated by a truncated series, the individual terms of which are functions of the independent variable x. This series then represents y to a given degree of accuracy, i.e., to the required number of significant figures, for a certain range of x. In some cases, this range may of necessity be quite small.

The second class of starting procedures is represented by the methods of Euler and those of Runge and Kutta. Here no attempt is made to obtain an expression of y in terms of x, but rather one calculates the change in y due to a given increment in x. That is, one finds the numerical value of Δy for a given Δx.

8.2. Taylor's Method. Consider the differential equation (8.1) with the initial condition $y = y_0$ when $x = x_0$. Let the required solution of this equation be

$$y = \phi(x) \qquad (8.5)$$

If $x = x_0$ is not a singular point of the function, the latter can be expanded in a Taylor's series about this point. Thus with

$$y_0^{(s)} = \phi^{(s)}(x_0) = \left[\frac{d^s y}{dx^s}\right]_{x=x_0}$$

$$y = y_0 + (x - x_0)y_0^{(1)} + \frac{1}{2!}(x - x_0)^2 y_0^{(2)}$$
$$+ \frac{1}{3!}(x - x_0)^3 y_0^{(3)} + \cdots \quad (8.6)$$

a power series in x that converges over some range $x_0 \leq x \leq x_b$.

There remains only the problem of determining the derivatives $y_0^{(1)}$, $y_0^{(2)}, \ldots$ that serve as coefficients in this equation. This is done by making use of the differential equation (8.1), which can be written

$$y^{(1)} = f(x,y) \equiv g_1(x,y) \quad (8.7)$$

Differentiating both sides with respect to x, we have

$$y^{(2)} = \frac{\partial g_1}{\partial x} + \frac{\partial g_1}{\partial y} y^{(1)} \equiv g_2(x,y,y^{(1)}) \quad (8.8)$$

and by repeated differentiation

$$y^{(s)} = g_s(x,y,y^{(1)},y^{(2)}, \ldots ,y^{(s-1)}) \quad s = 1, 2, 3, \ldots \quad (8.9)$$

where
$$g_s = \frac{\partial g_{s-1}}{\partial x} + \sum_{i=0}^{s-2} y^{(i+1)} \frac{\partial g_{s-1}}{\partial y^{(i)}} \quad (8.10)$$

Setting $x = x_0$ and $y = y_0$, one obtains the constants

$$y_0^{(s)} = g_s(x_0,y_0,y_0^{(1)},y_0^{(2)}, \ldots ,y_0^{(s-1)}) \quad s = 1, 2, 3, \ldots \quad (8.11)$$

required in Eq. (8.6), each constant $y_0^{(s)}$ being expressed in terms of x_0, y_0 and the constants preceding it in the series.

Example 1. Find an expansion for y over as large a range as feasible, given that

$$\frac{dy}{dx} = 0.1(x^3 + y^2)$$

and the initial condition $y = 1$ when $x = 0$. This expansion is to give y correct to four decimal places.

Since $x_0 = 0$ and $y_0 = 1$,

$$y_0^{(1)} = 0.1[x^3 + y^2]_0 = 0.1[x_0^3 + y_0^2] = 0.1$$
$$y_0^{(2)} = 0.1[3x^2 + 2yy^{(1)}]_0 = 0.1[3x_0^2 + 2y_0 y_0^{(1)}] = 0.02$$
$$y_0^{(3)} = 0.1[6x + 2yy^{(2)} + 2(y^{(1)})^2]_0 = 0.006$$
$$y_0^{(4)} = 0.1[6 + 2yy^{(3)} + 6y^{(1)}y^{(2)}]_0 = 0.6024$$
$$y_0^{(5)} = 0.2[yy^{(4)} + 4y^{(1)}y^{(3)} + 3(y^{(2)})^2]_0 = 0.1212$$
$$y_0^{(6)} = 0.2[yy^{(5)} + 5y^{(1)}y^{(4)} + 10y^{(2)}y^{(3)}]_0 = 0.08472$$
$$y_0^{(7)} = 0.2[yy^{(6)} + 6y^{(1)}y^{(5)} + 15y^{(2)}y^{(4)} + 10(y^{(3)})^2]_0 = 0.06770$$
$$y_0^{(8)} = 0.2[yy^{(7)} + 7y^{(1)}y^{(6)} + 21y^{(2)}y^{(5)} + 35y^{(3)}y^{(4)}]_0 = 0.06088$$
$$y_0^{(9)} = 0.2[yy^{(8)} + 8y^{(1)}y^{(7)} + 28y^{(2)}y^{(6)} + 56y^{(3)}y^{(5)} + 35(y^{(4)})^2]_0 = 2.58084$$

Making use of Eq. (8.6),

$$y = 1 + 0.1x + 0.01x^2 + 0.001x^3 + 0.0251x^4 + 0.00101x^5 + 0.000117x^6$$
$$+ 0.000013x^7 + 0.00000\ 15x^8 + 0.00000\ 71x^9 + \cdots \quad (8.12)$$

If the series is terminated with x^8, then for the approximation to be good to four decimal places we must have†

$$0.00000\ 71x^9 \leq 0.00005$$

or by the use of logarithms

$$x \leq 1.24$$

Thus the eighth-degree polynomial represented by the first nine terms of (8.12) probably gives y to four decimal places over the range $0 \leq x \leq 1.24$.

If the series is terminated at x^6, x must be small enough so that

$$0.000013x^7 \leq 0.00005$$

Hence the range is $0 \leq x \leq 1.21$. Finally, if the series is terminated at x^4, one has

$$0.00101x^5 < 0.00005$$

and the range is $0 \leq x \leq 0.55$.

Thus it is seen that while addition of the fifth- and sixth-degree terms doubles the range, only a small increase in the range is achieved by adding on additional terms beyond the sixth. It is therefore convenient to use the sixth-degree polynomial to approximate y in the range $0 \leq x \leq 1.2$ and then, if necessary, to take the value of y at $x = 1.2$ as a new initial value and repeat the expansion above.

Clearly the applicability of the method depends on how readily one can differentiate the function $f(x,y)$ repeatedly and on how large a range of x may be used.

8.3. Taylor's Method Applied to a System of Differential Equations.

The application of Taylor's method to the case of simultaneous equations of the first order, such as Eqs. (8.2), or to higher-order equations is illustrated by the following example.

Example 2. Obtain the power series in t for an x and y which satisfy the differential equations

$$\frac{dx}{dt} = x + y + t$$
$$\frac{d^2y}{dt^2} = x - t \quad (8.13)$$

subject to the initial conditions that $x = 0$, $y = 1$, $dy/dt = -1$, at $t = 0$.

Letting $z = dy/dt$ and denoting derivatives by a superscript these equations can be written in the form,

$$x^{(1)} = x + y + t$$
$$y^{(1)} = z \quad (8.14)$$
$$z^{(1)} = x - t$$

with the initial conditions $x = 0$, $y = 1$, $z = -1$, at $t = 0$.

† This assumes that the truncation is sufficiently well approximated by the first neglected term in the series.

Expanding x, y, and z in Taylor's series,

$$x = x_0 + x_0^{(1)}t + \frac{1}{2!}x_0^{(2)}t^2 + \frac{1}{3!}x_0^{(3)}t^3 + \cdots$$

$$y = y_0 + y_0^{(1)}t + \frac{1}{2!}y_0^{(2)}t^2 + \frac{1}{3!}y_0^{(3)}t^3 + \cdots \quad (8.15)$$

$$z = z_0 + z_0^{(1)}t + \frac{1}{2!}z_0^{(2)}t^2 + \frac{1}{3!}z_0^{(3)}t^3 + \cdots$$

To find derivatives $x_0^{(1)}$, $y_0^{(1)}$, $z_0^{(1)}$, $x_0^{(2)}$, $y_0^{(2)}$, $z_0^{(2)}$, ... needed in these equations, we repeatedly differentiate Eqs. (8.14).

By substituting the initial values directly in Eqs. (8.14), we have

$$x_0^{(1)} = (x + y + t)_0 = 1$$
$$y_0^{(1)} = z_0 = -1$$
$$z_0^{(1)} = (x - t)_0 = 0$$

Then by differentiating with respect to t

$$x_0^{(2)} = (x^{(1)} + y^{(1)} + 1)_0 = 1$$
$$y_0^{(2)} = z_0^{(1)} = 0$$
$$z_0^{(2)} = (x^{(1)} - 1)_0 = 0$$

Likewise by successive differentiations

$$x_0^{(3)} = (x^{(2)} + y^{(2)})_0 = 1$$
$$y_0^{(3)} = z_0^{(2)} = 0$$
$$z_0^{(3)} = x_0^{(2)} = 1$$
$$x_0^{(4)} = (x^{(3)} + y^{(3)})_0 = 1$$
$$y_0^{(4)} = z_0^{(3)} = 1$$
$$z_0^{(4)} = x_0^{(3)} = 1$$
$$x_0^{(5)} = (x^{(4)} + y^{(4)})_0 = 2$$
$$y_0^{(5)} = z_0^{(4)} = 1$$
$$z_0^{(5)} = x_0^{(4)} = 1$$

Substituting these results in Eqs. (8.15),

$$x = t + \frac{1}{2!}t^2 + \frac{1}{3!}t^3 + \frac{1}{4!}t^4 + \frac{2}{5!}t^5 + \cdots$$

$$y = 1 - t + \frac{1}{4!}t^4 + \frac{1}{5!}t^5 + \cdots$$

$$z = -1 + \frac{1}{3!}t^3 + \frac{1}{4!}t^4 + \frac{1}{5!}t^5 + \cdots$$

As a check we note that Eqs. (8.14) are indeed satisfied by these series.

8.4. Picard's Method. This method is also known as the method of *successive approximation*† or of *successive substitution*.

Let
$$\frac{dy}{dx} = f(x,y) \quad (8.16)$$

† For a more complete treatment of this method see Lester R. Ford, "Differential Equations," Chap. 5, McGraw-Hill Book Company, Inc., New York, 1955.

with the initial condition $y = y_0$ for $x = x_0$, and suppose that $f(x,y)$ is a single-valued continuous function of x and y in some region S of the xy plane including the point (x_0,y_0) (see Fig. 8.1). Furthermore, assume that $f(x,y)$ is bounded in S, so that

$$|f(x,y)| < M \qquad (8.17)$$

in S, where M is some positive constant.

Integrating both sides of Eq. (8.16), one has

$$\int_{x_0}^{x} \frac{dy}{dx} \, dx = y - y_0 = \int_{x_0}^{x} f(x,y) \, dx$$

or

$$y = y_0 + \int_{x_0}^{x} f(x,y) \, dx \qquad (8.18)$$

If one takes as a first approximation to the solution $y(x)$ of Eq. (8.16) a function† $u_1(x)$, then a better approximation is usually given by $u_2(x)$, obtained by substituting $u_1(x)$ for y in the right-hand side of Eq. (8.18); thus

$$u_2(x) = y_0 + \int_{x_0}^{x} f(x,u_1(x)) \, dx \qquad (8.19)$$

By generalizing this procedure one obtains the iterative formula

$$u_i(x) = y_0 + \int_{x_0}^{x} f(x,u_{i-1}(x)) \, dx \qquad (8.20)$$

for a set of functions $u_i(x)$ that, under suitable restrictions, converges to the solution $y(x)$ of the differential equation (8.16) (see Sec. 8.5).

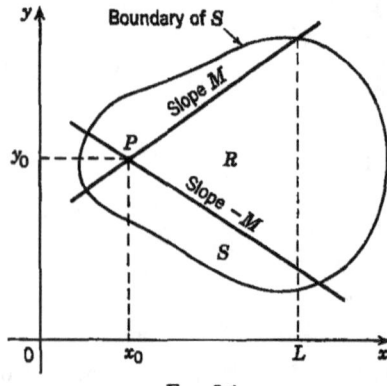

FIG. 8.1

Since $f(x,y)$ is bounded in S, all the functions $u_i(x)$, except possibly the initial function $u_1(x)$, are continuous for some interval I given by $x_0 \leq x \leq L$. This will now be shown explicitly. The constant L is determined by drawing two straight lines through the initial point (x_0,y_0) with slopes of M and $-M$ (see Fig. 8.1); L is then the largest value of x for which all points in a region R lying between these straight lines and to the left of $x = L$ are in S. We shall assume in the following discussion that x always lies in the interval I and that $u_1(x)$ is chosen so that it lies in R for all x in this interval. Then by Eqs. (8.17)

† Some of the restrictions on $u_1(x)$ will appear later.

and (8.19)

$$|u_2(x) - y_0| \leq \int_{x_0}^{x} |f(x,u_1(x))|\, dx \leq M(x - x_0)$$

and hence $u_2(x)$ lies in R. Likewise, since $u_2(x)$ lies in R, by Eq. (8.20) $|u_3(x) - y_0| \leq M(x - x_0)$, and hence $u_3(x)$ lies in R. Clearly, as one can easily prove by mathematical induction, all the other $u_i(x)$ must also lie in R. Thus, for x and $x + \Delta x$ in the interval I, by Eq. (8.20), (for $i \geq 2$)

$$|u_i(x + \Delta x) - u_i(x)| \leq \left| \int_{x}^{x+\Delta x} [f(x + \Delta x, u_{i-1}(x + \Delta x)) - f(x,u_{i-1}(x))]\, dx \right| \leq \left| \int_{x}^{x+\Delta x} 2M\, dx \right| = 2M|\Delta x|$$

since both $u_{i-1}(x)$ and $u_{i-1}(x + \Delta x)$ lie in the region R. Therefore $|u_i(x + \Delta x) - u_i(x)| \to 0$ as $\Delta x \to 0$, which requires that the $u_i(x)$, $i \geq 2$, be continuous functions.

8.5. Convergence of Picard's Method. The conditions of continuity and boundedness imposed on the single-valued function $f(x,y)$ in the last section are not sufficiently restrictive to ensure a unique solution of the differential equation for each given initial condition $y = y_0$ at $x = x_0$. For example, the differential equation

$$\frac{dy}{dx} = y^{\frac{2}{3}} \tag{8.21}$$

with initial condition $y = 0$ at $x = 0$ has both the solution $y = 0$ (everywhere) and the solution $y = x^3/27$.

One of the most common additional conditions to apply to $f(x,y)$ to ensure uniqueness of the solution is the *Lipschitz condition*, which requires that for any two vertically displaced points† (x,y) and (x,z) in S

$$|f(x,y) - f(x,z)| < K|y - z| \tag{8.22}$$

where K is some constant that depends on $f(x,y)$ and on the region S. For an $f(x,y)$ satisfying this condition the unique solution of Eq. (8.16) is given by the limit of the set of functions $u_i(x)$ defined by Eq. (8.20).

Since one may write $u_n(x)$ in the form

$$u_n(x) = u_1(x) + [u_2(x) - u_1(x)] + [u_3(x) - u_2(x)] + \cdots + [u_n(x) - u_{n-1}(x)] \tag{8.23}$$

it is clear that if a limiting function $u(x) = \lim\limits_{n \to \infty} u_n(x)$ exists, it is given by the infinite series

$$u(x) = u_1(x) + [u_2(x) - u_1(x)] + [u_3(x) - u_2(x)] + \cdots \tag{8.24}$$

† Strictly speaking, one must also require that the line segment joining (x,y) and (x,z) lie wholly in S.

It will now be shown that this series is absolutely and uniformly convergent. By Eqs. (8.20) and (8.22)

$$|u_{i+1}(x) - u_i(x)| \leq \int_{x_0}^{x} |f(x,u_i(x)) - f(x,u_{i-1}(x))|\, dx$$

$$\leq K \int_{x_0}^{x} |u_i(x) - u_{i-1}(x)|\, dx \qquad (8.25)$$

Since the $u_i(x)$, for interval I, were shown to lie in R, they can differ at most by $N = 2M(L - x_0)$ (see Fig. 8.1). Thus, in particular,

$$|u_2(x) - u_1(x)| \leq N$$

and, therefore, from Eq. (8.25) (for $i = 2$), letting $h = |L - x_0|$,

$$|u_3(x) - u_2(x)| \leq NK(x - x_0) \leq NKh$$

By repeated application of Eq. (8.25) we have

$$|u_{n+1}(x) - u_n(x)| \leq NK^{n-1} \frac{(x - x_0)^{n-1}}{(n - 1)!} \leq N \frac{(Kh)^{n-1}}{(n - 1)!} \qquad (8.26)$$

Starting with the second term in the series (8.24), each term of that series is less in absolute value than the corresponding terms of the series

$$N + NKh + N \frac{(Kh)^2}{2!} + N \frac{(Kh)^3}{3!} + \cdots$$

which converges to Ne^{Kh}. By the comparison test, therefore, the series (8.24) is absolutely and uniformly convergent. Since it is uniformly convergent and the individual terms [when $u_1(x)$ is canceled out from the first two terms] are continuous, the limit of the series (8.24) defines a continuous function $u(x)$.

By Eq. (8.20), for $i = n$,

$$\left| u - y_0 - \int_{x_0}^{x} f(x,u)\, dx \right| = \left| u - u_n - \int_{x_0}^{x} [f(x,u) - f(x,u_{n-1})]\, dx \right|$$

$$\leq |u - u_n| + K \int_{x_0}^{x} |u - u_{n-1}|\, dx$$

Since $u_n(x)$ and $u_{n-1}(x)$ converge uniformly to $u(x)$ as $n \to \infty$, the right-hand side approaches zero; therefore, the left-hand must be identically zero. Thus

$$u = y_0 + \int_{x_0}^{x} f(x,u)\, dx$$

and by differentiating this equation with respect to x, one finds that $u(x)$ satisfies the differential equation (8.16).

The uniqueness of the solution can also be proved by similar techniques, but we shall omit this part of the theory.

While the method is thus seen to converge to the desired solution for quite a general $f(x,y)$, it suffers from the very serious practical difficulty that one must be able to perform the integrations required. Since it is unlikely that one can perform more than the first few integrations, one uses $u_m(x)$, where m is usually fairly small, as an approximate solution of the differential equation. This permits one to obtain suitable starting values, which may, however, have to be improved by the method of Sec. 8.13. The latter method is nothing more than a method of successive approximation in which the integration is performed by means of some of the quadrature formulas of Chap. 7.

Example 3. Find the series expansion that gives y as a function of x in the neighborhood of $x = 0$ when

$$\frac{dy}{dx} = x^2 + y^2 \tag{8.27}$$

with boundary condition $y = 0$ for $x = 0$.

From Eq. (8.27) it is seen that $y' = 0$ at $x = 0$. Therefore, since $y = 0$ and its first derivative is zero, y will be quite small for x small. As a first approximation let us assume that $y = 0$. Then by Eq. (8.19), letting $u_1(x) = 0$,

$$u_2(x) = \int_0^x x^2 \, dx = \frac{1}{3} x^3$$

This is the second-order approximation to the solution of Eq. (8.27). The third-order approximation, by Eq. (8.20), is then

$$u_3(x) = \int_0^x \left(x^2 + \frac{x^6}{9} \right) dx = \frac{1}{3} x^3 + \frac{1}{63} x^7$$

Likewise the higher-order approximations are

$$u_4(x) = \int_0^x \left(x^2 + \frac{1}{9} x^6 + \frac{2}{189} x^{10} + \frac{1}{3,969} x^{14} \right) dx$$

$$= \frac{1}{3} x^3 + \frac{1}{63} x^7 + \frac{2}{2,079} x^{11} + \frac{1}{59,535} x^{15}$$

and

$$u_5(x) = \int_0^x \left(x^2 + \frac{1}{9} x^6 + \frac{2}{189} x^{10} + \frac{13}{14,553} x^{14} + \cdots \right) dx$$

$$= \frac{1}{3} x^3 + \frac{1}{63} x^7 + \frac{2}{2,079} x^{11} + \frac{13}{218,295} x^{15} + \cdots$$

If the series is terminated after the third term and used to approximate y to four decimal places, then using the first neglected term as an approximation of the truncation error,

$$\frac{13}{218,295} x^{15} \leq 0.00005$$

or

$$15 \log x \leq \log \frac{(0.00005)(218,295)}{13}$$

$$x \leq 0.988$$

Since negative values can be used as well as positive ones, the polynomial

$$\frac{1}{3} x^3 + \frac{1}{63} x^7 + \frac{2}{2,079} x^{11} \tag{8.28}$$

represents y correct to four decimal places for the range $-0.988 \leq x \leq 0.988$.

8.6. Euler's Method. Euler's method, as such, is of very little practical importance, but it illustrates in simple form the basic idea of those numerical methods which seek to determine the change Δy in y corresponding to a small increase in the argument x.

Again take the differential equation

$$\frac{dy}{dx} = f(x,y) \tag{8.29}$$

with the boundary condition $y = y_0$ at $x = x_0$. By definition

$$\frac{dy}{dx} = \lim_{\Delta x \to 0} \frac{\Delta y}{\Delta x}$$

therefore, for small values of Δx

$$\Delta y \cong \frac{dy}{dx} \Delta x$$

Thus, the increase in $y(x)$, the solution of the differential Eq. (8.29), when x increases to $x + \Delta x$, is approximately

$$\Delta y = f(x,y) \, \Delta x \tag{8.30}$$

Starting from x_0, where $y = y_0$, one may construct a table of y for given steps of $\Delta x = h$ in x. If x_0 is not the initial value of x, one needs to take Δx negative to obtain those values of y to the left of x_0.

Example 4. Given

$$\frac{dy}{dx} = \frac{y - x}{y + x}$$

with the initial condition $y = 1$ when $x = 0$, find y for $x = 0.1$.

If one takes $\Delta x = 0.1$, then by Eq. (8.30)

$$\Delta y = \frac{1 - 0}{1 + 0} \Delta x = 0.1$$

and therefore
$$y = y_0 + \Delta y = 1.1$$

Much better accuracy is achieved by breaking up the interval 0 to 0.1 into five steps so that $\Delta x = h = 0.02$; then for the first step

$$\Delta y = h = 0.02$$

and therefore
$$y = 1.02 \quad \text{at } x = 0.02$$

For the second step

$$\Delta y = \frac{1.02 - 0.02}{1.02 + 0.02} h = 0.0192$$

therefore
$$y = 1.02 + 0.0192 = 1.0392 \quad \text{at } x = 0.04$$

It is convenient to arrange the work in the following tabular form, the calculations being made from left to right across the table.

x	y	$y + x$	$y - x$	$f = \dfrac{y-x}{y+x}$	$\Delta y = hf$
0.00	1.0000	1.0000	1.0000	1.000	0.0200
0.02	1.0200	1.0400	1.0000	0.962	0.0192
0.04	1.0392	1.0792	0.9992	0.926	0.0185
0.06	1.0577	1.1177	0.9977	0.893	0.0179
0.08	1.0756	1.1556	0.9956	0.862	0.0172
0.10	1.0928	1.1928	0.9928	0.832	0.0166

The analytical solution of the differential equation in this example is the implicit function defined by the relation

$$g(x,y) = \tfrac{1}{2} \ln (x^2 + y^2) + \arctan \frac{y}{x} = C$$

where C is an arbitrary constant. Differentiating both sides of this equation, it is found that $g_x + y'g_y = 0$, where the subscripts denote partial derivatives; thus, since

$$g_x = \frac{x-y}{x^2+y^2} \qquad g_y = \frac{x+y}{x^2+y^2}$$

one verifies that y satisfies the differential equation

$$y' = -\frac{g_x}{g_y} = \frac{y-x}{y+x}$$

To find C we substitute the initial condition $x = 0$, $y = 1$ and have

$$C = \arctan \infty + \tfrac{1}{2} \ln 1 = \frac{\pi}{2}$$

To find y corresponding to a given x, this value of x must be substituted in the equation $g(x,y) - \pi/2 = 0$ and the resulting equation for y solved by some method of successive approximation, such as the method of false position. Since the function $g(x,y)$ involves natural logarithms and arctangents, it is necessary to have good tables of these functions, or their inverse functions e^x and $\tan x$, to find y accurately. Thus the work involved in constructing the table above from the analytical solution is much greater than that involved in the numerical solution, even after the analytical solution is given.

8.7. The Accumulation of Errors in Euler's Method. It is well before passing on to more practical methods to study the error made in the numerical procedure above. This study should serve as an introduction to the more difficult error analyses of Chap. 9. It will be assumed in this analysis that the accumulated errors are not large enough to cause the values of y computed, which we shall label Y_i, to differ sizably from the true y_i.

Suppose we have integrated Eq. (8.29) from x_0 to $x_{n+1} = x_0 + (n + 1)h$

using formula (8.30), which now will be written

$$\Delta Y_i = Y_{i+1} - Y_i = f(x_i, Y_i)h \tag{8.31}$$

Let the errors ϵ_i be defined by the equation

$$Y_i + \epsilon_i = y_i = y(x_i) \tag{8.32}$$

where $y(x)$ is the true solution of the differential equation for the boundary condition $y = y_0$ at $x = x_0$.

By Taylor's series

$$y_{i+1} = y_i + hf(x_i, y_i) + \frac{h^2}{2} y''(x_i) + O(h^3) \tag{8.33}$$

where $f(x_i, y_i)$ replaces $y'(x_i)$ and $O(h^3)$ stands for terms of order h^3 or higher. Subtracting Y_{i+1} as given by Eq. (8.31) from this equation, we have

$$\epsilon_{i+1} = \epsilon_i + h[f(x_i, y_i) - f(x_i, Y_i)] + h^2 a_{i+1}$$

where $h^2 a_{i+1}$ represents the last two terms of Eq. (8.33). Since Y_i is assumed to differ from y_i by a small amount ϵ_i

$$\epsilon_{i+1} = (1 + hK_i)\epsilon_i + h^2 a_{i+1} \tag{8.34}$$

where†

$$K_i \equiv \frac{f(x_i, y_i) - f(x_i, Y_i)}{y_i - Y_i} \cong \frac{\partial f(x_i, y_i)}{\partial y} \tag{8.35}$$

Since there is no error in the initial value, $\epsilon_0 = 0$, and by Eq. (8.34) (for $i = 0$) $\epsilon_1 = h^2 a_1$. Likewise for $i = 1$

$$\epsilon_2 = (1 + hK_1)\epsilon_1 + h^2 a_2 = h^2[(1 + hK_1)a_1 + a_2]$$

and for $i = 2$

$$\begin{aligned}\epsilon_3 &= (1 + hK_2)\epsilon_2 + h^2 a_3 \\ &= h^2[(1 + hK_1)(1 + hK_2)a_1 + (1 + hK_2)a_2 + a_3]\end{aligned} \tag{8.36}$$

Proceeding in this way, one obtains for the error at the nth step

$$\epsilon_n = h^2 \left[a_1 \prod_{j=1}^{n-1} (1 + hK_j) + a_2 \prod_{j=2}^{n-1} (1 + hK_j) + \cdots + a_n \right] \tag{8.37}$$

or

$$\epsilon_n = h^2 \sum_{i=1}^{n} A_i a_i \tag{8.38}$$

where

$$A_i = \prod_{j=i}^{n-1} (1 + hK_j) \qquad i \neq n \tag{8.39}$$
$$A_n = 1$$

† By reference to Sec. 5.7 we see that K_i is the first divided difference with respect to y of $f(x,y)$.

Thus the errors $h^2 a_i$ introduced at each step because of the inaccuracy of formula (8.31) are to be multiplied by the *amplification factors* A_i before being summed.

It can be seen from Eqs. (8.39) that two cases may arise. If the K_j are positive, the A_i increase with n, and an error $h^2 a_i$ introduced at the ith step grows as the integration continues. On the other hand, if the K_j are negative, such an error dies out.† The first case, $K_j > 0$, corresponds to a differential equation in which the plots of the individual solutions for various boundary conditions, in the neighborhood of the

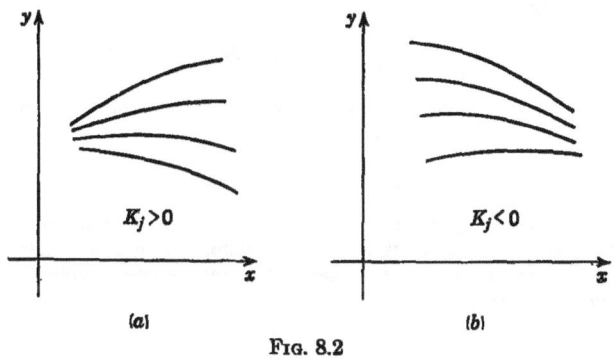

$K_j > 0$

$K_j < 0$

(a)

(b)

FIG. 8.2

desired solution, diverge as one moves to the right (see Fig. 8.2a). For the second case, $K_j < 0$, they converge.

Suppose the K_j are all equal to some value K and let α be determined by the equation

$$e^{\alpha h} = 1 + hK \tag{8.40}$$

Then from Eqs. (8.39)

$$A_i = (1 + hK)^{n-i} = e^{\alpha h (n-i)} = e^{\alpha (x_n - x_i)} \tag{8.41}$$

This shows that in the region where $K \cong \partial f / \partial y$ is nearly constant the errors grow (or decrease) exponentially.

The $h^2 a_i$, thus far, have been the errors introduced at each step because of the inaccuracy of the integration formula, the so-called *truncation errors* (see Appendix A). No account has thus been taken of the *round-off errors* due to the use of a limited number of significant figures to represent the numbers, it being assumed that a sufficient number of figures is retained to keep these errors small. Nevertheless, no harm would be done in letting $h^2 a_i$ represent *formally* the sum of both the *truncation error* and the *round-off error* introduced at the ith step in the actual evaluation of Y_i by Eq. (8.31).

† Note that a reversal of the direction of integration would cause h to be replaced by $-h$ and would thus reverse these statements.

By Eq. (8.33) one sees that the truncation error due to the integration formula is approximately given by $\frac{1}{2}h^2y''$. Therefore, by (8.29) this error is estimated as

$$\frac{1}{2}h^2y'' = \frac{1}{2}(f_x + y'f_y)h^2 = \frac{1}{2}(f_x + ff_y)h^2 \qquad (8.42)$$

where f_x and f_y are the partial derivatives of $f(x,y)$ with respect to x and y, respectively.

Example 5. Find the accumulation of errors in Example 4.

This analysis is best made at the same time as the computation in Example 4. It consists in computing

$$f_x = \frac{-2y}{(y + x)^2} \quad \text{and} \quad f_y = \frac{2x}{(y + x)^2}$$

and from these determining the approximate truncation error $\frac{1}{2}h^2y_i''$ [by Eq. (8.42)] and the amplification factor for each step

$$1 + hK_i \cong 1 + hf_y \qquad (8.43)$$

[see Eq. (8.35)]. Having an approximate value of the truncation error and the amplification factor $1 + hK_i$, one may use the difference equation (8.34) to compute the errors ϵ_i. Since we wish to take account of the round-off errors $\bar{\epsilon}_i$ made in the computation, we replace h^2a_{i+1} by

$$h^2(f_x + ff_y) + \bar{\epsilon}_i$$

The round-off errors in $y + x$ and $y - x$ in the table of Example 4 are in the same direction; hence the round-off error in f is just the final rounding error of 5×10^{-4}. However, since Δy is obtained by multiplying f by 0.02 and dropping the last figure, one need only consider the error in rounding off the last figure. In this way we obtain the round-off error $\bar{\epsilon}$ for each step. The computation is shown in the table below.

x	$\frac{1}{2}h^2y''$	$\bar{\epsilon}$	$\frac{1}{2}h^2y'' + \bar{\epsilon}$	$1 + hf_y$	ϵ
0.00	-4.0×10^{-4}	0	-4.0×10^{-4}	1.0000	-4.0×10^{-4}
0.02	-3.7	0.4×10^{-4}	-3.3	1.0007	-7.3
0.04	-3.4	0.2	-3.2	1.0014	-10.5
0.06	-3.2	-0.4	-3.6	1.0019	-14.1
0.08	-3.0	0.4	-2.6	1.0024	-16.7
0.10	-2.8×10^{-4}	0.4×10^{-4}	-2.4×10^{-4}	......	-19.1×10^{-4}

Thus the error term for $x = 0.10$ is -19×10^{-4}, and the correct value should therefore be, by Eq. (8.32), $y = 1.0928 - 0.0019 = 1.0909$. Note that the amplification factor $1 + hf_y$ stays so near unity that one may simply add the errors at each step, $\frac{1}{2}h^2y'' + \bar{\epsilon}$, to obtain the accumulated error ϵ. The correct value of y for $x = 0.1$, as found from the analytical solution, is $y = 1.0911$; thus the actual error is -17×10^{-4}. We see that the above error analysis is sufficiently accurate for our purpose.

It is clear by the above analysis that the errors accumulate too rapidly for one to obtain suitable accuracy over many steps. One could improve the accuracy by reducing the interval h; but since the round-off error per step would remain the same, one would arrive at an interval such that no further improvement in accuracy could be obtained by such a reduction, unless, of course, one increased the number of figures carried along in the computation. Note that both the reduction of interval and an increase in the number of figures retained add greatly to the work required. It seems logical, therefore, to attempt to reduce the truncation error by using a more accurate formula for the integration.

8.8. Modified Method of Euler. By treating y' as a new variable one may replace the differential equation (8.29) by the two equations

$$\frac{dy}{dx} = y'$$
$$y' = f(x,y)$$

Integrating the first equation from x_i to x_{i+1}, where $x_i = x_0 + ih$, we have

$$y_{i+1} = y_i + \int_{x_i}^{x_{i+1}} y' \, dx$$
$$y'_i = f(x_i, y_i) \tag{8.44}$$

If one uses the rectangular rule $Q_{01}(0)$ (see Chap. 7) to perform the integration in the first equation, one has

$$y_{i+1} = y_i + hy'_i = y_i + hf(x_i, y_i) \tag{8.45}$$

which is equivalent to Eq. (8.30), the basic formula in Euler's method.

If, instead, one uses the trapezoidal rule Q_{11}, one obtains in place of Eq. (8.45)

$$\begin{aligned} y_{i+1} &= y_i + \tfrac{1}{2}h(y'_i + y'_{i+1}) \\ &= y_i + \tfrac{1}{2}h[f(x_i, y_i) + f(x_{i+1}, y_{i+1})] \end{aligned} \tag{8.46}$$

This equation may be solved by the method of iteration (see Chap. 1) for y_{i+1}. For this purpose one substitutes a given approximation for y_{i+1} in the right-hand side to obtain the next approximation, the process being repeated until no significant change occurs in y_{i+1}. This procedure constitutes the *modified method of Euler*. For the first approximation, one uses the y_{i+1} given by Eq. (8.45), i.e., that which would be given by the unmodified method. Note that Eq. (8.46) differs from Eq. (8.45) only in having $f(x_i, y_i)$ replaced by the average of $f(x_i, y_i)$ and $f(x_{i+1}, y_{i+1})$, a sort of average slope for the interval.

8.9. Convergence of the Iterations Involved in the Modified Method. If one lets $y_{i+1}^{(k)}$ be the kth approximation to y_{i+1}, the value to which the modified method converges, one may write the iteration equation used in the modified method of Euler as

$$y_{i+1}^{(k)} = y_i + \tfrac{1}{2}h[f(x_i, y_i) + f(x_{i+1}, y_{i+1}^{(k-1)})] \tag{8.47}$$

where $k \geq 2$ and $y_{i+1}^{(1)}$ is obtained from Eq. (8.45).

From this equation and Eq. (8.46) by subtraction

$$|y_{i+1}^{(k)} - y_{i+1}| = \tfrac{1}{2}h|f(x_{i+1},y_{i+1}^{(k-1)}) - f(x_{i+1},y_{i+1})| \qquad (8.48)$$

Assuming that $f(x,y)$ satisfies the Lipschitz condition [see Eq. (8.22)],

$$|f(x_{i+1},y_{i+1}^{(k-1)}) - f(x_{i+1},y_{i+1})| < K_i|y_{i+1}^{(k-1)} - y_{i+1}| \qquad (8.49)$$

where K_i is a suitable finite constant, one has

$$|y_{i+1}^{(k)} - y_{i+1}| < \tfrac{1}{2}hK_i|y_{i+1}^{(k-1)} - y_{i+1}|$$

As a consequence,

$$|y_{i+1}^{(k)} - y_{i+1}| < (\tfrac{1}{2}hK_i)^{k-1}|y_{i+1}^{(1)} - y_{i+1}| \qquad (8.50)$$

Thus if $\tfrac{1}{2}hK_i$ is less than unity, which will be the case if h is taken sufficiently small, the right-hand side of Eq. (8.50) approaches zero as k approaches infinity, and therefore $y_{i+1}^{(k)}$ converges to y_{i+1}. We shall refer to $\tfrac{1}{2}hK_i$ as the *convergence factor* for the modified method.

Example 6. Given that

$$\frac{dy}{dx} = \log (x + y)$$

with the initial condition that $y = 1$ when $x = 0$, find y for $x = 0.2$ and $x = 0.5$.

Let $x_0 = 0$, $x_1 = 0.2$, and $x_2 = 0.5$; then $y_0 = 1$, while y_1 and y_2 are to be computed. By Eq. (8.45), for the step from x_0 to x_1,

$$\Delta y = [\log (0 + 1)] \Delta x = 0$$

Therefore, as a first approximation, $y_1 = 1$. Using Eq. (8.46), with $h = 0.2$,

$$\Delta y = \tfrac{1}{2}[\log 1 + \log (0.2 + 1)](0.2) = 0.0079$$

and

$$y_1 = 1.0079$$

On repeating the process,

$$\Delta y = \tfrac{1}{2}[\log 1 + \log (0.2 + 1.0079)](0.2) = 0.0082$$
$$y_1 = 1.0082$$

On the second repetition

$$\Delta y = \tfrac{1}{2}[\log 1 + \log (0.2 + 1.0082)](0.2) = 0.0082$$

which is the same as before; hence we take

$$y_1 = 1.0082$$

To obtain y_2, the value of y at $x = 0.5$, we again make use of Eq. (8.45) for obtaining a first estimate of Δy. Thus, since $\Delta x = 0.3$,

$$\Delta y = [\log (0.2 + 1.0082)](0.3) = 0.0246$$

and hence

$$y_2 = 1.0082 + 0.0246 = 1.0328$$

To improve this estimate of y_2 we use Eq. (8.46), from which

$$\Delta y = \tfrac{1}{2}[\log 1.2082 + \log (0.5 + 1.0328)](0.3) = 0.0401$$

and

$$y_2 = 1.0483$$

Repeating the process,

$$\Delta y = \tfrac{1}{2}[\log 1.2082 + \log (0.5 + 1.0483)](0.3) = 0.0408$$
and
$$y_2 = 1.0082 + 0.0408 = 1.0490$$

Another time around will not change this value, and hence it is taken as the value of y at $x = 0.5$.

8.10. Third-order Approximation Formulas of Runge, Heun, and Kutta.†

All of the third-order approximation formulas of this kind, including those of Runge and of Heun, can be obtained by the following general method, which is due to Kutta.

The true value of Δy for a given Δx is given by the Taylor's series of Eq. (8.6), provided only that this series converges. The constants $y_0^{(1)}$, $y_0^{(2)}$, $y_0^{(3)}$ can be expressed in terms of $f(x,y)$ and its various partial derivatives by the following procedure. Starting with the differential equation

$$y^{(1)} \equiv \frac{dy}{dx} = f(x,y) \tag{8.51}$$

with the initial condition $y = y_0$ for $x = x_0$, we have

$$y_0^{(1)} = f(x_0,y_0) \tag{8.52}$$

Differentiating Eq. (8.51) with respect to x, we have

$$y^{(2)} = \frac{\partial f}{\partial x} + \frac{\partial f}{\partial y}\frac{dy}{dx} = f_x + f_y f \tag{8.53}$$

where the subscripts x and y denote, respectively, differentiation with respect to x and y. Differentiating once more,

$$y^{(3)} = \frac{\partial y^{(2)}}{\partial x} + \frac{\partial y^{(2)}}{\partial y} f = (f_{xx} + f_{xy}f + f_x f_y) + (f_{xy} + f_{yy}f + f_y^2)f$$
or
$$y^{(3)} = f_{xx} + 2f_{xy}f + f_{yy}f^2 + (f_x + f_y f)f_y \tag{8.54}$$

Continuing this procedure yields $y^{(4)}$, $y^{(5)}$, etc. Upon substituting $x = x_0$ and $y = y_0$ in these expressions for the derivatives, we obtain the values of the constants $y_0^{(1)}$, $y_0^{(2)}$, $y_0^{(3)}$, etc. This procedure is, of course, very similar to that used in Taylor's method.

Putting in these values of the constants and letting $x - x_0 = h$ and $y - y_0 = \Delta y$, Eq. (8.6) becomes

$$\Delta y = f_0 h + \tfrac{1}{2}(f_x + f_y f)_0 h^2 + \tfrac{1}{6}[f_{xx} + 2f_{xy}f + f_{yy}f^2 + (f_x + f_y f)f_y]_0 h^3 + \cdots \tag{8.55}$$

where the zero subscript designates that the functions are to be evaluated at the point (x_0,y_0).

† See Levy and Baggott, *op. cit.*, pp. 96–103.

Consider increments in y defined by the equations

$$\Delta'y = f(x_0, y_0)\,\Delta x = f_0 h$$
$$\Delta''y = f(x_0 + mh, y_0 + m\,\Delta'y)h \qquad (8.56)$$
$$\Delta'''y = f(x_0 + \lambda h, y_0 + \rho\,\Delta''y + (\lambda - \rho)\,\Delta'y)h$$

where m, λ, and ρ are constants. Taylor's series for two independent variables† applied to $f(x,y)$ may be written

$$f(x_0 + p, y_0 + q) = f_0 + (f_x)_0 p + (f_y)_0 q + \tfrac{1}{2}[(f_{xx})_0 p^2 + 2(f_{xy})_0 pq + (f_{yy})_0 q^2] + \cdots \qquad (8.57)$$

where the subscript 0 has the meaning indicated above. Therefore

$$\Delta'y = hf_0$$
$$\Delta''y = h\{f_0 + mh(f_x)_0 + m\,\Delta'y(f_y)_0 + \tfrac{1}{2}[(mh)^2(f_{xx})_0 + 2m^2 h\,\Delta'y(f_{xy})_0 + (m\,\Delta'y)^2(f_{yy})_0] + \cdots\}$$
$$\Delta'''y = h\{f_0 + \lambda h(f_x)_0 + [\rho\,\Delta''y + (\lambda - \rho)\,\Delta'y](f_y)_0 \qquad (8.58)$$
$$+ \tfrac{1}{2}\{(\lambda h)^2(f_{xx})_0 + 2\lambda h[\rho\,\Delta''y + (\lambda - \rho)\,\Delta'y](f_{xy})_0 + [\rho\,\Delta''y + (\lambda - \rho)\,\Delta'y]^2(f_{yy})_0\} + \cdots\}$$

Using the first of these equations, one can eliminate $\Delta'y$ from the right-hand side of the second equation and obtain the first two equations below. Next, by the use of these two equations, we may eliminate $\Delta'y$ and $\Delta''y$ from the third equation of Eqs. (8.58). Thus to terms in h^3

$$\Delta'y = f_0 h$$
$$\Delta''y = f_0 h + m(f_x + f_y f)_0 h^2 + \tfrac{1}{2}m^2(f_{xx} + 2f_{xy}f + f_{yy}f^2)_0 h^3$$
$$\Delta'''y = f_0 h + \lambda(f_x + f_y f)_0 h^2 + \tfrac{1}{2}[\lambda^2(f_{xx} + 2f_{xy}f + f_{yy}f^2) \qquad (8.59)$$
$$+ 2m\rho(f_x + f_y f)f_y]_0 h^3$$

If the first equation in Eq. (8.59) is multiplied by a, the second by b, and the third by c, and the three equations are added, we obtain the first four terms of the Taylor's expansion given in Eq. (8.55), provided the following relations hold among the constants

$$a + b + c = 1$$
$$bm + c\lambda = \tfrac{1}{2}$$
$$bm^2 + c\lambda^2 = \tfrac{1}{3} \qquad (8.60)$$
$$c\rho m = \tfrac{1}{6}$$

Since there are six constants to be determined, namely, a, b, c, m, λ, and ρ, and only four conditions, we can choose two constants and express all the other constants in terms of them. If we choose m and λ as the two

† See F. S. Wood, "Advanced Calculus," p. 85, Ginn & Company, Boston, 1926.

independent constants,

$$\rho = \frac{\lambda(\lambda - m)}{m(2 - 3m)}$$

$$a = \frac{6m\lambda - 3(m + \lambda) + 2}{6m\lambda}$$

$$b = \frac{2 - 3\lambda}{6m(m - \lambda)}$$ (8.61)

$$c = \frac{2 - 3m}{6\lambda(\lambda - m)}$$

When these conditions are met, the increment

$$\Delta y = a\,\Delta'y + b\,\Delta''y + c\,\Delta'''y$$ (8.62)

agrees with the Taylor's expansion for Δy out to terms in h^3. Hence, the error in applying (8.62) to a determination of Δy is of the order of h^4. Particular values of m and λ yield the third-order formulas of Heun and Kutta. Kutta thus shows that any individual formula belongs to a twofold infinity of possible formulas. Heun's work on this subject is equivalent to assuming that $\rho = \lambda$.

Runge's original third-order formula, although it approximates the Taylor's series out to and including terms in h^3, is not obtainable from the above formulas because of the fact that it requires the calculation of a $\Delta^{iv}y$. This original formula is

$$\Delta y = \Delta^{iv}y + \tfrac{1}{3}[\tfrac{1}{2}(\Delta'y + \Delta'''y) - \Delta^{iv}y]$$ (8.63)

where

$$\Delta'y = f(x,y)h$$
$$\Delta''y = f(x + h, y + \Delta'y)h$$
$$\Delta'''y = f(x + h, y + \Delta''y)h$$ (8.64)
$$\Delta^{iv}y = f(x + \tfrac{1}{2}h, y + \tfrac{1}{2}\Delta'y)h$$

Since it requires four evaluations of the function $f(x,y)$ instead of three, it is only of historical interest.

From the last equation of Eqs. (8.60), it is clear that c cannot be zero. Moreover, a and b cannot both be zero, since in that case the first three equations of Eqs. (8.60) cannot be satisfied. Thus we must have at least two of the constants a, b, and c different from zero in Eq. (8.62) for Δy.

If $b = 0$, then by Eqs. (8.60)

$$\lambda = \tfrac{2}{3} \quad c = \tfrac{3}{4} \quad a = \tfrac{1}{4} \quad \text{and} \quad \rho m = \tfrac{2}{9}$$ (8.65)

When $a = 0$, by the second equation of Eqs. (8.61)

$$\lambda = \frac{2 - 3m}{3 - 6m}$$ (8.66)

and the other constants are obtained from the remaining equations.

Turning now to Eqs. (8.56), one might seek to simplify these equations by having one or more of the constants m, λ, ρ, and $\lambda - \rho$ vanish. From the last equation of Eqs. (8.60) it is at once clear that neither m nor ρ may vanish. If $\lambda = 0$, then from these same equations

$$m = \tfrac{2}{3} \qquad b = \tfrac{3}{4} \qquad c = \frac{1}{4\rho} \qquad \text{and} \qquad a = \frac{\rho - 1}{4\rho} \qquad (8.67)$$

Finally, if $\lambda = \rho$, corresponding to Heun's work on the subject, we have, in place of Eqs. (8.61), the equations

$$\lambda = 3m(1 - m)$$
$$a = \frac{2 - 12m + 27m^2 - 18m^3}{18m^2(1 - m)}$$
$$b = \frac{3m - 1}{6m^2} \qquad\qquad (8.68)$$
$$c = \frac{1}{18m^2(1 - m)}$$

Formulas of Particular Simplicity. It is thus seen that there are four possible simplifications in the formulas, namely, (1) $a = 0$, (2) $b = 0$, (3) $\lambda = 0$, and (4) $\lambda - \rho = 0$. We shall next seek to determine those formulas for which two or more of these simplifications hold. Starting with $b = 0$, we have by Eq. (8.65) that $a \neq 0$ and $\lambda \neq 0$. The only further simplification is to make $\rho = \lambda = \tfrac{2}{3}$; thus

1. $\lambda = \rho = \tfrac{2}{3}$, $m = \tfrac{1}{3}$, $a = \tfrac{1}{4}$, $b = 0$, $c = \tfrac{3}{4}$.

Substituting these values in Eqs. (8.62) and (8.56), we have *Heun's formula*

$$\Delta y = \tfrac{1}{4}(\Delta'y + 3\Delta'''y) \qquad\qquad (8.69)$$
where
$$\Delta'y = f(x,y)h$$
$$\Delta''y = f(x + \tfrac{1}{3}h,\ y + \tfrac{1}{3}\Delta'y)h \qquad\qquad (8.70)$$
$$\Delta'''y = f(x + \tfrac{2}{3}h,\ y + \tfrac{2}{3}\Delta''y)h$$

If $\lambda = 0$, then by Eq. (8.67) $b \neq 0$ and $\rho \neq 0$. The only additional simplification of the above type is obtained by setting $a = 0$, for which $\rho = 1$. Thus we have the particular set of values

2. $\lambda = 0$, $\rho = 1$, $m = \tfrac{2}{3}$, $a = 0$, $b = \tfrac{3}{4}$, $c = \tfrac{1}{4}$.

From Eqs. (8.62) and (8.56), therefore,

$$\Delta y = \tfrac{1}{4}(3\Delta''y + \Delta'''y) \qquad\qquad (8.71)$$
where
$$\Delta'y = f(x,y)h$$
$$\Delta''y = f(x + \tfrac{2}{3}h,\ y + \tfrac{2}{3}\Delta'y)h \qquad\qquad (8.72)$$
$$\Delta'''y = f(x,\ y + \Delta''y - \Delta'y)h$$

Since the possibilities of $b = 0$ and of $\lambda = 0$ have been fully investigated, there remains only the possibility of $a = 0$ simultaneously with $\lambda = \rho$. For $\lambda = \rho$ Eqs. (8.68) hold. Thus, if also $a = 0$, by the second equation of this set

$$18m^3 - 27m^2 + 12m - 2 = 0$$

Since the right-hand side is equal to $-\tfrac{2}{3}$ at $m = \tfrac{2}{3}$ and 1 at $m = 1$, there is a real root of the equation between $\tfrac{2}{3}$ and 1. This is an irrational root, and, therefore, some of the simplicity of the resulting formula would be lost.

Besides the simplifications investigated above, there is an advantage in having $\lambda = m$ in Eqs. (8.56). The second and third equations of Eqs. (8.60) then become

$$m(b + c) = \tfrac{1}{2}$$
$$m^2(b + c) = \tfrac{1}{3}$$

from which $m = \lambda = \tfrac{2}{3}$ and $b + c = \tfrac{3}{4}$. Using the other equations, we find that

$$\lambda = m = \tfrac{2}{3} \qquad a = \tfrac{1}{4} \qquad c = \frac{1}{4\rho} \qquad b = \frac{3\rho - 1}{4\rho} \tag{8.73}$$

Thus letting $\rho = \tfrac{1}{3}$ results in the special case
3. $\lambda = m = \tfrac{2}{3}, \rho = \tfrac{1}{3}, a = \tfrac{1}{4}, b = 0, c = \tfrac{3}{4}$.
By Eqs. (8.62) and (8.56), therefore,

$$\Delta y = \tfrac{1}{4}(\Delta'y + 3\Delta'''y) \tag{8.74}$$

where
$$\Delta'y = f(x,y)h$$
$$\Delta''y = f(x + \tfrac{2}{3}h, y + \tfrac{2}{3}\Delta'y)h \tag{8.75}$$
$$\Delta'''y = f(x + \tfrac{2}{3}h, y + \tfrac{1}{3}(\Delta'y + \Delta''y))h$$

On the other hand, if $\rho = \tfrac{2}{3}$,
4. $\lambda = m = \rho = \tfrac{2}{3}, a = \tfrac{1}{4}, b = c = \tfrac{3}{8}$.

Therefore,
$$\Delta y = \tfrac{1}{8}(2\Delta'y + 3\Delta''y + 3\Delta'''y) \tag{8.76}$$
where
$$\Delta'y = f(x,y)h$$
$$\Delta''y = f(x + \tfrac{2}{3}h, y + \tfrac{2}{3}\Delta'y)h \tag{8.77}$$
$$\Delta'''y = f(x + \tfrac{2}{3}h, y + \tfrac{2}{3}\Delta''y)h$$

Also worthy of notice is the set of constants
5. $\lambda = 1, \rho = 2, m = \tfrac{1}{2}, a = \tfrac{1}{6}, b = \tfrac{2}{3}, c = \tfrac{1}{6}$

for which
$$\Delta y = \tfrac{1}{6}(\Delta'y + 4\Delta''y + \Delta'''y) \tag{8.78}$$
and
$$\Delta'y = f(x,y)h$$
$$\Delta''y = f(x + \tfrac{1}{2}h, y + \tfrac{1}{2}\Delta'y)h \tag{8.79}$$
$$\Delta'''y = f(x + h, y + 2\Delta''y - \Delta'y)h$$

This is analogous to Simpson's one-third rule.† It is referred to as *Kutta's third-order rule*.

8.11. Kutta's Fourth-order Approximations.

Kutta extended this analysis to the case where the error allowable is of the order of h^5 instead of h^4. The formulas then have the form

$$\Delta y = a \Delta'y + b \Delta''y + c \Delta'''y + d \Delta^{iv}y \tag{8.80}$$

† In fact if $f(x,y)$ is independent of y, $f(x,y) = f(x)$, and this rule gives

$$\Delta y = \frac{h}{6}\left[f(x) + 4f\left(x + \frac{h}{2}\right) + f(x + h) \right]$$

Since from the differential equation (8.51)

$$\Delta y = \int_x^{x + h} f(x,y)\, dx = \int_x^{x+h} f(x)\, dx$$

the above expression is seen to be just the integral evaluated by Simpson's rule.

where

$$\Delta'y = f(x,y)h$$
$$\Delta''y = f(x + mh, y + m\,\Delta'y)h$$
$$\Delta'''y = f(x + \lambda h, y + \rho\,\Delta''y + (\lambda - \rho)\,\Delta'y)h \qquad (8.81)$$
$$\Delta^{iv}y = f(x + \mu h, y + \sigma\,\Delta'''y + \tau\,\Delta''y + (\mu - \sigma - \tau)\,\Delta'y)h$$

The treatment is very similar to that given in Sec. 8.10. Again there is a twofold infinity of possible formulas since all the other constants are expressible in terms of the arbitrary constants m and λ.†

We shall mention only two particular formulas:

Kutta-Simpson One-third Rule

$$\Delta y = \tfrac{1}{6}(\Delta'y + 2\Delta''y + 2\Delta'''y + \Delta^{iv}y) \qquad (8.82)$$

where
$$\Delta'y = f(x,y)h$$
$$\Delta''y = f(x + \tfrac{1}{2}h, y + \tfrac{1}{2}\Delta'y)h$$
$$\Delta'''y = f(x + \tfrac{1}{2}h, y + \tfrac{1}{2}\Delta''y)h \qquad (8.83)$$
$$\Delta^{iv}y = f(x + h, y + \Delta'''y)h$$

By an argument similar to that used in discussing Kutta's third-order rule, Eq. (8.82) is seen to reduce to Simpson's one-third rule whenever $f(x,y)$ is a function of x alone.

Kutta-Simpson Three-eighths Rule

$$\Delta y = \tfrac{1}{8}(\Delta'y + 3\Delta''y + 3\Delta'''y + \Delta^{iv}y) \qquad (8.84)$$

where
$$\Delta'y = f(x,y)h$$
$$\Delta''y = f(x + \tfrac{1}{3}h, y + \tfrac{1}{3}\Delta'y)h$$
$$\Delta'''y = f(x + \tfrac{2}{3}h, y + \Delta''y - \tfrac{1}{3}\Delta'y)h \qquad (8.85)$$
$$\Delta^{iv}y = f(x + h, y + \Delta'''y - \Delta''y + \Delta'y)h$$

As one might guess, this formula reduces to Simpson's three-eighths rule whenever $f(x,y)$ is a function of x only.

Both of the rules above are correct to the first four powers of $h = \Delta x$ and have, therefore, errors of the order of h^5.

8.12. Generalization of the Runge-Kutta Formulas to a Set of Simultaneous Differential Equations. The way in which the Runge-Kutta formulas may be generalized to a set of simultaneous differential equations of the first order is illustrated by writing down the Kutta-Simpson one-third rule for the pair of differential equations

$$\frac{dy}{dx} = f(x,y,z)$$
$$\frac{dz}{dx} = g(x,y,z) \qquad (8.86)$$

† See Levy and Baggott, *op. cit.*, p. 103, and *Z. Math. u. Phys.*, **46**: 435–453 (1901).

This rule becomes in this case,

$$\Delta y = \tfrac{1}{6}(\Delta'y + 2\Delta''y + 2\Delta'''y + \Delta^{iv}y)$$
$$\Delta z = \tfrac{1}{6}(\Delta'z + 2\Delta''z + 2\Delta'''z + \Delta^{iv}z)$$

(8.87)

where (for $\Delta x = h$)

$$\Delta'y = f(x_0,y_0,z_0)h$$
$$\Delta'z = g(x_0,y_0,z_0)h$$
$$\Delta''y = f(x_0 + \tfrac{1}{2}h,\ y_0 + \tfrac{1}{2}\Delta'y,\ z_0 + \tfrac{1}{2}\Delta'z)h$$
$$\Delta''z = g(x_0 + \tfrac{1}{2}h,\ y_0 + \tfrac{1}{2}\Delta'y,\ z_0 + \tfrac{1}{2}\Delta'z)h$$
$$\Delta'''y = f(x_0 + \tfrac{1}{2}h,\ y_0 + \tfrac{1}{2}\Delta''y,\ z_0 + \tfrac{1}{2}\Delta''z)h$$
$$\Delta'''z = g(x_0 + \tfrac{1}{2}h,\ y_0 + \tfrac{1}{2}\Delta''y,\ z_0 + \tfrac{1}{2}\Delta''z)h$$
$$\Delta^{iv}y = f(x_0 + h,\ y_0 + \Delta'''y,\ z_0 + \Delta'''z)h$$
$$\Delta^{iv}z = g(x_0 + h,\ y_0 + \Delta'''y,\ z_0 + \Delta'''z)h$$

(8.88)

If these increments in y and z are computed in the order given, only previously computed increments will be needed at each step in computation.

8.13. Correction of the Starting Values. The methods of this chapter are suitable for finding a few starting values (three, five, or more) needed for the methods of Chap. 9. Some of these methods for starting a solution, such as Euler's modified method, may not yield sufficient accuracy. In this case, before going further, one employs a numerical form of Picard's method of successive approximation to improve these starting values.

Picard's method may be summarized by the iteration equations

$$v_i(x) = f(x,u_i(x))$$

(8.89)

$$u_{i+1}(x) = y_0 + \int_{x_0}^{x} v_i(x)\,dx$$

(8.90)

where the boundary condition is $y = y_0$ at $x = x_0$. These equations are at once seen to be equivalent to the single equation (8.20). The introduction of $v(x)$, however, is helpful in tying this discussion in with the methods employed in the next chapter. From a given approximation $u_i(x)$ to the solution y of the differential equation $y' = f(x,y)$ one obtains from Eq. (8.89) a function $v_i(x)$ which is an approximation to $y' = dy/dx$. Then from Eq. (8.90) one obtains a better approximation $u_{i+1}(x)$ to y. If $f(x,y)$ satisfies the Lipschitz condition of (8.22) for some region including (x_0,y_0), then the series of functions $u_i(x)$ converges to y, and the series of functions $v_i(x)$ converges to y' as $i \to \infty$, at least in the neighborhood of the point (x_0,y_0).

The starting values y_k and y'_k, $k = 0, 1, \ldots, m$ may be considered to be a tabulation of the functions $u_1(x)$ and $v_1(x)$, respectively. They should already satisfy Eq. (8.89). Improved values of the y_k corresponding to a tabulation of $u_2(x)$ may, therefore, be found by the use of Eq.

(8.90). Since, however, the integration must be performed numerically making use only of the tabulated values, one replaces each integral in Eq. (8.90) by an appropriate quadrature formula and writes

$$\bar{y}_k = y_0 + \int_{x_0}^{x_k} v_1(x)\,dx = y_0 + h \sum_{j=0}^{m} w_{kj}y'_j \qquad (8.91)$$

Here $\bar{y}_k$ are the improved values of the y_k, and w_{kj} are constants giving the weights to be assigned to the y'_j by virtue of the quadrature formula. Some of these weights may, of course, be zero. Improved values of y'_j may then be obtained from the equation

$$\bar{y}'_j = f(x,\bar{y}_j) \qquad (8.92)$$

corresponding to Eq. (8.89).

Instead of integrating over the entire range x_0 to x_k in Eq. (8.91), one may make use of a previously improved value $\bar{y}_r$ to reduce the range of integration. Thus Eq. (8.91) may be replaced by the more general equation

$$\bar{y}_k = \bar{y}_r + \int_{x_r}^{x_k} v_1(x)\,dx = \bar{y}_r + h \sum_{j=0}^{m} w_{kj}y'_j \qquad (8.93)$$

The weights, of course, are different from those in Eq. (8.91).

Having obtained a set of improved starting values $\bar{y}_k$ and $\bar{y}'_k$, one may repeat the improvement procedure using these in place of the original starting values. When no further improvement can be obtained in this way, the values are taken as correct. Since there remains, in such a numerical application of Picard's method, the truncation error due to the approximate nature of the quadrature formula, one must take the interval h between tabulated values small enough to make this truncation error negligible.

Illustration of this correction procedure employing specific quadrature formulas will be given in Sec. 9.6.

PROBLEMS

1. Use Taylor's method to find a solution to the differential equation

$$\frac{dy}{dx} = x + y$$

with boundary condition $y = 1$ at $x = 0$. Take the first five terms of the series, and determine roughly over what range this solution will hold to four decimal places.

2. Find a solution of the set of simultaneous equations

$$x' = x + y + t$$
$$y' = 2x - t$$

subject to the initial conditions $x = 0$, $y = 1$, at $t = 1$. Obtain this solution in the form of series expansions for x and y using (a) Taylor's method, (b) Picard's method.

3. Use Euler's method and an interval of $h = 0.1$ to find a solution of the differential equation (8.27) in the range $0 \leq x \leq 1$. Check the value at $x = 1$ against that given by (8.28).

4. Determine the approximate error to be expected in Prob. 3.

5. Obtain a solution of the set of simultaneous equations

$$\frac{dx}{dt} = y \qquad \frac{dy}{dt} = -x$$

with boundary conditions $y = 1$ and $x = 0$ at $t = 0$, for the range $0 \leq x \leq \pi/4$. Use Euler's modified method and an interval of 0.2. Check results against the analytical solutions.

6. Perform the integration of the differential equation in Example 6 for the same two steps but use Heun's formula [Eq. (8.69)].

7. Solve the set of simultaneous equations given in Prob. 5 using the Kutta-Simpson one-third rule.

8. (a) Using Euler's modified method, obtain a solution of the differential equation

$$\frac{dy}{dx} = x + |\sqrt{y}|$$

with boundary condition $y = 1$ at $x = 0$ for the range $0 \leq x \leq 1$ in steps of 0.2.

(b) Correct the starting values so obtained by the method of Sec. 8.13.

CHAPTER 9

SOLUTION OF ORDINARY DIFFERENTIAL EQUATIONS: METHODS FOR CONTINUING THE SOLUTION

The methods of Chap. 8, although necessary for obtaining the initial values in the solution of a differential equation, generally involve too much labor to be used for obtaining a numerical solution over an extended range. Fortunately, a general class of methods involving extrapolation or integration ahead by open-type quadrature formulas is available for continuing a solution with much less work once a few initial values have been obtained. These methods will be discussed in this chapter.

The above step-by-step methods are applicable only if all the boundary conditions apply to a single value of the independent variable. Besides these *one-point* boundary conditions we shall consider also the more difficult *two-point* and *multipoint* boundary conditions† involving conditions imposed at two or more values of the independent variable. The latter generally require some method of successive approximation for obtaining a solution. Either the differential equation is assumed to hold at all points and the boundary conditions are achieved by successive approximations, or the boundary conditions are assumed initially and the difference equation, corresponding to the differential equation, is satisfied at each point by some iterative method. Richardson‡ refers to the one-point boundary problems as *marching problems* and to the two-point problems as *jury problems*. Included in the latter are *periodic boundary* conditions such as the requirement that $y(b) = y(a)$ and *normalization* conditions such as $\int_a^b y^2\, dx = 1$.

Sometimes the above *two-point* boundary-value problems are referred to simply as *boundary-value* problems, in distinction to *initial-value* problems. They play the same role with respect to ordinary differential equations as that played by elliptic-type equations in partial differential equations (see Chap. 13).

---

† These arise only for equations of the second or higher order or for a set of two or more equations of the first order.

‡ L. F. Richardson, *Phil. Trans. Roy. Soc. (London)*, **A226**: 300 (1927).

9.1. Milne's Method.† Suppose that one is integrating the differential equation

$$y' = f(x,y) \tag{9.1}$$

with the boundary condition $y = y_0$ at $x = x_0$ and that by some starting process described in Chap. 8 one has already obtained sufficiently accurate numerical values for y_0, y_1, y_2, and y_3 and for y_0', y_1', y_2', y_3'.‡ Then it is possible, by a step-by-step process, to find the values of all subsequent y_k and y_k' by making use of suitable numerical integration formulas, such as those discussed in Chap. 7. Suppose, in particular, that all values of the y_k and y_k' out to $k = n$ have been obtained; then by the open-type quadrature formula Q_{24} of Chap. 7,

$$y_{n+1} - y_{n-3} = \int_{x_{n-3}}^{x_{n+1}} y' \, dx = \frac{4h}{3} (2y_{n-2}' - y_{n-1}' + 2y_n')$$
$$+ \tfrac{14}{45} h^5 y^{(5)}(\xi_1) \tag{9.2}$$

where $y = y(x)$ is the required solution of the differential equation and $x_{n-3} < \xi_1 < x_{n+1}$. If one takes h small enough to ensure that the remainder term involving $y^{(5)}(\xi_1)$ is small, then y_{n+1} given by the formula

$$y_{n+1} = y_{n-3} + \frac{4h}{3} (2y_{n-2}' - y_{n-1}' + 2y_n') \tag{9.3}$$

is a good approximation to the next ordinate. This formula is termed a *predictor*.

Having an approximate value of y_{n+1}, one may obtain a first approximation to y_{n+1}' by substitution in the differential equation (9.1). Having this approximation, it is possible to obtain a better value of y_{n+1} by means of Simpson's rule, quadrature formula Q_{22},

$$y_{n+1} - y_{n-1} = \int_{x_{n-1}}^{x_{n+1}} y' \, dx = \frac{h}{3} (y_{n-1}' + 4y_n' + y_{n+1}') \tag{9.4}$$

The remainder term $-h^5 y^{(5)}(\xi_2)/90$ for this formula, where $x_{n-1} < \xi_2 < x_{n+1}$, is now ordinarily much smaller than that in Eq. (9.2), and hence this formula is preferable to Eq. (9.3) as soon as a good estimate of y_{n+1}' has been obtained. This formula is therefore termed a *corrector*.

Having an improved value of y_{n+1}', one may again apply the corrector [Eq. (9.4)] to find a still better value of y_{n+1}. However, this correction can be kept negligible, so that only a single application of the corrector is necessary, by keeping the interval h sufficiently small. Once y_{n+1} and

† See W. E. Milne, "Numerical Calculus," chap. V, Princeton University Press, Princeton, N.J., 1949.

‡ Here y_i and y_i' are the values of y and its derivative y' corresponding to $x_i = x_0 + ih$, $i = 0, 1, 2, 3$, etc.

y'_{n+1} are obtained to this accuracy, the next pair of values y_{n+2} and y'_{n+2} may be obtained by a repetition of the process. Since an automatic electronic computer is well adapted to any repetition procedure, such a step-by-step method is readily coded for automatic computation.

9.2. Rate of Convergence. Let the correction in y_{n+1} introduced by the corrector be†

$$\delta = \bar{y}_{n+1} - y_{n+1} \tag{9.5}$$

Then the change in y'_{n+1} due to this change in y_{n+1}, by Eq. (9.1), is

$$\bar{y}'_{n+1} - y'_{n+1} = f(x_{n+1}, \bar{y}_{n+1}) - f(x_{n+1}, y_{n+1}) = K(\bar{y}_{n+1} - y_{n+1}) = K\delta \tag{9.6}$$

where‡

$$K = \frac{f(x_{n+1}, \bar{y}_{n+1}) - f(x_{n+1}, y_{n+1})}{\bar{y}_{n+1} - y_{n+1}} \tag{9.7}$$

If $f(x,y)$ possesses a continuous first derivative $f_y(x,y)$ with respect to y, then by the *mean-value theorem*

$$K = f_y(x_{n+1}, \eta_{n+1}) \tag{9.8}$$

where η_{n+1} lies between y_{n+1} and $\bar{y}_{n+1}$. The change in y'_{n+1} given by Eq. (9.6) on a second application of the corrector formula (9.4) would, therefore, cause a change of $\frac{1}{3}hK\delta$ in y_{n+1}. Thus the first correction δ to y_{n+1} is multiplied by the *convergence factor*

$$\Theta = \tfrac{1}{3}hK \tag{9.9}$$

in obtaining the second correction to the value of y_{n+1}. To get a third correction, one would again multiply by Θ.§

If $|\Theta|$ is less than unity, the values obtained for y_{n+1} will converge to a limit. This limit $\bar{y}_{n+1}$ may be found by adding to the first value all the corrections produced by repeated applications of the corrector formula; thus

$$\bar{y}_{n+1} = y_{n+1} + \delta + \Theta\delta + \Theta^2\delta + \cdots$$
$$= y_{n+1} + \frac{\delta}{1 - \Theta} \tag{9.10}$$

If Θ is small compared to 1, a sufficiently accurate value of $\bar{y}_{n+1}$ will be given by $\bar{y}_{n+1} = y_{n+1} + \delta$ [see Eq. (9.5)].

As a practical rule one should keep h small enough so that the difference $\bar{y}_{n+1} - \bar{y}_{n+1} = \Theta\delta/(1 - \Theta)$ does not affect the last figure retained in y. Thus, if N decimal places are retained in the y column, one should make sure that $|\Theta| |\delta|/|1 - \Theta| < 10^{-N}$ or satisfy the equivalent requirement

---

† Here $\bar{y}_{n-1}$ is the new value of y_{n-1} obtained from the corrector.
‡ K is the first divided difference of $f(x,y)$ with respect to y (see Sec. 5.7).
§ We shall neglect the change in K, which is usually quite small.

that

$$|D| < \frac{1 - \Theta}{|\Theta|} \qquad (9.11)$$

where $D = \delta \times 10^N$ is the change in the last figure retained in y. This assumes, of course, that h has been chosen small enough so that $|\Theta| < 1$. For $|\Theta| \ll 1$, this condition reduces to

$$|D| < \frac{1}{|\Theta|} = \frac{3}{h|K|} \qquad (9.12)$$

9.3. Truncation Error in the Corrector Formula. The use of the sum of the first n terms of an infinite series as the sum of the series introduces a so-called *truncation* error. It is conveniently represented by a remainder term. More generally, any approximate formula may be thought of as being the first few terms of a series, and hence the term truncation error is applied quite generally to the error in any approximation formula. We shall take the sign of such an error to be the same as that of the remainder term. Thus the error should be added to the computed value to obtain the true value.

The remainder term for the corrector formula (9.4), as stated previously, is $R \equiv -h^5 y^{(5)}(\xi_2)/90$, where $x_{n-1} < \xi_2 < x_{n+1}$, while the error term of the predictor formula (9.3) is given in Eq. (9.2). Since the difference $\delta = \bar{y}_{n+1} - y_{n+1}$ is due only to the difference between these error terms, one has

$$\delta = \frac{h^5}{90} [y^{(5)}(\xi_2) + 28 y^{(5)}(\xi_1)]$$

Assuming R does not change too rapidly from step to step, $y^{(5)}(\xi_1) \cong y^{(5)}(\xi_2)$, and therefore

$$R \cong -\frac{\delta}{29} \qquad (9.13)$$

In terms of the last figure retained, the error term is thus $-D/29$, where, as before, D is the change in the last figure of y_{n+1} on applying the corrector. To reduce the truncation error per step to less than one unit in the last figure retained h must, therefore, be kept small enough so that

$$|D| < 29 \qquad (9.14)$$

Thus, by means of a column giving D, one can determine both whether a single application of the corrector is sufficient [inequality (9.12)] and whether the truncation error per step is small [inequality (9.14)]. Such a column also provides one with a check on accidental errors in the computation, since these will usually show up as sudden changes in the size of D. It will not provide, of course, a check against systematic errors due to an improper computational procedure.

No account has here been taken of two other types of errors, namely, (1) *round-off* errors, which are due to the use of only a finite number of decimal places in the computation, and (2) *inherited* errors introduced through Eq. (9.4), which are due to incorrect values for the' previously computed y'_i. Discussion of these errors will be undertaken in Sec. 9.5, where the accumulation of error in the numerical computation is considered.

9.4. Changing the Interval. If the truncation error $R = -h^5 y^{(5)}/90$ as indicated by $-D/29$ is too large, it can be reduced to approximately $\frac{1}{32}$ of its value by halving the interval h. This necessitates finding intermediate values of y' by interpolation. If the last values obtained using the interval h are y_n and y'_n, one needs the intermediate values $y'_{n-\frac{3}{2}}$ and $y'_{n-\frac{1}{2}}$. Convenient formulas for this purpose are

$$y'_{n-\frac{3}{2}} = \tfrac{1}{16}[9(y'_{n-2} + y'_{n-1}) - (y'_{n-3} + y'_n)] + \frac{3h^4}{128} y^{(5)}(\xi) \qquad (9.15)$$

and $\quad y'_{n-\frac{1}{2}} = \tfrac{1}{16}[15y'_{n-1} + 5(y'_{n-2} - y'_n) + y'_{n-3}] - \frac{5h^4}{128} y^{(5)}(\xi) \qquad (9.16)$

where for each formula $x_{n-3} < \xi < x_n$. On the other hand, if D is only 0 or 1, it may be possible to reduce the work to about half by doubling the interval. In this case, one picks out every other value and continues the computation using the double interval. Note that the truncation error may be expected to increase to about 32 times its magnitude when the interval is doubled.

Example 1. Given that $y = 1$ when $x = 0$ and that y satisfies the differential equation

$$\frac{dy}{dx} = \frac{x^3 + y^2}{10} \qquad (9.17)$$

tabulate y as a function of x to four decimal places over the range $0 \le x \le 2$.

Table 9.1 gives the numerical solution of this problem. To ensure the reliability of the four decimal places required, one carries along five decimal places in the calculation of the y. Since, however, the derivatives are multiplied by h in obtaining y, four decimal places for y' are sufficient. In the y column only the final numbers obtained from the corrector formula are given. The initial values obtained from the predictor formula (9.3) were smaller than these by the amount D given in the last column. These numbers in the y column rounded off to four decimal places constitute the required table.

Some of the details of the calculations are as follows. A Taylor's-series expansion for y about $x = 0$ was obtained in Chap. 8 [see Eq. (8.12)]. From this series and the differential equation one may obtain the values lying above the line. The first estimate of y_4 corresponding to $x = 0.4$ may then be obtained from Eq. (9.3) with the substitutions $n = 3$ and $h = 0.1$. It is then possible to calculate $0.1x^3$ and $0.1y^2$ and by adding them to obtain $y'_4 = 0.1150$. Simpson's rule in the form of Eq. (9.4), again with $n = 3$ and $h = 0.1$, is next used to recalculate y_4. No change will be found in this value, so one enters a zero in the D column.

TABLE 9.1. INTEGRATION OF THE DIFFERENTIAL EQUATION $y' = 0.1(x^3 + y^2)$

x	y	$0.1x^3$	$0.1y^2$	y'	D
0.0	1.00000	0.0000	0.1000	0.1000	
0.1	1.01010	0.0001	0.1020	0.1021	
0.2	1.02045	0.0008	0.1041	0.1049	
0.3	1.03113	0.0027	0.1063	0.1090	
0.4	1.04231	0.0064	0.1086	0.1150	0
0.5	1.05422	0.0125	0.1111	0.1236	1
0.6	1.06714	0.0216	0.1139	0.1355	0
0.8	1.09755	0.0512	0.1205	0.1717	1
1.0	1.13724	0.1000	0.1293	0.2293	1
1.2	1.19113	0.1728	0.1419	0.3147	4
1.4	1.26541	0.2744	0.1601	0.4345	1
1.6	1.36775	0.4096	0.1870	0.5966	10
1.8	1.50772	0.5832	0.2272	0.8104	32
2.0	1.69613	0.8000	0.2877	1.0877	14

Repetition of this procedure for $n = 4$ and $n = 5$ yields y_5 and y_6. By now it is clear that D is so small that the interval h may be safely doubled. One therefore drops out the values at $x = 0.1$, 0.3, and 0.5 and applies the above procedure to the remaining entries, h now being taken as 0.2. Although D grows as one approaches $x = 2$, the estimated truncation error $D/29$ in y for each step of the computation is not large enough to affect the fourth decimal place. However, the over-all error in y at $x = 2$ is not ordinarily just the sum of the truncation errors in the individual steps, and one must investigate the amplification of these errors that takes place in a step-by-step application of Eq. (9.4).

9.5. Accumulation of Error in Milne's Method.†   By keeping track of the difference between the value of y given by the predictor [Eq. (9.3)] and that given by the corrector [Eq. (9.4)] it is possible to adjust the interval h so as to keep the *truncation* error less than one in the last decimal place retained. Moreover, one can make the *round-off* error negligible with respect to the truncation error by carrying along one or two *guarding figures*, i.e., extra decimal places. There remains, however, the need to analyze *inherited* error, the error in the determination of y_{n+1} by Eq. (9.4) due to the errors in y'_{n-1} and y'_n. These latter errors in turn include errors inherited from still earlier values.

The predictor formula is ignored in this analysis since it is assumed that the interval is kept small enough for the corrector to have wiped out any initial error from the predictor. If this were not the case, another application of the corrector would be called for. Since the corrector here is Simpson's rule, the analysis given in this section is applicable to any method employing Simpson's rule as the final formula.

† An approach similar to that used here was suggested to the author by Garrett Birkhoff.

Let the true values of the y_i and y_i' be designated by z_i and z_i', respectively. These would correspond to the values obtained from an exact analytical solution of the differential equation. The errors in the y_i and y_i' will then be defined by the equations

$$z_i = y_i + \epsilon_i$$
$$z_i' = y_i' + \epsilon_i' \qquad (9.18)$$

Note that the errors thus defined are opposite in sign to the usual convention and represent the corrections that need to be added to the y_i and y_i' to give the true values. They are in this respect similar to remainder (or error) terms in quadrature formulas.

From the differential equation (9.1) by application of the mean-value theorem

$$\epsilon_i' = z_i' - y_i' = f(x_i,z_i) - f(x_i,y_i) = K_i(z_i - y_i)$$

hence $\qquad \epsilon_i' = K_i \epsilon_i \qquad\qquad\qquad\qquad\qquad\qquad (9.19)$

where [see Eqs. (9.6) to (9.8)]

$$K_i = f_y(x_i, \eta_i) \qquad (9.20)$$

and η_i lies somewhere between y_i and z_i.

The z_i and z_i' satisfy the equation

$$z_{n+1} - z_{n-1} = \frac{h}{3}(z_{n-1}' + 4z_n' + z_{n+1}') - h^5 A_n \qquad (9.21)$$

where $A_n = z^{(5)}(\xi)/90$ and $x_{n-1} < \xi < x_{n+1}$.
Subtracting Eq. (9.4), one has

$$\epsilon_{n+1} - \epsilon_{n-1} = \frac{h}{3}(\epsilon_{n-1}' + 4\epsilon_n' + \epsilon_{n+1}') - h^5 A_n \qquad (9.22)$$

and, on making use of Eq. (9.19), this difference equation may be written

$$(1 - \tfrac{1}{3}hK_{n+1})\epsilon_{n+1} = (\tfrac{4}{3}hK_n)\epsilon_n + (1 + \tfrac{1}{3}hK_{n-1})\epsilon_{n-1} - h^5 A_n \qquad (9.23)$$

The K_i, given by Eq. (9.20), are characteristic of the differential equation. Over a sufficiently restricted range they can usually be represented by a single value K. Likewise, the A_i can be replaced by a single number A. Consequently, by introducing the convergence factor $\Theta = \tfrac{1}{3}hK$ of Eq. (9.9), Eq. (9.23) may be replaced by

$$\epsilon_{n+1} = a\epsilon_n + b\epsilon_{n-1} - \frac{h^5 A}{1 - \Theta} \qquad (9.24)$$

where $\qquad\qquad a = \dfrac{4\Theta}{1 - \Theta} \qquad b = \dfrac{1 + \Theta}{1 - \Theta} \qquad (9.25)$

A particular solution of the difference equation (9.24) may be obtained by taking $\epsilon_i = c = $ constant. This constant must satisfy the equation

$$c = (a + b)c - \frac{h^2 A}{1 - \theta}$$

therefore, from Eq. (9.25)

$$c = \frac{h^2 A}{(a + b - 1)(1 - \theta)} = \frac{h^2 A}{6\theta} \tag{9.26}$$

To this must be added the general solution of the homogeneous equation

$$\epsilon_{n+1} = a\epsilon_n + b\epsilon_{n-1} \tag{9.27}$$

Such a solution can be obtained by letting $\epsilon_n = \lambda^n$. This is a solution of Eq. (9.27) provided λ satisfies the quadratic equation

$$\lambda^2 - a\lambda - b = 0$$

The two solutions of this quadratic are

$$\begin{aligned} \lambda_1 &= \tfrac{1}{2}(a + \sqrt{a^2 + 4b}) \\ \lambda_2 &= \tfrac{1}{2}(a - \sqrt{a^2 + 4b}) \end{aligned} \tag{9.28}$$

The general solution of Eq. (9.27) is then

$$\epsilon_n = c_1\lambda_1^n + c_2\lambda_2^n$$

and, therefore, the general solution of Eq. (9.24) is

$$\epsilon_n = c_1\lambda_1^n + c_2\lambda_2^n + \frac{h^5 A}{6\theta} \tag{9.29}$$

In terms of θ, by virtue of Eqs. (9.25),

$$\begin{aligned} \lambda_1 &= \frac{2\theta + \sqrt{1 + 3\theta^2}}{1 - \theta} \cong 1 + 3\theta \\ \lambda_2 &= \frac{2\theta - \sqrt{1 + 3\theta^2}}{1 - \theta} \cong -(1 - \theta) \end{aligned} \tag{9.30}$$

where the approximations shown involve the neglect of terms in θ^2.

The constants c_1 and c_2 in Eq. (9.29) may be expressed in terms of the first two errors ϵ_i in the chosen range, that in which K_i and A_i are taken to be independent of i. Suppose, for the purpose of this discussion, that these first errors are ϵ_0 and ϵ_1. Since ϵ_0 is the error in the initial value y_0, it should ordinarily be zero, and for this case it is then easily shown that

$$\epsilon_n = \frac{\lambda_1^n - \lambda_2^n}{\lambda_1 - \lambda_2} \epsilon_1 + \frac{h^5 A}{6\theta} \left(1 - \frac{1 - \lambda_2}{\lambda_1 - \lambda_2} \lambda_1^n + \frac{1 - \lambda_1}{\lambda_1 - \lambda_2} \lambda_2^n \right) \tag{9.31}$$

Making use of the approximate values for λ_1 and λ_2 of Eq. (9.30), one has $1 - \lambda_1 \cong -3\Theta$, $1 - \lambda_2 \cong 2$, $\lambda_1 - \lambda_2 \cong 2$, and thus

$$\epsilon_n \cong \frac{\epsilon_1}{2} (\lambda_1^n - \lambda_2^n) + \frac{h^5 A}{6\Theta} (1 - \lambda_1^n - \tfrac{3}{2}\Theta\lambda_2^n) \tag{9.32}$$

It is important to note that, whether K is positive or negative, one of the λ's will be greater than 1 in absolute magnitude. This means that in every case the accumulated error grows exponentially. This is in contrast to the analysis of error accumulation made for Euler's method (see Sec. 8.7), where, for K negative, the errors died out. In fact, from Eqs. (9.30) one sees that

$$\begin{aligned} \lambda_1 &\cong e^{3\Theta} \\ \lambda_2 &\cong -e^{-\Theta} \end{aligned} \tag{9.33}$$

Since $x_n = x_0 + nh$ and $\Theta = \tfrac{1}{3}hK$,

$$\begin{aligned} \lambda_1^n &\cong e^{nhK} = \exp [K(x_n - x_0)] \\ \lambda_2^n &\cong (-1)^n \exp [-\tfrac{1}{3}K(x_n - x_0)] \end{aligned} \tag{9.34}$$

In Eq. (9.32) ϵ_1 should be of the order of the remainder term $h^5 A$ in Simpson's rule and hence should be considerably smaller than $h^5 A/6\Theta$. Making use of Eqs. (9.34) and as an approximation setting $\epsilon_1 = 0$, one may write in place of Eq. (9.32) that

$$\epsilon(x_n) \cong \frac{h^4 A}{2K} [1 - e^{K(x_n - x_0)} + (-1)^{n+1} \tfrac{1}{2}hKe^{-\frac{1}{3}K(x_n - x_0)}] \tag{9.35}$$

If K is positive, the last term dies out as x_n increases. In this case the error is always of the same sign. Since $e^{2.3} = 10$, the error is multiplied by 10 each time x_n increases by $2.3/K$.

If K is negative, the second term dies out, and the error is alternately positive and negative. The magnitude of the error, in this case, becomes 10 times as large every time x_n increases by about $7/K$.

9.6. Obtaining Starting Values for Milne's Method. Any of the methods of Chap. 8 may be used to obtain the four starting values needed for Milne's method; however, the following method of successive approximation using the three "starter" formulas

$$\begin{aligned} y_1 &= y_0 + \frac{h}{24} (7y_1' + 16y_0' + y_{-1}') + \frac{h^2 y_0''}{4} - \tfrac{1}{180}h^5 y^{(5)}(\xi) \\ y_{-1} &= y_0 - \frac{h}{24} (y_1' + 16y_0' + 7y_{-1}') + \frac{h^2 y_0''}{4} + \tfrac{1}{180} h^5 y^{(5)}(\xi) \\ y_2 &= y_0 + \frac{2h}{3} (5y_1' - y_0' - y_{-1}') - 2h^2 y_0'' + \tfrac{7}{45}h^5 y^{(5)}(\xi) \end{aligned} \tag{9.36}$$

is particularly adapted to this purpose. Note that one uses an ordinate y_{-1} to the left of the starting point x_0. The last term in each formula represents, of course, the error term.

Since y_0 is given, y_0' may be found directly from the differential equation (9.1). By differentiating such an equation one obtains an expression for y_0'' in terms of x_0, y_0, and y_0'. Thus the terms in y_0, y_0', and y_0'' of Eqs. (9.36) may be calculated once and for all.

One obtains initial values for y_1' and y_{-1}' from the equations

$$y_1' = y_0' + hy_0''$$
$$y_{-1}' = y_0' - hy_0'' \tag{9.37}$$

and uses the first two equations of (9.36) to find trial values of y_1 and y_{-1}. The differential equation is then used to calculate new values of y_1' and y_{-1}'. These latter two steps may be repeated until no further change occurs in the values of y_1, y_{-1}, y_1', and y_{-1}'. Since the coefficients of the error terms in the first two equations of (9.36) are one-half that for Simpson's rule, the final accuracy of the values is usually sufficient.

A trial value of y_2 is obtained from the third equation of (9.36), and it is used to obtain a trial value of y_2' from the differential equation. All values may then be checked and perhaps improved by making use of Simpson's rule.

9.7. Milne's Method for Second-order Equations. A second-order differential equation for y in terms of the independent variable x may ordinarily be solved for $y'' = d^2y/dx^2$. Consider, therefore, the solution of the differential equation

$$y'' = f(x,y,y') \tag{9.38}$$

with boundary conditions

$$y = y_0 \qquad y' = y_0' \qquad \text{at } x = x_0 \tag{9.39}$$

Assuming that we have already obtained y_i, y_i', y_i'' for $i = 0, 1, 2, \ldots$, n, we may obtain a prediction for y_{n+1}' from the equation

$$y_{n+1}' - y_{n-3}' = \frac{4h}{3}(2y_n'' - y_{n-1}'' + 2y_{n-2}'') \tag{9.40}$$

and then obtain y_{n+1} by the use of Simpson's rule.

$$y_{n+1} - y_{n-1} = \frac{h}{3}(y_{n+1}' + 4y_n' + y_{n-1}') \tag{9.41}$$

The differential equation (9.38) permits us to calculate y_{n+1}''. Using this value and Simpson's rule applied to the y_i'', which may be written

$$y_{n+1}' - y_{n-1}' = \frac{h}{3}(y_{n+1}'' + 4y_n'' + y_{n-1}'') \tag{9.42}$$

one obtains an improved value of y'_{n+1}. If this correction is sizable, one recalculates y_{n+1}, using Eq. (9.41). The procedure is thus a natural extension of the use of the predictor and corrector formulas of Eqs. (9.3) and (9.4).

A very important special case of a second-order differential equation is that in which the first derivative does not occur. For this case Eq. (9.38) becomes

$$y'' = f(x,y) \tag{9.43}$$

If four starting values have been found, one may proceed with the calculation by using the quadrature formula

$$y_{n+1} = y_n + y_{n-2} - y_{n-3} + \frac{h^2}{4}\left(5y''_n + 2y''_{n-1} + 5y''_{n-2}\right)$$
$$+ \tfrac{17}{240}h^6 y^{(6)}(\xi) \tag{9.44}$$

as a predictor and

$$y_n = 2y_{n-1} - y_{n-2} + \frac{h^2}{12}\left(y''_n + 10y''_{n-1} + y''_{n-2}\right) - \tfrac{1}{240}h^6 y^{(6)}(\xi) \tag{9.45}$$

as a corrector. These formulas will be derived in the next section.

9.8. Quadrature Formulas for Second-order Equations.† Newton's backward-interpolation formula [see Eq. (3.61)] applied to

$$y''(x) = d^2 y/dx^2$$

may be written

$$y''(x) = y''_n + u\,\nabla y''_n + \frac{u^{[2]}}{2!}\nabla^2 y''_n + \cdots + \frac{u^{[k]}}{k!}\nabla^k y''_n + R \tag{9.46}$$

where $u = (x - x_n)/h$, R is the remainder term, and

$$u^{[k]} = u(u + 1)\cdots(u + k - 1)$$

Integrating each side of Eq. (9.46) twice with respect to x, one has, for $j \neq 0$,

$$\frac{1}{j}\int_{x_n}^{x_{n+j}} dx \int_{x_n}^{x} y''(x)\,dx$$
$$= \frac{h^2}{j}\int_0^j du \int_0^u \left(y''_n + u\,\nabla y''_n + \cdots + \frac{u^{[k]}}{k!}\nabla^k y''_n + R\right) du$$

The left-hand side of this equation, when the indicated integrations are performed, becomes

† This treatment is based on the article by W. E. Milne, On the Numerical Integration of Certain Differential Equations of the Second Order, *Am. Math. Monthly*, **40**: 322–327 (1933).

$$\frac{1}{j} \int_{x_n}^{x_{n+j}} [y'(x) - y_n'] \, dx = \frac{1}{j} [(y_{n+j} - y_n) - (x_{n+j} - x_n)y_n']$$

$$= \frac{1}{j} (y_{n+j} - y_n) - hy_n'$$

where $y_s = y(x_s)$.   Thus one obtains the general formula

$$\frac{1}{j} (y_{n+j} - y_n) - hy_n'$$
$$= h^2[B_0(j)y_n'' + B_1(j) \nabla y_n'' + \cdots + B_k(j) \nabla^k y_n''] + R_j \quad (9.47)$$

where
$$B_s(j) = \frac{1}{j} \int_0^j du \int_0^u \frac{u^{(s)}}{s!} \, du \quad (9.48)$$

and R_j is the corresponding double integral of R.

A table of the coefficients $B_s(j)$ for $s = 0, 1, 2, \ldots, 7$ and for $j = -5, \ldots, -1, 0, 1$ is given in Table 9.2.

TABLE 9.2

j	$B_0(j)$	$B_1(j)$	$B_2(j)$	$B_3(j)$	$B_4(j)$	$B_5(j)$	$B_6(j)$	$B_7(j)$
-5	-5	25	-75	250	-125	275	1,375	8,125
-4	-4	16	-32	64	0	128	1,280	7,936
-3	-3	9	-9	9	9	135	1,269	7,776
-2	-2	4	0	4	8	128	1,216	7,456
-1	-1	1	1	4	7	107	995	6,031
0								
1	1	1	3	19	45	863	9,625	67,906
Common denominator for the column......	2	6	24	180	480	10,080	120,960	907,200

Let w_j be any set of numbers for $j = -k$ to 1 such that $w_0 = 0$ and

$$\sum_{j=-k}^{1} w_j = 0 \quad (9.49)$$

If one now multiplies Eq. (9.47) on both sides by w_j and sums over j, leaving out $j = 0$, one obtains the general quadrature formula

$$\sum_{j=-k}^{1}{}' \frac{1}{j} w_j(y_{n+j} - y_n)$$
$$= h^2(c_0 y_n'' + c_1 \nabla y_n'' + \cdots + c_k \nabla^k y_n'') + \bar{R} \quad (9.50)$$

where the prime on the summation indicates that the term $j = 0$ is to be omitted and

$$c_s = \sum_{j=-k}^{1}{}' w_j B_s(j) \tag{9.51}$$

This formula expresses the y_s in terms of their second derivative and is, therefore, very useful in the integration of second-order differential equations.

For example, let $w_{-1} = -1$, $w_0 = 0$, $w_1 = 1$; then from Table 9.2

$$c_0 = 1 \qquad c_1 = 0 \qquad c_2 = \tfrac{1}{12} \qquad c_3 = \tfrac{1}{12} \qquad c_4 = \tfrac{19}{240} \qquad \text{etc.}$$

Since the left-hand side of Eq. (9.50) for this choice of the w_j becomes $(y_{n-1} - y_n) + (y_{n+1} - y_n)$, one has

$$y_{n+1} - 2y_n + y_{n-1}$$
$$= h^2(y_n'' + \tfrac{1}{12}\nabla^2 y_n'' + \tfrac{1}{12}\nabla^3 y_n'' + \tfrac{19}{240}\nabla^4 y_{n}'' + \cdots) \tag{9.52}$$

a formula that was employed by Störmer† (see Sec. 9.12).

On the other hand, if $w_{-2} = -2$, $w_{-1} = 2$, and $w_0 = w_1 = 0$, one has

$$y_n - 2y_{n-1} + y_{n-2}$$
$$= h^2(y_n'' - \nabla y_n'' + \tfrac{1}{12}\nabla^2 y_n'' - \tfrac{1}{240}\nabla^4 y_n'' + \cdots) \tag{9.53}$$

Note that no third difference occurs. By expressing the first and second differences on the right in terms of y_n'', y_{n-1}'', and y_{n-2}'' and neglecting the fourth and all higher differences we arrive at Eq. (9.45). The remainder term follows immediately if one is able to justify the assumption that the latter formula is simplex (see Chap. 6) since it must then yield the same error for y a sixth-order polynomial as that predicted by the fourth difference in Eq. (9.53).

Taking $w_{-3} = -3$, $w_{-2} = 2$, $w_{-1} = w_0 = 0$, $w_1 = 1$, one has from Eqs. (9.50) and (9.51) that

$$y_{n+1} - y_n - y_{n-2} + y_{n-3}$$
$$= h^2(3y_n'' - 3\nabla y_n'' + \tfrac{5}{4}\nabla^2 y_n'' + \tfrac{17}{240}\nabla^4 y_n'' + \cdots) \tag{9.54}$$

By the same procedure as above this formula can be made to yield Eq. (9.44).

9.9. The Numerov-Manning-Millman Method.‡ Consider the linear differential equation

$$\frac{d^2y}{dx^2} = f(x)y \tag{9.55}$$

This equation may be integrated step by step using Eq. (9.44) as a predictor and Eq. (9.45) as a corrector. If, however, one has a method of

† *Arch. sci. phys. et nat.*, juillet–octobre, 1907, pp. 63ff.

‡ B. Numerov, *Publs. Observ. Astrophys. Central Russie*, **2**: 188 (1933); M. F. Manning and J. Millman, *Phys. Rev.*, **53**: 673 (1938).

estimating y_n'', one may use Eq. (9.45) alone. For the linear differential equation (9.55) this may be done by using the differential equation to express y_n'' in terms of y_n; thus Eq. (9.45) without the remainder term may be written

$$y_n - \frac{h^2}{12} y_n'' = \left[1 - \frac{h^2}{12} f(x_n) \right] y_n = 2y_{n-1} - y_{n-2} + \frac{h^2}{12} (10y_{n-1}'' + y_{n-2}'')$$

or solving for y_n

$$y_n = \frac{2y_{n-1} - y_{n-2} + (h^2/12)y_{n-2}'' + 10(h^2/12)y_{n-1}''}{1 - (h^2/12)f(x_n)} \qquad (9.56)$$

The method is based on this quadrature formula. One observes that only two starting values are needed and that $1 - (h^2/12)f(x_n)$ may be tabulated before the integration is begun.

One may obviously extend the idea of replacing y_n'' by means of the differential equation to other types of equations.

Example 2. Integrate the differential equation

$$\frac{d^2y}{dx^2} = xy \qquad (9.57)$$

over the range $x = 0$ to $x = 1$, given that at $x = 0$, $y = 1$, and $y' = 0$.

One finds very readily by Taylor's method that y for small values of x is given approximately by

$$y = 1 + \frac{x^3}{6} + \frac{x^6}{180}$$

therefore, at $x = 0.1$, $y = 1.00017$. Having filled in the last two columns of Table 9.3, one uses the first two values of y to obtain from the differential equation the first

TABLE 9.3 SOLUTION OF $y'' = xy$

$$\frac{h^2}{12} = 0.00083\ 33$$

x	y	y''	$\frac{h^2}{12} y''$	$\frac{h^2}{12} x$	$1 - \frac{h^2}{12} x$
0.0	1.00000	0	0	0	1.000000
0.1	1.00017	0.10002	0.000083	0.000083	0.999917
0.2	1.00133	0.20027	0.000167	0.000167	0.999833
0.3	1.00449	0.30135	0.000251	0.000250	0.999750
0.4	1.01066	0.40426	0.000336	0.000333	0.999667
0.5	1.02087	0.51044	0.000425	0.000417	0.999583
0.6	1.03618	0.62171	0.000518	0.000500	0.999500
0.7	1.05771	0.74040	0.000617	0.000583	0.999417
0.8	1.08665	0.86932	0.000724	0.000667	0.999333
0.9	1.12429	1.01186	0.000843	0.000750	0.999250
1.0	1.17206	1.17206	0.000976	0.000833	0.999167

two values of y'' and $(h^2/12)y''$ of columns 3 and 4. One next applies formula (9.56) to find the third value of y. Except for a multiplication by 2 and a shift of the decimal point (multiplication by 10) the numerator in this formula involves only the addition of previously calculated numbers. The denominator, of course, is the number appearing in the last column. Having the third value of y, one again obtains columns 3 and 4 by means of the differential equation, and one is ready for the next step. The values of y shown here were checked by differencing. It was found that the fifth differences did not exceed four units in the fifth decimal place.

9.10. Method of Clippinger and Dimsdale. A very interesting method is that recently devised by R. F. Clippinger and B. Dimsdale. Its principal advantage is that it does not involve a separate method for starting a solution. This permits one to code a single routine on an automatic electronic computer and to change the size of the interval as often as such a change is indicated by some suitable criterion.

We saw in Chap. 7 that, neglecting terms of the higher order,

$$f'(x_1) = \frac{f(x_2) - f(x_0)}{2h} \tag{9.58}$$

and

$$y''(x_1) = \frac{y(x_0) - 2y(x_1) + y(x_2)}{h^2} \tag{9.59}$$

Letting $f(x) = y'(x)$ in the first of these equations permits one to equate the right-hand sides and thereby obtain the equation

$$y(x_1) = \tfrac{1}{2}[y(x_0) + y(x_2)] + \frac{h}{4}[y'(x_0) - y'(x_2)] \tag{9.60}$$

By the use of Taylor's series

$$\tfrac{1}{2}[y(x_1 - h) + y(x_1 + h)] = y(x_1) + \frac{h^2}{2!}y^{(2)}(x_1) + \frac{h^4}{4!}y^{(4)}(x_1) + \cdots$$

$$\frac{h}{4}[y'(x_1 - h) - y'(x_1 + h)] = \frac{h}{2}\left[-hy^{(2)}(x_1) - \frac{h^3}{3!}y^{(4)}(x_1) - \cdots\right]$$

and hence

$$\tfrac{1}{2}[y(x_0) + y(x_2)] + \frac{h}{4}[y'(x_0) - y'(x_2)] = y(x_1) - \frac{h^4}{12}y^{(4)}(x_1) - \cdots$$

Thus the error term in Eq. (9.60) is given approximately by

$$\frac{h^4}{12}y^{(4)}(x_1)$$

Equation (9.60) is used together with Simpson's rule,

$$y(x_2) = y(x_0) + \frac{h}{3}[y'(x_0) + 4y'(x_1) + y'(x_2)] \tag{9.61}$$

to obtain $y(x_1)$ and $y(x_2)$ from a known $y(x_0)$ by a method of successive approximation. To obtain an initial approximation to $y(x_2)$ for use in

Eq. (9.60) one may take

$$y(x_2) = y(x_0) + 2hy'(x_0) \qquad (9.62)$$

This equation, which is just that used in Euler's method, may be obtained at once from Eq. (9.58). Having obtained $y(x_1)$ from Eq. (9.60), one may use Eq. (9.61) to find a better value of $y(x_2)$ and then return to Eq. (9.60), etc.

The application of this method is illustrated by the following example.

Example 3. Determine the solution of the differential equation†

$$y' = 0.1(x^3 + y^2)$$

over the interval $x = 0$ to $x = 1$ given that $y = 1$ at $x = 0$.

The computation is shown in Table 9.4. One first fills in the columns for x and $0.1x^3$, leaving room for several repetitions of the values of y and y' in the other columns. Next, having found that $y'(0) = 0.1000$ from the differential equation, one finds the first value listed for $y(0.2)$, namely, 1.02000, by applying Eq. (9.62). Here, as throughout the calculation, once y is known, one fills in the corresponding y' by

TABLE 9.4. SOLUTION OF $y' = 0.1(x^3 + y^2)$

$$2h = 0.2 \qquad \frac{h}{4} = 0.025$$

x	y	y'	$0.1x^3$	x	y	y'	$0.1x^3$
0.0	1.00000	0.1000	0.0000	0.6	1.06530	0.1351	0.0216
0.1	1.00988	0.1021	0.0001		1.06708	0.1355	
	1.01010	0.1021			1.06713	0.1355	
0.2	1.02000	0.1048	0.0008	0.7	1.07980	0.1509	0.0343
	1.02044	0.1049			1.08139	0.1512	
0.3	1.03023	0.1088	0.0027		1.08142	0.1512	
	1.03110	0.1090		0.8	1.09423	0.1709	0.0512
	1.03111	0.1090			1.09746	0.1716	
0.4	1.04096	0.1148	0.0064		1.09753	0.1717	
	1.04227	0.1150		0.9	1.11329	0.1968	0.0729
	1.04230	0.1150			1.11588	0.1974	
0.5	1.05330	0.1234	0.0125		1.11594	0.1974	
	1.05418	0.1236		1.0	1.13187	0.2281	0.1000
	1.05420	0.1236			1.13710	0.2293	
					1.13722	0.2293	

means of the differential equation. The next step is to use Eq. (9.60) to obtain a first approximation to $y(0.1)$, which is

$$y(0.1) = (0.5)(1.00000 + 1.02000) - (0.025)(0.1048 - 0.1000) = 1.00988$$

Having now available a set of approximate values for y and y' for $x = 0.0, 0.1,$ and 0.2, one may use Simpson's rule, given by Eq. (9.61), to get a better value of 1.02044 for $y(0.2)$. Using this value in place of the old value of 1.02000 and the new value

† This same differential equation was solved by another method in Example 1.

$y'(0.2) = 0.1049$, one recomputes $y(0.1)$ by Eq. (9.60), obtaining the value 1.01010. This calculation may be checked by noting that the change in $y(0.1)$ is given by

$$\delta y(0.1) = \tfrac{1}{2}\delta y(0.2) - \frac{h}{4}\,\delta y'(0.2)$$

$$= 0.00022 - 0.00000\,25 = 0.00022$$

[Here $\delta y(0.1)$ designates the change in $y(0.1)$ rather than a central difference.] One terminates the successive approximations to $y(0.1)$ and $y(0.2)$ represented by the above procedure when no further change takes place.

The next two values $y(0.3)$ and $y(0.4)$ are obtained in exactly the same way from the knowledge of only $y(0.2)$. Thus, each step involves a calculation of two new values and entails exactly the same procedure. It is clear, therefore, that one may change, at any time, to a new interval by simply substituting a new value of h in the formulas.

9.11. Adams-Bashforth Method.† This method is based on the following quadrature formula involving ascending differences

$$y_{n+1} - y_n = h(y'_n + \tfrac{1}{2}\nabla y'_n + \tfrac{5}{12}\nabla^2 y'_n + \tfrac{3}{8}\nabla^3 y'_n + \tfrac{251}{720}\nabla^4 y'_n) \quad (9.63)$$

obtained by integrating Newton's backward-interpolation formula applied to y'_n [see Eq. (9.46)] between the limits of $u = 0$ and $u = 1$, where now $u = (x - x_n)/h$. The open-type quadrature formula (9.63) may also be obtained immediately from the lozenge diagram of Fig. 7.3.

Given a first-order differential equation

$$y' = f(x,y)$$

with the initial value $y = y_0$ at $x = x_0$, one may use any of the methods of Chap. 8 to find the five starting values y_0, y_1, y_2, y_3, and y_4, together with their derivatives. It is customary, however, to associate Taylor's method for starting a solution with this method. Having $y'_0, y'_1, \cdots, y'_4$, one forms a difference table, as shown in Table 9.5, using these values. Thus, the starting method provides one with all the numbers shown in that table.

TABLE 9.5

x	y	y'				
x_0	y_0	y'_0				
x_1	y_1	y'_1	$\nabla y'_1$	$\nabla^2 y'_2$		
x_2	y_2	y'_2	$\nabla y'_2$	$\nabla^2 y'_3$	$\nabla^3 y'_3$	$\nabla^4 y'_4$
x_3	y_3	y'_3	$\nabla y'_3$	$\nabla^2 y'_4$	$\nabla^3 y'_4$	
x_4	y_4	y'_4	$\nabla y'_4$			

The ordinate y_5 may be obtained directly from Eq. (9.63) (if n is set equal to 4). It is, of course, desirable to convert the coefficients in that equation to decimal form. One may, indeed, choose to include h

† F. Bashforth and J. C. Adams, "An Attempt to Test the Theories of Capillary Action . . . ," pp. 15–62, Cambridge University Press, Cambridge, 1883.

in these coefficients.† Once y_5 is found, y_5' may be obtained from the differential equation, and having y_5', a new line of differences can be added to the table. The process can then be repeated to find y_6.

Thus the method employs only a predictor, the open-type formula (9.63). The interval h must therefore be kept small enough so that such a formula is sufficiently accurate. Since the next term in the formula is approximately $(h/5) \nabla^5 y_n'$, one should carry along fifth differences in the table in order to estimate this truncation error. A mistake in the computation will usually show up as a sudden change in size or trend of the highest difference kept.

One may elect to improve the accuracy per step and check the computation by the use of a closed-type formula as a corrector. A convenient formula for this purpose is the formula

$$y_{n+1} - y_n = h(y_{n+1}' - \tfrac{1}{2}\nabla y_{n+1}' - \tfrac{1}{12}\nabla^2 y_{n+1}' - \tfrac{1}{24}\nabla^3 y_{n+1}'$$
$$- \tfrac{19}{720}\nabla^4 y_{n+1}') - \frac{3h^6}{160} y^{(6)}(\xi) \quad (9.64)$$

where $x_{n-3} < \xi < x_{n+1}$.

A method employing a single quadrature formula, such as the Adams-Bashforth method, is well adapted to programming for automatic computation. In this case one usually checks the computation by differencing the y.

Clearly, one could use an open-type quadrature formula employing the tabulated values rather than one employing the differences. Suitable formulas for this purpose are formulas $Q_{n,n+2}$ of Chap. 7. Thus for $n = 2$, we would have the predictor formula of Eq. (9.3). The use of such a formula for the numerical integration of differential equations was suggested by Steffensen.‡

9.12. Adams-Störmer Method for Second-order Equations. The Adams method may be extended to a second-order differential equation by first reducing the differential equation to two simultaneous first-order equations. Thus, suppose the differential equation is

$$y'' = f(x,y,y') \quad (9.65)$$

then letting $g(x,y)$ be defined by the equation

$$y' = g(x,y) \quad (9.66)$$

one has in place of (9.65)

$$g' = f(x,y,g) \quad (9.67)$$

---

† It is customary in this method to tabulate $q_n = hy_n'$ in place of y_n' in order to get rid of h in formula (9.63). However, the h will then reappear in the calculation of $q_{n+1} = hf(x_{n+1},y_{n+1})$ from y_{n+1}.

‡ J. F. Steffensen, "Interpolation," 2d ed., p. 170, Chelsea Publishing Company, New York, 1950.

Thus the pair of first-order equations (9.66) and (9.67) may be solved instead of the single second-order equation (9.65).

Suppose y_i, y'_i, g_i, and g'_i have been computed out to $i = n$. Then the pair of formulas (9.63) and the equivalent formula for g_{n+1}

$$g_{n+1} = g_n + h(g'_n + \tfrac{1}{2}\nabla g'_n + \tfrac{5}{12}\nabla^2 g'_n + \tfrac{3}{8}\nabla^3 g'_n + \tfrac{251}{720}\nabla^4 g'_n) \qquad (9.68)$$

permit one to calculate y_{n+1} and g_{n+1} using only the difference tables of y' and g' built up from the known values. Having y_{n+1} and g_{n+1}, one may use the differential equations (9.66) and (9.67) to find y'_{n+1} and g'_{n+1}. This extends the computation another step. Clearly higher-order equations may be similarly reduced to a set of first-order equations and these equations solved numerically by the quadrature formula represented by Eq. (9.68).

If the first derivative is absent in Eq. (9.65), one may integrate this differential equation directly using the quadrature formula of Eq. (9.52). One finds the new value y_{n+1} in terms of the values already calculated by applying the latter equation and then finds y''_{n+1} from the differential equation.

9.13. Extrapolation Techniques. In the Adams-Bashforth method the predictor formula of Milne's method is used, but the corrector formula is dispensed with. One may, on the other hand, eliminate the predictor formula and retain the more accurate closed-type corrector formula by extrapolating the difference table of y'. Thus, in Table 9.5 one may assume that the next fourth difference $\nabla^4 y'_5$ is the same as $\nabla^4 y'_4$ and extend the differences back to find y'_5. Thus $\nabla^4 y'_5$ added to $\nabla^3 y'_4$ gives $\nabla^3 y'_5$, and this added to $\nabla^2 y'_4$ gives $\nabla^2 y'_5$, which added to $\nabla y'_4$ gives $\nabla y'_5$, etc. Having these ascending differences of y'_5, one may apply the closed-type formula of Eq. (9.64) (with $n = 4$) to find y_5.

If one is not using a difference table, as, for example, in the Milne method, the extrapolation may be done by again assuming that the fifth difference of y', or perhaps some other difference, is constant. Thus to find y'_{n+1} one may set $\nabla^5 y'_{n+1} = 0$ and obtain the extrapolation formula

$$y'_{n+1} = 5y'_n - 10y'_{n-1} + 10y'_{n-2} - y'_{n-3} \qquad (9.69)$$

Having an estimate of y'_{n+1} from this formula, one may apply directly a closed-type formula such as Simpson's rule. It is important, however, to bear in mind that formula (9.69) tends to amplify greatly any errors in the previously calculated values of y'_i, and care must be exercised to see that the quadrature formula employed washes out any error from Eq. (9.69).

9.14. Methods of Successive Approximation. Instead of reducing the interval h over that needed for Milne's method to permit the use of only the predictor formula, one may increase the interval at the cost of repeated application of the corrector formula. Thus one first applies a predictor formula, such as Eq. (9.3), and then a corrector formula, such as Eq. (9.4). Since, however, the interval is kept somewhat larger than in Milne's method, it is necessary to apply the corrector formula again to obtain an improved value of y_{n+1} and from the differential equation to obtain a new value of y'_{n+1}. These two steps are repeated until no further change occurs in y_{n+1}.

Whether this method entails more or less work than Milne's method depends on how much of an increase in h is permitted within the restriction that the convergence factor $\Theta = \frac{1}{3}hK$ of Eq. (9.9) be less than unity and on how much work is involved in the repeated substitution into the differential equation. One thing, however, is clear. Since the interval is larger in this method, the truncation error should ordinarily be greater than for Milne's method.

One advantage of the method of successive approximation is that an accidental error is likely to be caught in subsequent iterations. Levy and Baggott† suggest the use of

$$y_{n+1} = y_{n-1} + h(2y'_n + \tfrac{1}{3}\nabla^2 y'_n) \tag{9.70}$$

as a predictor and

$$y_{n+1} = y_{n-1} + h(2y'_n + \tfrac{1}{3}\nabla^2 y'_{n+1}) \tag{9.71}$$

as a corrector. The latter is just Simpson's rule [Eq. (9.4)] expressed in difference form, and the first can be looked upon as this same formula with the initially unknown difference $\nabla^2 y'_{n+1}$ replaced by an extrapolated value $\nabla^2 y'_n$. The extrapolation, in this case, amounts to assuming that the third difference is zero.

9.15. Solution by Interpolation in Two-point Boundary-value Problems. Consider the differential equation

$$y'' = f(x,y,y') \tag{9.72}$$

with boundary conditions $y = y_a$ at $x = a$ and $y = y_b$ at $x = b$. One may assume a value $y' = N_0$ for the derivative at $x = a$ and integrate the differential equation by a step-by-step method out to $x = b$. This will yield some value $y = M_0$ for the dependent variable at $x = b$. Since this will undoubtedly be different from the required value of y_b, one needs to reintegrate using a different value $y' = N_1$ of the initial derivative. Suppose this yields a value $y = M_1$ at $x = b$. It is then possible to interpolate for the proper value of y' at $x = a$ by assuming that y at $x = b$ varies linearly with the value taken for y' at $x = a$. Suppose this interpolation yields a value of $y' = N_2$. Then one again integrates by some step-by-step method to find a new value $y = M_2$ of the depend-

---

† H. Levy, and E. A. Baggott, "Numerical Studies in Differential Equations," vol. I, pp. 132–133, C. A. Watts & Co., Ltd., London, 1934.

ent variable at $x = b$. At this point there are solutions for three values of the initial derivative. This may permit a rather accurate determination of the proper choice for the initial value of y'. If so, one performs the integration using this value to obtain the desired solution. If not, one must interpolate for a better value of y'. For the proper method of using the results of each trial to predict the next value to try, see the method of false position of Chap. 1.

It is clear from the above description that this method is very laborious, requiring usually some half dozen step-by-step integrations over the range $a \leq x \leq b$. However, it is clear that the first integrations can be very rough, so that the work is greatly reduced. This is done by using a simple quadrature formula or a large interval h or both. An approximate analytical solution, obtained perhaps by using an approximate expression for $f(x,y,y')$ or perhaps by means of a series solution carried out to only a few terms, is also admirably suited to locating the proper y' roughly. Such a solution may contain y_b or y' at $x = a$ as a parameter and permit one to determine at once, within the approximation, the best choice of y'. However, the best all-round method of fixing y' approximately is to use some method of graphical integration.

In the method of this section one works always with solutions of the differential equation and seeks to fit the boundary conditions by successive approximation. In the following section, the boundary conditions are met at once, and one obtains successively better approximations to the solution of differential equation.

9.16. Solution by Successive Integrations. In the solution of the differential equation (9.72) the range $a \leq x \leq b$ may be broken up into n equal parts and the $n + 1$ values $y_0, y_1, \ldots , y_n$ (where y_i is the value of the dependent variable corresponding to $x_i = a + ih$) used to represent the required solution $y(x)$. The boundary conditions may then be met at once by setting $y_0 = y_a$ and $y_n = y_b$. Finally, the remaining y_i are assigned any likely values, and these values are adjusted by successive approximation to meet more and more closely the requirements of the differential equation.

Suppose for simplicity that the first derivative is absent from the differential equation (9.72). Then from the assigned values of y_i one determines from the differential equation

$$y'' = f(x,y) \tag{9.73}$$

the corresponding values of y_i''. From these y_i'' by numerical integration the set of numbers

$$\lambda_i = y_i' - y_0' = \int_{x_0}^{x_i} y'' \, dx \tag{9.74}$$

may be computed. The derivatives y_i' are then given by

$$y_i' = \lambda_i + y_0' \tag{9.75}$$

and the y_i are obtained from these by a second integration. Thus

$$y_i = \int_{x_0}^{x_i} (y' - y_0') \, dx + ihy_0' + y_0 \tag{9.76}$$

and one chooses y_0' so that $y_n = y_b$. Using these new values of y_i, by the differential equation (9.74) one can recompute the y_i'' and repeat the process. Just as in Picard's method (see Chap. 8) such a method tends to integrate out the initial errors in the assigned values of the y_i, and the successive sets of values of the y_i, obtained after each repetition of the process, generally converge.

If y' occurs in the differential equation, one needs to assign at the beginning not only the y_i but also the y_i'. This may be done best by using the finite-difference approximation

$$y_i' = \frac{1}{2h} (y_{i+1} - y_{i-1}) \tag{9.77}$$

The rest of the process is the same as above.

9.17. Replacement of the Differential Equation by a Difference Equation. A third method of handling two-point boundary conditions is to replace the differential equation by a difference equation, i.e., an equation relating just the y_i. This may be done by letting y_i'' be approximated by the finite-difference formula

$$y_i'' = \frac{1}{h^2} \delta^2 y_i = \frac{1}{h^2} (y_{i+1} - 2y_i + y_{i-1}) \tag{9.78}$$

and y_i' by Eq. (9.77). The differential equation (9.72) then requires that

$$y_{i+1} - 2y_i + y_{i-1} = h^2 f\left(x_i\, y_i,\, \frac{y_{i+1} - y_{i-1}}{2h}\right) \tag{9.79}$$

where $i = 1, 2, \ldots, n - 1$.

The difference equation (9.79) represents a set of $n - 1$ equations for the $n - 1$ unknowns y_i, $i = 1, 2, \ldots, n - 1$, in terms of the boundary values $y_0 = y_a$ and $y_n = y_b$. Since many problems in numerical analysis may thus be reduced to the problem of finding the solution of a set of simultaneous equations, the methods available for solving such a set of equations will be discussed in the next chapter.

PROBLEMS

1. Use Milne's method to find a solution of the differential equation

$$\frac{dy}{dx} = x + ye^{-y}$$

in the range $0 \leq x \leq 1$ for the boundary condition $y = 0$ at $x = 0$.

2. Estimate the accumulated error in the integration of Prob. 1.

3. Use Milne's method on the second-order equation $y'' = (1 + x^2)y$ with boundary conditions $y = 0$, $y' = 0.5$ at $x = 0$. Integrate over the range $0 \leq x \leq 2$, and attempt to retain three significant figures of accuracy.

4. Solve the differential equation in Prob. 3 using the Numerov-Manning-Millman method of Sec. 9.9.

5. Given that $x' = \cos x + \sin t$ with $x = 0$ at $t = 0$, find x at $t = 1$. Use Adams's method and an interval of $h = 0.1$.

6. Solve Prob. 5 using any method other than Adams's.

7. Find by numerical integration the maximum value of x if $x'' + xx' + 1 = 0$ and $x = 0$ and $x' = 1$ at $t = 0$.

8. If the drag on a rocket is assumed proportional to the velocity and the rate of burning, and if the thrust is assumed constant, the differential equation to be satisfied by $x = x(t)$, until the fuel is consumed, is $(m - rt)x'' = -kx' - (m - rt)g + T$, where T is the constant giving the thrust. Consider then the problem of finding the maximum value of x which for $0 < t < 1$ satisfies the differential equation $(2 - t)x'' + x' + (2 - t) - 4 = 0$ and for $t > 1$ the differential equation $x'' + x' + 1 = 0$. As initial conditions take $x = 0$, $x' = 0$ at $t = 0$.

9. Find what initial value of x' would be required for the maximum of x to occur at $t = 2$.

10. Find the solution of the differential equation in Prob. 7 for the two-point boundary condition $x = 0$ at $t = 0$ and $x = 0$ at $t = 1$.

CHAPTER 10

SIMULTANEOUS EQUATIONS AND DETERMINANTS

Although both linear and nonlinear sets of algebraic equations will be considered in this chapter, by far the greatest emphasis will be placed on finding the numerical solution of sets of simultaneous linear equations. The solution of such a set of linear equations can, of course, be expressed by means of determinants; however, such a solution is prohibitively long for sets of 10 or more equations. Thus for 10 equations in 10 unknowns, the method of determinants requires some 70,000,000 multiplications, neglecting occurrence of 0 or 1 in the elements of the determinant. This is to be compared to 410 multiplications required by some of the methods of this chapter.

Until recent years the amount of labor involved in a solution of a large number of equations, say 50 or more, was completely prohibitive, except in special cases. The advent of automatic electronic calculators has made such a solution feasible.

Not only, however, must one consider the time required to make the large number of multiplications, divisions, and additions required in any method, but because of the large number of operations involved, the accumulation of round-off errors is also a very serious problem.†

10.1. Simultaneous Linear Equations.‡ Simultaneous linear equations of high orders arise in many fields of applied mathematics. Thus, in statistics the normal equations may involve many unknowns, and in electrical-network theory the number of equations, being equal to the number of meshes, may again be quite large. In particular, whenever one approximates the solution of a linear differential equation or an integral equation by making use of lattice points and difference equations,

---

† V. Bargmann, D. Montgomery, and J. von Neumann, "Solution of Linear Systems of High Order," a report prepared in accordance with Contract Nord-9596 by the Bureau of Ordnance and the Institute of Advanced Study, Princeton University, Princeton, N.J., 1946.

‡ It is assumed in this discussion that the student is already acquainted with the classical theory of simultaneous equations and the properties of determinants such as given in Louis A. Pipes, "Applied Mathematics for Engineers and Physicists," chap. 4, McGraw-Hill Book Company, Inc., New York, 1946. See also M. Bôcher, "Introduction to Higher Algebra," The Macmillan Company, New York, 1907.

the number of unknown values—one for each interior point of the lattice —must be taken large in order for the approximations to be sufficiently accurate.

Consider the set of n linear equations in n unknowns

$$\begin{aligned} a_{11}x_1 + a_{12}x_2 + \cdots + a_{1n}x_n &= b_1 \\ a_{21}x_1 + a_{22}x_2 + \cdots + a_{2n}x_n &= b_2 \\ \cdots\cdots\cdots\cdots\cdots\cdots\cdots\cdots \\ a_{n1}x_1 + a_{n2}x_2 + \cdots + a_{nn}x_n &= b_n \end{aligned} \tag{10.1}$$

If the determinant

$$\Delta = \begin{vmatrix} a_{11} & a_{12} & a_{13} & \cdots & a_{1n} \\ a_{21} & a_{22} & a_{23} & \cdots & a_{2n} \\ a_{31} & a_{32} & a_{33} & \cdots & a_{3n} \\ \cdots\cdots\cdots\cdots\cdots\cdots \\ a_{n1} & a_{n2} & a_{n3} & \cdots & a_{nn} \end{vmatrix} \tag{10.2}$$

is not zero, then the solution of Eqs. (10.1) is given by Cramér's rule, e.g.,

$$x_1 = \frac{D_1}{\Delta}$$

$$x_2 = \frac{D_2}{\Delta} \tag{10.3}$$

$$\cdots\cdots$$

$$x_n = \frac{D_n}{\Delta}$$

where

$$D_s = \begin{vmatrix} a_{11} & a_{12} & \cdots & a_{1,s-1} & b_1 & a_{1,s+1} & \cdots & a_{1n} \\ a_{21} & a_{22} & \cdots & a_{2,s-1} & b_2 & a_{2,s+1} & \cdots & a_{2n} \\ \cdots\cdots\cdots\cdots\cdots\cdots\cdots\cdots\cdots\cdots \\ a_{n1} & a_{n2} & \cdots & a_{n,s-1} & b_n & a_{n,s+1} & \cdots & a_{nn} \end{vmatrix} \tag{10.4}$$

Thus D_s is the same as the Δ but with the b's replacing the a's in the sth column.

This solution, however, requires the evaluation of $n + 1$ determinants of the nth order, and each of the nth-order determinants, in general, involves $\lambda n!$ multiplications if expanded in terms of minors. Here $1 \leq \lambda < e - 1 = 1.718$, and λ approaches $e - 1$ as n becomes large.†

† Let $G(n)$ be the number of multiplications needed to expand an nth-degree determinant by minors. Since such an expansion involves the multiplication of n elements by their minors and there are n minors requiring $G(n - 1)$ multiplications each,

$$G(n) = n + nG(n - 1)$$

This is a difference equation for $G(n)$ that can be simplified by letting $G(n) = n!\lambda(n)$

Thus the solution requires $\lambda(n + 1)!$ multiplications and n divisions plus approximately $(n + 1)!$ additions (and/or subtractions). Since the addition time is usually small compared to the multiplication or division time, for large values of n the time needed for determining the x's is indicated by the approximate figure of

$$2(n + 1)! \text{ multiplications}$$

Even for $n = 10$, the number of multiplications would be about 70,000,000.

10.2. Method of Chiò for Evaluating Determinants. The $\lambda n!$ multiplications $(1 \leq \lambda < e - 1)$ necessary to evaluate an nth-order determinant (see Sec. 10.1) can be greatly reduced by using a method due to F. Chiò.†

Consider the determinant

$$D \equiv \begin{vmatrix} a_1 & a_2 & a_3 & a_4 & a_5 \\ b_1 & b_2 & b_3 & b_4 & b_5 \\ c_1 & c_2 & c_3 & c_4 & c_5 \\ d_1 & d_2 & d_3 & d_4 & d_5 \\ e_1 & e_2 & e_3 & e_4 & e_5 \end{vmatrix} \qquad (10.5)$$

Suppose that some element, say c_4, is equal to unity. If there are no unit elements in the determinant, we can always divide a given row (or

and hence $G(n - 1) = (n - 1)!\lambda(n - 1)$; then

$$\lambda(n) - \lambda(n - 1) = \frac{1}{(n - 1)!}$$

Since a second-order determinant requires two multiplications, $G(2) = 2$, $\lambda(2) = 1$ and the solution is

$$\lambda(n) = \sum_{p=2}^{n} \frac{1}{(p - 1)!}$$

Note that $\lim_{n \to \infty} \lambda(n) = e - 1$.

If one expands a determinant as a sum of products formed by taking an element from each row and each column, the number of multiplications needed is $n!(n - 1)$. This can be seen as follows. There are n ways of choosing an element from the first row, and after that is done, there are $n - 1$ ways of choosing an element from the second row, $n - 2$ ways from the third row, etc. Thus, there are $n!$ products of n numbers each or $n!$ times $n - 1$ multiplications. Thus the expansion by minors is preferable.

All of these calculations, of course, assume that all the elements of the determinant are different from zero. Any zeros greatly reduce the number of multiplications.

† F. Chiò, "Mémoire sur les fonctions connues sous le nom de résultantes ou de dèterminants," Turin, 1853. See also, E. Whittaker and G. Robinson, "The Calculus of Observations," 4th ed., chap. V, Blackie & Son, Ltd., Glasgow, 1944.

column) by one of its elements and take that element out in front of the determinant. We shall, however, for the moment assume that, as the determinant stands, $c_4 = 1$. Having chosen the pivotal element to be c_4, we make all the other elements in that row 0 by multiplying all members of the fourth column by c_1 and subtracting the resulting numbers from those of the first column,† by multiplying next by c_2 and subtracting the results from the second column, by multiplying this same fourth column of D by c_3 and subtracting from the third column, and finally, by multiplying the fourth column by c_5 and subtracting from the fifth column. The result, since $c_4 = 1$, is

$$D = \begin{vmatrix} a_1 - c_1a_4 & a_2 - c_2a_4 & a_3 - c_3a_4 & a_4 & a_5 - c_5a_4 \\ b_1 - c_1b_4 & b_2 - c_2b_4 & b_3 - c_3b_4 & b_4 & b_5 - c_5b_4 \\ 0 & 0 & 0 & 1 & 0 \\ d_1 - c_1d_4 & d_2 - c_2d_4 & d_3 - c_3d_4 & d_4 & d_5 - c_5d_4 \\ e_1 - c_1e_4 & e_2 - c_2e_4 & e_3 - c_3e_4 & e_4 & e_5 - c_5e_4 \end{vmatrix}$$

or expanding by minors in terms of the elements of the third row,

$$D = (-1)^{3+4} \begin{vmatrix} a_1 - c_1a_4 & a_2 - c_2a_4 & a_3 - c_3a_4 & a_5 - c_5a_4 \\ b_1 - c_1b_4 & b_2 - c_2b_4 & b_3 - c_3b_4 & b_5 - c_5b_4 \\ d_1 - c_1d_4 & d_2 - c_2d_4 & d_3 - c_3d_4 & d_5 - c_5d_4 \\ e_1 - c_1e_4 & e_2 - c_2e_4 & e_3 - c_3e_4 & e_5 - c_5e_4 \end{vmatrix} \quad (10.6)$$

Thus a determinant of the fifth order [Eq. (10.5)] is reduced to one of fourth order [Eq. (10.6)].

The fourth-order terms are seen to be obtainable from the original determinant by the following steps:

1. Choose a pivotal element, such as c_4, which must either be equal to unity in advance or be made unity by division of the row (or column) by that element.

2. Cross out the row and column belonging to this element.

3. Subtract from each of the remaining elements the product of the elements found at the base of perpendiculars drawn from that element to the crossed-out row and column.

4. Multiply the determinant by $(-1)^{r+s}$, where r is the row and s is the column of the pivotal element.

This constitutes *Chio's rule*, which can be applied repeatedly to reduce a given determinant eventually to a first-order determinant, i.e., to a number. Each application of the rule reduces the order by one.

If a unit element can be found, the reduction of an nth-order determinant to an $(n - 1)$th-order determinant requires $(n - 1)^2$ multiplications, as can be seen from Eq. (10.6). We cannot count on a pivotal element's

† By a property of determinants, such operations do not change the value of the determinant.

being available before factoring out a number from one of the rows. This involves $n - 1$ divisions and requires a subsequent multiplication by the value of the pivotal element. Thus the above reduction requires

$$(n - 1)^2 + 1 \text{ multiplications}$$
$$n - 1 \text{ divisions}$$

By finding the reciprocal of the pivotal element one can replace the $n - 1$ divisions by 1 division and $n - 1$ multiplications. The reduction of the $(n - 1)$th-degree determinant requires similarly

$$(n - 2)^2 + 1 \text{ multiplications}$$
$$n - 2 \text{ divisions}$$

and so forth. Finally, the reduction from a third-order to a second-order determinant requires

$$2^2 + 1 \text{ multiplications}$$
$$2 \text{ divisions}$$

and the evaluation of the second-order determinant requires

$$1^2 + 1 \text{ multiplications}$$
$$1 \text{ division} \dagger$$

Therefore, the total number of multiplications and divisions is

$$\sum_{s=1}^{n-1} (s^2 + 1) = \frac{(n - 1)(2n^2 - n + 6)}{6} \text{ multiplications}$$

$$\sum_{s=1}^{n-1} s = \frac{n(n - 1)}{2} \text{ divisions} \tag{10.7}$$

For large values of n, assuming that a division time is at most only a few times longer than a multiplication time, the time required for evaluating a determinant by Chiò's method is indicated by the approximate figure of

$$\frac{n^3}{3} \text{ multiplications}$$

Thus the solution of a set of n linear equations by determinants could be done by using approximately

$$\frac{n^4}{3} \text{ multiplications} \tag{10.8}$$

For $n = 10$, this amounts to about 3,000 multiplications as against 70,000,000 without Chiò's method. However, for $n = 100$ the number

---

† This division can, of course, be eliminated by using the usual expansion of the second-order determinant.

of multiplications again reaches a large figure, namely, 30,000,000 (approximately).†

10.3. Method of Elimination (Gauss' Method). The process of eliminating one unknown at a time from the set of equations is perhaps the simplest approach to their solution and at the same time one of the shortest methods known. Many adaptations of the method have been proposed, and many schemes for organizing the required computations have been devised.‡ Some care must be exercised in the order of elimination of the x's, especially if they are of widely different magnitudes. When possible, it is advisable to begin with the smallest one, proceeding in order of increasing magnitude. If this is not done, the cumulative errors in the calculations may produce unreliable values of the unknowns.

The first equation of Eqs. (10.1) can be multiplied through by the reciprocal of a_{11} and written

$$x_1 = a_{11}^{-1}b_1 - \sum_{i=2}^{n} (a_{11}^{-1}a_{1i})x_i \tag{10.9}$$

In the other equations, which can be written in the form

$$a_{j1}x_1 + \sum_{i=2}^{n} a_{ji}x_i = b_j \qquad j = 2, 3, \ldots, n$$

x_1 can be eliminated by the use of Eq. (10.9), and one obtains the $n - 1$ equations in $n - 1$ unknowns $x_2, x_3, \ldots, x_n$

$$\sum_{i=2}^{n} (a_{ji} - a_{j1}a_{11}^{-1}a_{1i})x_i = b_j - a_{j1}a_{11}^{-1}b_1 \tag{10.10}$$

where again $j = 2, 3, \ldots, n$.

We can next proceed to eliminate x_2. Writing Eqs. (10.10) as

$$\sum_{i=2}^{n} a'_{ji}x_i = b'_j \qquad j = 2, 3, \ldots, n \tag{10.11}$$

the process can be repeated. Thus from the first equation

$$x_2 = (a'_{22})^{-1}b'_2 - \sum_{i=3}^{n} (a'_{22})^{-1}a'_{2i}x_i \tag{10.12}$$

---

† This corresponds to about 19 days of running time for Mark III (built by the Harvard Computation Laboratory). The approximate time required for additions, transfer of numbers, checks, etc., is included in this estimate. If the elimination method is used, this time is reduced to $4\frac{2}{3}$ hours (see Sec. 10.3). These times would be greatly reduced in the more recent machines.

‡ An example of such organization of the computation is D. E. Richardson, "Electrical Network Calculations," D. Van Nostrand Company, Inc., Princeton, N.J., 1946.

and substituting this expression for x_2 in the remaining equation, one has, corresponding to Eqs. (10.10), the equations

$$\sum_{i=3}^{n} [a'_{ji} - a'_{j2}(a'_{22})^{-1}a'_{2i}]x_i = b'_j - a'_{j2}(a'_{22})^{-1}b'_2 \qquad (10.13)$$

where now $j = 3, 4, \ldots, n$.

By repeated elimination we arrive at a single equation in the unknown x_n, which can then be solved by a single division. Having x_n, we can substitute to find x_{n-1}; and having x_{n-1}, we can substitute in the appropriate equation to find x_{n-2}, etc.; and finally having $x_n, x_{n-1}, \ldots, x_3$, we can substitute in Eq. (10.12) to find x_2 and in Eq. (10.9) to find x_1. The number of multiplications required to find the a'_{ji} and b'_j of Eqs. (10.11) [see Eqs. (10.10)] is seen to be n^2. The total number of multiplications and divisions needed is

$$\frac{n(n + 1)(2n + 1)}{6} \text{ multiplications}$$
$$n \text{ divisions} \qquad (10.14)$$

This number of multiplications is just the sum of squares up to n^2. Therefore, for large n we can take as our figure

$$\frac{n^3}{3} \text{ multiplications} \qquad (10.15)$$

Thus we see that there is a factor of n to our advantage in this method over the determinant method (using Chiò's rule). For $n = 100$ the number of multiplications is now roughly 300,000. A little consideration of the mathematical operations performed reveals that Chiò's method is essentially the method of elimination applied to the evaluation of a determinant.

10.4. Gauss-Jordan Method.† The purpose of this modification of the Gauss method is to make the approach to a solution more uniform by eliminating the need for working backward from the knowledge of x_n to find x_{n-1}, x_{n-2}, etc. This is done at the expense of somewhat more calculation, but it permits the procedure to be adapted more readily to automatic machine computation.

The first step of eliminating x_1 is performed as before, and the second step of eliminating x_2 is modified only to the extent that, after it is performed (as given in Sec. 10.3), one also eliminates x_2 from Eq. (10.9). Similarly after the elimination of x_3 from the $n - 3$ equations given by Eqs. (10.13), one also eliminates x_3 from Eqs. (10.9) and (10.12). Thus,

† W. Jordan, "Handbuch der Vermessungskunde," 7th ed., vol. I, p. 36, Stuttgart, 1920.

by the time one has reached a single equation in x_n in the above elimination process, one has also reduced Eq. (10.9) to an equation in x_1 alone and Eq. (10.12) to an equation in x_2. In fact, we end up with n equations, each involving only one of the unknowns $x_1, x_2, \ldots, x_n$ and requiring only a single division for its solution.

10.5. The Square-root Method of Banachiewicz and Dwyer.† The set of linear equations (10.1) can be written in the form

$$AX = B \qquad (10.16)$$

where A is the matrix of the coefficients a_{ij}, and X and B are the column matrices $\{x_i\}$ and $\{b_i\}$. The method is applicable only if A is symmetric, i.e., $a_{ij} = a_{ji}$.

Let the triangular matrix

$$S \equiv \begin{bmatrix} s_{11} & s_{12} & s_{13} & \cdots & s_{1n} \\ 0 & s_{22} & s_{23} & \cdots & s_{2n} \\ 0 & 0 & s_{33} & \cdots & s_{3n} \\ \cdots\cdots\cdots\cdots\cdots\cdots \\ 0 & 0 & 0 & \cdots & s_{nn} \end{bmatrix} \qquad (10.17)$$

be defined by the equation

$$S'S = A \qquad (10.18)$$

where S' is the transpose of S, i.e., $s'_{ij} = s_{ji}$. Then

$$S'(SX) = B$$

which can be expressed as two equations, namely,

$$\begin{aligned} S'K &= B \\ SX &= K \end{aligned} \qquad (10.19)$$

where K is a column matrix $\{k_i\}$. Once S is determined from Eq. (10.18), Eqs. (10.19) can easily be solved in succession, the first for K and the second for X. The ease of solving Eqs. (10.19) lies in the fact that S' and S are triangular matrices.

First Step. Obtain S from the Relation S'S = A. Equating individual elements of A and $S'S$,

$$\sum_{r=1}^{n} s'_{ir} s_{rj} = \sum_{r=1}^{n} s_{ri} s_{rj} = \sum_{r=1}^{i} s_{ri} s_{rj} = a_{ij} \qquad j \geq i \qquad (10.20)$$

† This treatment of the method is based on Jack Laderman, The Square Root Method for Solving Simultaneous Linear Equations, *Math. Tables and Other Aids to Computation*, **3**(21): 13–16 (1948). See also D. B. Duncan, and J. F. Kenney, "On the Solution of Normal Equations and Related Topics," J. W. Edwards, Publisher, Inc., Ann Arbor, Mich., 1946.

Writing out these equations, we have

$$s_{11}s_{11} = a_{11}$$
$$s_{11}s_{1j} = a_{1j}$$
$$s_{12}s_{12} + s_{22}s_{22} = a_{22}$$
$$s_{12}s_{1j} + s_{22}s_{2j} = a_{2j}$$
$$\cdot \cdot \cdot \cdot \cdot \cdot \cdot \cdot \cdot$$
$$\sum_{r=1}^{i-1} s_{ri}s_{ri} + s_{ii}s_{ii} = a_{ii}$$
$$\sum_{r=1}^{i-1} s_{ri}s_{rj} + s_{ii}s_{ij} = a_{ij}$$

and solving, we have

$$s_{11} = \sqrt{a_{11}}$$
$$s_{1j} = \frac{a_{1j}}{s_{11}}$$
$$s_{22} = (a_{22} - s_{12}^2)^{\frac{1}{2}}$$
$$s_{2j} = \frac{1}{s_{22}}(a_{2j} - s_{12}s_{1j})$$
$$\cdot \cdot \cdot \cdot \cdot \cdot \cdot \cdot \cdot \cdot \cdot \quad (10.21)$$
$$s_{ii} = \left(a_{ii} - \sum_{r=1}^{i-1} s_{ri}^2\right)^{\frac{1}{2}}$$
$$s_{ij} = \frac{1}{s_{11}}\left(a_{ij} - \sum_{r=1}^{i-1} s_{ri}s_{rj}\right)$$

If computed in this order, all the quantities on the right-hand sides will be known.

Second Step. Obtain K from S'K = B. Equating individual terms of this matrix equation,

$$\sum_{r=1}^{n} s'_{ir}k_r = \sum_{r=1}^{i} s_{ri}k_r = b_i \qquad i = 1, 2, \ldots, n$$

Writing out these equations,

$$s_{11}k_1 = b_1$$
$$s_{12}k_1 + s_{22}k_2 = b_2$$
$$\cdot \cdot \cdot \cdot \cdot \cdot \cdot \cdot \cdot$$
$$\sum_{r=1}^{i-1} s_{ri}k_r + s_{ii}k_i = b_i$$

therefore

$$k_1 = \frac{b_1}{s_{11}}$$

$$k_2 = \frac{1}{s_{22}}(b_2 - s_{12}k_1)$$

(10.22)

$$\cdots \cdots \cdots \cdots \cdots$$

$$k_i = \frac{1}{s_{ii}}\left(b_i - \sum_{r=1}^{i-1} s_{ri}k_r\right)$$

Third Step. *Obtain* X *from* $SX = K$. Again equating individual elements,

$$\sum_{r=1}^{n} s_{ir}x_r = \sum_{r=i}^{n} s_{ir}x_r = k_i \qquad i = 1, 2, \ldots, n$$

These equations written out are

$$s_{nn}x_n = k_n$$

$$s_{n-1,n-1}x_{n-1} + s_{n-1,n}x_n = k_{n-1}$$

$$\cdots \cdots \cdots \cdots \cdots \cdots$$

$$s_{ii}x_i + \sum_{r=i+1}^{n} s_{ir}x_r = k_i$$

therefore

$$x_n = \frac{k_n}{s_{nn}}$$

$$x_{n-1} = \frac{1}{s_{n-1,n-1}}(k_{n-1} - s_{n-1,n}x_n)$$

(10.23)

$$\cdots \cdots \cdots \cdots \cdots \cdots$$

$$x_i = \frac{1}{s_{ii}}\left(k_i - \sum_{r=i+1}^{n} s_{ir}x_r\right)$$

A check can be applied to each part of the calculation by solving simultaneously with the above set of linear equations that set obtained by replacing x_i by $\bar{x}_i - 1$. This second set may be written in matrix notation as

$$A\bar{X} = \bar{B}$$

(10.24)

Since by definition

$$\sum_{r=1}^{n} a_{ir}(\bar{x}_r - 1) = b_i$$

it is clear from Eq. (10.24) that the components of $\bar{B}$ are

$$\bar{b}_i = \sum_{r=1}^{n} a_{ir} + b_i$$

(10.25)

Equation (10.24) is to be solved for $\bar{x}_i$, the elements of $\bar{X}$, in the same way Eq. (10.16) was solved for the x_i. Since A is the same as before, S is the same; only the second and third steps are different.

The computation is checked by the two equations

$$\bar{x}_i = x_i + 1 \tag{10.26}$$

$$\bar{k}_i = k_i + \sum_{r=i}^{n} s_{ir} \tag{10.27}$$

The first equation is from the definition of the $\bar{x}_i$, and the second is derived from the equation†

$$S\bar{X} = \bar{K} \tag{10.28}$$

This matrix equation written in terms of the elements is

$$\bar{k}_i = \sum_{r=i}^{n} s_{ir}\bar{x}_r = \sum_{r=i}^{n} s_{ir}x_r + \sum_{r=i}^{n} s_{ir} = k_i + \sum_{r=i}^{n} s_{ir}$$

For the checking to be efficient, the computation should be made in the following order:

1. Obtain all the elements of $\bar{B}$ from Eq. (10.25).

2. Obtain the entire first row of S, K, and $\bar{K}$, and check $\bar{k}_1$ by Eq. (10.27). Then obtain the entire second row and check $\bar{k}_2$, etc., through to the nth row.

3. Obtain x_n and $\bar{x}_n$ and check by Eq. (10.26); then obtain x_{n-1} and $\bar{x}_{n-1}$ and check by the same equation, etc., until x_1 and $\bar{x}_1$ are obtained and checked.

In this manner each row is checked before proceeding to the next row.

If a desk calculator is used, one needs to write down only those quantities indicated in the following scheme:

a_{11}	a_{12}	a_{13}	a_{14}	$\cdots$	a_{1n}	b_1	$\bar{b}_1$
	a_{22}	a_{23}	a_{24}	$\cdots$	a_{2n}	b_2	$\bar{b}_2$
		a_{33}	a_{34}	$\cdots$	a_{3n}	b_3	$\bar{b}_3$
			a_{44}	$\cdots$	a_{4n}	b_4	$\bar{b}_4$
				$\cdots$		$\cdots$	$\cdots$
					a_{nn}	b_n	$\bar{b}_n$
s_{11}	s_{12}	s_{13}	s_{14}	$\cdots$	s_{1n}	k_1	$\bar{k}_1$
	s_{22}	s_{23}	s_{24}	$\cdots$	s_{2n}	k_2	$\bar{k}_2$
		s_{33}	s_{34}	$\cdots$	s_{3n}	k_3	$\bar{k}_3$
			s_{44}	$\cdots$	s_{4n}	k_4	$\bar{k}_4$
				$\cdots$		$\cdots$	$\cdots$
					s_{nn}	k_n	$\bar{k}_n$
x_1	x_2	x_3	x_4	$\cdots$	x_n		
$\bar{x}_1$	$\bar{x}_2$	$\bar{x}_3$	$\bar{x}_4$	$\cdots$	$\bar{x}_n$		

† This equation is the complete analogue of the second equation of Eqs. (10.19).

10.6. The Crout Method. Crout[†] modified the elimination method of Gauss to take advantage of the fact that modern desk-type calculators are able to find the sum of a series of products and, if desired, to make a final division in one continuous machine operation. The procedure is based on the following variation of the method of Sec. 10.3. We write Eq. (10.9) as

$$x_1 = b_1' - \sum_{i=2}^{n} a_{1i}' x_i \tag{10.29}$$

where b_1' and a_{1i}' are calculated from the relation

$$b_1' = \frac{b_1}{a_{11}} \quad \text{and} \quad a_{1i}' = \frac{a_{1i}}{a_{11}} \quad i \geq 2 \tag{10.30}$$

Letting
$$\begin{aligned} a_{j1}' &= a_{j1} \\ a_{j2}' &= a_{j2} - a_{j1}' a_{12}' \quad j \geq 2 \end{aligned} \tag{10.31}$$

Eqs. (10.10) can be written

$$a_{j2}' x_2 + \sum_{i=3}^{n} (a_{ji} - a_{j1}' a_{1i}') x_i = b_j - a_{j1}' b_1' \tag{10.32}$$

where again $j = 2, 3, \ldots, n$. The first equation of this set can be solved for x_2, yielding in place of Eq. (10.12) the equation

$$x_2 = b_2' - \sum_{i=3}^{n} a_{2i}' x_i \tag{10.33}$$

where now, however,

$$\begin{aligned} b_2' &= (b_2 - a_{21}' b_1') \frac{1}{a_{22}'} \\ a_{2i}' &= (a_{2i} - a_{21}' a_{1i}') \frac{1}{a_{22}'} \quad i \geq 2 \end{aligned} \tag{10.34}$$

Using this expression for x_2 to eliminate it from the other equations of Eqs. (10.32), one has in place of Eqs. (10.13) the equations

$$a_{j3}' x_3 + \sum_{i=4}^{n} (a_{ji} - a_{j1}' a_{1i}' - a_{j2}' a_{2i}') x_i = b_j - a_{j1}' b_1' - a_{j2}' b_2' \tag{10.35}$$

where $j = 3, 4, \ldots, n$ and

$$a_{j3}' = a_{j3} - a_{j1}' a_{13}' - a_{j2}' a_{23}' \quad j \geq 3 \tag{10.36}$$

Once again the first equation of (10.35) can be solved for x_3, yielding the

[†] P. D. Crout, A Short Method of Evaluating Determinants and Solving Systems of Linear Equations, *Trans. AIEE*, **60**: 1235 (1941).

equation

$$x_3 = b_3' - \sum_{i=4}^{n} a_{3i}' x_i \tag{10.37}$$

where

$$b_3' = (b_3 - a_{31}' b_1' - a_{32}' b_2') \frac{1}{a_{33}'}$$

$$a_{3i}' = (a_{31} - a_{3i}' a_{1i}' - a_{32}' a_{2i}') \frac{1}{a_{33}'} \qquad i \geq 4 \tag{10.38}$$

It is clear that this procedure differs from the elimination method only in the fact that the constants are reevaluated not at each step of the process but only at the stage where they are needed for subsequent calculations. One thus arrives at the same set of equations as in the elimination method, namely, the equations

$$x_j = b_j' - \sum_{i=j+1}^{n} a_{ji}' x_i \qquad j = 1, 2, \ldots, n \tag{10.39}$$

[see Eqs. (10.29), (10.33), and (10.37)], which clearly have as an *augmented*† matrix the triangular matrix

$$\begin{pmatrix} 1 & a_{12}' & a_{13}' & a_{14}' & \cdots & a_{1n}' & b_1' \\ 0 & 1 & a_{23}' & a_{24}' & \cdots & a_{2n}' & b_2' \\ 0 & 0 & 1 & a_{34}' & \cdots & a_{3n}' & b_3' \\ 0 & 0 & 0 & 1 & \cdots & a_{4n}' & b_4' \\ \cdots & \cdots & \cdots & \cdots & \cdots & \cdots & \cdots \\ 0 & 0 & 0 & 0 & \cdots & 1 & b_n' \end{pmatrix} \tag{10.40}$$

The method of determining these constants is indicated by Eqs. (10.30), (10.31), (10.34), (10.36), and (10.38) and is summarized by the equations

$$a_{ji}' = a_{ji} - \sum_{k=1}^{i-1} a_{jk}' a_{ki}' \qquad i \leq j$$

$$a_{ji}' = \frac{1}{a_{jj}'} \left(a_{ji} - \sum_{k=1}^{j-1} a_{jk}' a_{ki}' \right) \qquad i > j \tag{10.41}$$

$$b_j' = \frac{1}{a_{jj}'} \left(b_j - \sum_{k=1}^{j-1} a_{jk}' b_k' \right)$$

† An augmented matrix represents a set of linear equations in condensed form. It is composed of the matrix of the coefficients of the x_i augmented on the right by a column giving the constant terms. Note that equality signs may be visualized just ahead of the elements of this last column.

All these primed constants can be thought of as belonging to the matrix

$$
\begin{pmatrix}
a'_{11} & a'_{12} & a'_{13} & \cdots & a'_{1n} & b'_1 \\
a'_{21} & a'_{22} & a'_{23} & \cdots & a'_{2n} & b'_2 \\
a'_{31} & a'_{32} & a'_{33} & \cdots & a'_{3n} & b'_3 \\
\multicolumn{6}{c}{\cdots\cdots\cdots\cdots\cdots\cdots} \\
a'_{n1} & a'_{n2} & a'_{n3} & \cdots & a'_{nn} & b'_n
\end{pmatrix}
\tag{10.42}
$$

which is termed the *auxiliary matrix*. Of these constants, only those found in the matrix (10.40) are of significance; the others are used only in the intermediate calculations. Rules for finding the elements of the auxiliary matrix directly will be given later.

Having the elements of the auxiliary matrix (10.42), and hence also the elements of the triangular matrix (10.40), one can solve Eqs. (10.39) in reverse order for $x_n, x_{n-1}, \ldots, x_1$. This procedure is exactly the same as in the elimination method. The solutions can be represented by the equations

$$
\begin{aligned}
x_n &= b'_n \\
x_{n-1} &= b'_{n-1} - a'_{n-1,n} x_n \\
x_{n-2} &= b'_{n-2} - a'_{n-2,n-1} x_{n-1} - a'_{n-2,n} x_n \\
\multicolumn{2}{c}{\cdots\cdots\cdots\cdots\cdots\cdots\cdots\cdots}
\end{aligned}
$$

$$
x_j = b'_j - \sum_{i=j+1}^{n} a'_{ji} x_i
\tag{10.43}
$$

$$
\cdots\cdots\cdots\cdots\cdots
$$

$$
x_1 = b'_1 - \sum_{i=2}^{n} a'_{2i} x_i
$$

which are merely Eqs. (10.39) written in reverse order, since all the x's occurring on the right of any one equation have been evaluated in the preceding equations. These values of $x_1, x_2, \ldots, x_n$, respectively, written in a column are termed by Crout the *final matrix*.

Crout gives the following working rules for obtaining the *auxiliary matrix* (10.42) from the given *augmented matrix* [representing the equations (10.1)],

$$
\begin{pmatrix}
a_{11} & a_{12} & a_{13} & \cdots & a_{1n} & b_1 \\
a_{21} & a_{22} & a_{23} & \cdots & a_{2n} & b_2 \\
a_{31} & a_{32} & a_{33} & \cdots & a_{3n} & b_3 \\
\multicolumn{6}{c}{\cdots\cdots\cdots\cdots\cdots\cdots} \\
a_{n1} & a_{n2} & a_{n3} & \cdots & a_{nn} & b_n
\end{pmatrix}
\tag{10.44}
$$

1. The various numbers, or elements, are determined in the following order: elements of the first column; then elements of the first row to the right of the first column; elements of the second column below the first

row; then elements of the second row to the right of the second column; elements of the third column below the second row; then elements of the third row to the right of the third column; and so on until all elements are determined.

2. The first column is identical with the first column of the given matrix. Each element of the first row except the first is obtained by dividing the corresponding element of the given matrix by that first element.

3. Each element on or below the principal diagonal is equal to the corresponding element of the given matrix minus the sum of those products of elements in its row and corresponding elements in its column (in the auxiliary matrix) which involve only previously computed elements.

4. Each element to the right of the principal diagonal is given by a calculation which differs from rule 3 only in that there is a final division by its diagonal element (in the auxiliary matrix).

Crout gives the following working rules for obtaining the one-column *final matrix* from the *auxiliary matrix:*

1. The elements are determined in the following order: last, next to the last, second from the last, third from the last, and so forth.

2. The last element is equal to the corresponding element in the last column of the auxiliary matrix.

3. Each element is equal to the corresponding element of the last column of the auxiliary matrix minus the sum of those products of elements in its row in the auxiliary matrix and corresponding elements in the final matrix which involve only previously computed elements.

Since the Crout method is basically the elimination method, it is clear that the number of multiplications is approximately

$$\frac{n^3}{3} \text{ multiplications}$$

10.7. Method of Postmultiplication.† The elimination process can be brought about by multiplication by suitably chosen matrices. Consider the set of linear equations

$$\sum_{j=1}^{n} x_j a_{ji} = b_i \qquad (10.45)$$

which can be written in matrix form as

$$XA = B \qquad (10.46)$$

---

† Postmultiplication of a matrix A by a matrix M means that A is to be multiplied on the right by M. Premultiplication used in same way signifies that A is to be multiplied on the left by M. Only if A and M commute, i.e., $AM = MA$, can one ignore the order of the multiplication.

where X and B are row matrices and A is the matrix

$$A \equiv \begin{pmatrix} a_{11} & a_{12} & a_{13} & \cdots & a_{1n} \\ a_{21} & a_{22} & a_{23} & \cdots & a_{2n} \\ a_{31} & a_{32} & a_{33} & \cdots & a_{3n} \\ \cdots\cdots\cdots\cdots\cdots\cdots\cdots \\ a_{n1} & a_{n2} & a_{n3} & \cdots & a_{nn} \end{pmatrix} \tag{10.47}$$

Let it be assumed in this discussion that diagonal elements of A are all different from zero.

If one postmultiplies Eq. (10.46) on both sides by the matrix

$$M_1 \equiv \begin{pmatrix} 1 & -\dfrac{a_{12}}{a_{11}} & -\dfrac{a_{13}}{a_{11}} & \cdots & -\dfrac{a_{1n}}{a_{11}} \\ 0 & 1 & 0 & \cdots & 0 \\ 0 & 0 & 1 & \cdots & 0 \\ \cdots\cdots\cdots\cdots\cdots\cdots\cdots \\ 0 & 0 & 0 & \cdots & 1 \end{pmatrix} \tag{10.48}$$

one obtains the matrix equation

$$X A_1 = B_1 \tag{10.49}$$

where $A_1 = A M_1$, $B_1 = B M_1$, and in particular

$$A_1 = \begin{pmatrix} a'_{11} & 0 & 0 & \cdots & 0 \\ a'_{21} & a'^{i}_{22} & a'_{23} & \cdots & a'_{2n} \\ a'^{i}_{31} & a'_{32} & a'_{33} & \cdots & a'_{3n} \\ \cdots\cdots\cdots\cdots\cdots\cdots\cdots \\ a'_{n1} & a'_{n2} & a'_{n3} & \cdots & a'_{nn} \end{pmatrix} \tag{10.50}$$

The a'_{ij} are functions of the elements of A. If next Eq. (10.49) is postmultiplied by

$$M_2 \equiv \begin{pmatrix} 1 & 0 & 0 & \cdots & 0 \\ 0 & 1 & -\dfrac{a'_{23}}{a'_{22}} & \cdots & -\dfrac{a'_{2n}}{a'_{22}} \\ 0 & 0 & 1 & \cdots & 0 \\ \cdots\cdots\cdots\cdots\cdots\cdots\cdots \\ 0 & 0 & 0 & \cdots & 1 \end{pmatrix} \tag{10.51}$$

the matrix becomes $X A_2 = B_2$, where now A_2 has zeros in the first and second row to the right of the diagonal.

By continuing in this way the coefficients of the unknowns become a triangular matrix. Let

$$A_k = A M_1 M_2 \cdots M_k$$
$$B_k = B M_1 M_2 \cdots M_k \tag{10.52}$$

where $k = 1, 2, \ldots, n - 1$, and where M_k is the unit matrix except for elements to the right of the diagonal in the kth row. For these elements one substitutes the corresponding elements of A_{k-1} divided by the negative of the diagonal term [see Eqs. (10.48) and (10.51)].

By this procedure A_{n-1} in Eq. (10.52) turns out to be triangular. One may thus write for the transformed equation

$$XA' = B' \tag{10.53}$$

where $A' = A_{n-1}$ and $B' = B_{n-1}$.

The linear equations given by the matrix equation (10.53) are readily solved for the unknowns $x_n, x_{n-1}, \ldots, x_1$, in that order, by successive substitution in these equations. Alternatively one can construct a set of matrices $W_1, W_2, \ldots, W_{n-1}$ similar to the M matrices above, which, when used as postmultipliers of Eq. (10.53), convert that equation into the equation†

$$XA'' = B'' \tag{10.54}$$

where
$$A'' \equiv A'W_1 \cdots W_{n-1} = \text{diagonal matrix}$$
$$B'' \equiv B'W_1 \cdots W_{n-1} \tag{10.55}$$

Since A'' is diagonal, each of the linear equations of Eq. (10.54) contains only one unknown, and hence each of these is obtainable by a single division.

This method of applying the elimination process involves $2(n - 1)$ multiplications of n-by-n matrices to obtain A'' and $2(n - 1)$ multiplications of a row matrix of n elements by an n-by-n matrix. Since each of the first requires n^3 multiplications and the latter requires n^2, the total for large n is

$$2n^2(n^2 - 1) \cong 2n^4 \text{ multiplications} \tag{10.56}$$

Although a large percentage of these are multiplications by 0 or by 1, a large-scale calculator must still spend a certain amount of nonproductive time on them. Hence, the method in this form is useful only in hand computing, where one can speedily pass over such multiplications.

One may reduce the number of multiplications somewhat by working with the augmented matrix $\mathfrak{A}$ formed by adding a top row to A [see Eq. (10.47)], the elements of this row being the given constants $b_1, b_2, \ldots, b_n$, and by eliminating the superfluous columns in the postmultipliers.

An example of this technique for the special case of four equations in four unknowns ($n = 4$) is outlined below. Here the restriction that all the diagonal elements are different from zero is removed. The bold-

---

† See R. A. Frazer, W. J. Duncan, and A. R. Collar, "Elementary Matrices and Some Applications to Dynamics and Differential Equations," p. 102, Cambridge University Press, London, 1952.

faced elements shown here are assumed, however, to be chosen from the nonzero elements of the matrix.

$$(\mathfrak{a})\quad
\begin{array}{c}
\begin{array}{cccc} b_1 & b_2 & b_3 & b_4 \end{array}\\
\begin{array}{cccc}
a_{11} & a_{12} & a_{13} & a_{14}\\
a_{21} & a_{22} & a_{23} & a_{24}\\
a_{31} & a_{32} & a_{33} & a_{34}\\
a_{41} & a_{42} & a_{43} & a_{44}
\end{array}
\end{array}
\left|\;
\begin{array}{ccc}
-\dfrac{a_{22}}{a_{21}} & -\dfrac{a_{23}}{a_{21}} & -\dfrac{a_{24}}{a_{21}}\\[4pt]
1 & 0 & 0\\
0 & 1 & 0\\
0 & 0 & 1
\end{array}
\right.\quad(\mathfrak{M}_1)$$

$$(\mathfrak{a}')\quad
\begin{array}{c}
\begin{array}{ccc} b_2' & b_3' & b_4' \end{array}\\
\begin{array}{ccc}
a_{12}' & a_{13}' & a_{14}'\\
a_{32}' & a_{33}' & a_{34}'\\
a_{42}' & a_{43}' & a_{44}'
\end{array}
\end{array}
\left|\;
\begin{array}{cc}
1 & 0\\
-\dfrac{a_{32}'}{a_{33}'} & -\dfrac{a_{34}'}{a_{33}'}\\[4pt]
0 & 1
\end{array}
\right.\quad(\mathfrak{M}_2)\quad(10.57)$$

$$(\mathfrak{a}'')\quad
\begin{array}{c}
\begin{array}{cc} b_2'' & b_4'' \end{array}\\
\begin{array}{cc}
a_{12}'' & a_{14}''\\
a_{42}'' & a_{44}''
\end{array}
\end{array}
\left|\;
\begin{array}{c}
1\\
-\dfrac{a_{12}''}{a_{14}''}
\end{array}
\right.\quad(\mathfrak{M}_3)$$

$$(\mathfrak{a}''')\quad
\begin{array}{c}
b_2'''\\
a_{42}'''
\end{array}$$

The four given equations are expressed by the augmented matrix $\mathfrak{a}$. When $\mathfrak{a}$ is postmultiplied by the matrix $\mathfrak{M}_1$, we obtain a matrix which, when a row of zeros is crossed out, is the matrix $\mathfrak{a}'$. Thus, one now has three equations in three unknowns. Likewise $\mathfrak{a}'$ multiplied by $\mathfrak{M}_2$ leads to $\mathfrak{a}''$; and this, in turn, when multiplied by $\mathfrak{M}_3$ becomes $\mathfrak{a}'''$. Now $\mathfrak{a}'''$ represents the single equation

$$x_4 a_{42}''' = b_2'''$$

and thus it yields the value of x_4. Substituting this value in the two equations represented by $\mathfrak{a}''$, we find x_1. Likewise, we find from $\mathfrak{a}$ and $\mathfrak{a}'$, respectively, the values of x_3 and x_2.

10.8. Evaluation of Determinants by Postmultiplication. Since the determinants of the postmultipliers used in Sec. 10.7 are all unity, one can employ the technique for finding A' from the given matrix A to evaluate the determinant of A. Consider the first matrix equation of (10.52). Since the determinant of a product of two square matrices is the product of their determinants and $A' = A_{n-1}$,

$$\det A' = \det A_{n-1}$$
$$= (\det A)(\det M_1)(\det M_2) \cdots (\det M_{n-1}) \quad (10.58)$$

From the form of the M_k [see Eqs. (10.48) and (10.51)] it is clear that

$$\det M_1 = \det M_2 = \cdots = \det M_{n-1} = 1 \qquad (10.59)$$

Moreover, since A' is a triangular matrix, it is easily seen that its determinant is just the product of its diagonal terms. Therefore, from Eqs. (10.58) and (10.59)

$$\det A = \det A' = \prod_{i=1}^{n} (A')_{ii} \qquad (10.60)$$

where $(A')_{ii}$ are the diagonal terms of A'.

10.9. Solution of Linear Equations by the Inversion of the Matrix of the Coefficients. If a set of linear equations expressed by the matrix equation

$$XA = B \qquad (10.61)$$

is to be solved for several choices of B using the same coefficients of x as contained in the matrix A, then it is usually expedient to find first the inverse of A, designated A^{-1}, which has the property

$$AA^{-1} = A^{-1}A = I \qquad (10.62)$$

Here I is the *unit matrix*. Its diagonal elements are all ones and the rest of its elements are all zeros. It plays the part of unity in matrix multiplication, since for any arbitrary matrix M

$$MI = IM = M$$

Having obtained A^{-1}, we can postmultiply both sides of Eq. (10.61) by it and obtain an explicit expression of X in terms of B,

$$X = BA^{-1} \qquad (10.63)$$

Since B is a row matrix, we need only n^2 additional multiplications to find X for each new choice B.

We shall find that A can be inverted by as few as n^3 multiplications. This is to be compared to $n^3/3$ for finding a single X for a single specified B. Therefore, if four or more values of the n constants in B are to be assigned and $n > 12$, it requires fewer multiplications to invert A and use Eq. (10.63) to find the x's than to solve each of four, or more, sets of equations separately. Also, it is obviously necessary to use matrix inversion in those problems in which the elements of B are to be treated as variables.

10.10. Inversion of a Matrix by the Elimination Method. If in the elimination method of Sec. 10.3 the b's are treated as unspecified letters instead of numbers, it will be impossible to evaluate numerically some of the additions indicated in that section. This does not change the processes there described, since such additions can still be indicated by plus signs. The x's, of course, will now be expressed not as numbers but as sums of terms involving the individual b's, i.e.,

$$x_j = \sum_{k=1}^{n} a'_{jk} b_k \qquad j = 1, 2, \ldots, n$$

By comparing these equations with the matrix equation (10.63) it becomes clear that the numbers a'_{jk} are the elements of A^{-1}.

Another way of applying the elimination method to the inversion of a matrix is to write the augmented matrix

$$\begin{bmatrix} a_{11} & a_{12} & a_{13} & \cdots & a_{1n} & 1 & 0 & 0 & \cdots & 0 \\ a_{21} & a_{22} & a_{23} & \cdots & a_{2n} & 0 & 1 & 0 & \cdots & 0 \\ a_{31} & a_{32} & a_{33} & \cdots & a_{3n} & 0 & 0 & 1 & \cdots & 0 \\ \cdots & \cdots & \cdots & & \cdots & \cdots & \cdots & \cdots & & \cdots \\ a_{n1} & a_{n2} & a_{n3} & \cdots & a_{nn} & 0 & 0 & 0 & & 1 \end{bmatrix} \qquad (10.64)$$

associated with the given equations

$$\sum_{k=1}^{n} a_{ik} x_k = b_i \qquad i = 1, 2, \ldots, n \qquad (10.65)$$

The first n columns of this matrix represent, respectively, the coefficients of $x_1, x_2, \ldots, x_n$; the last n columns represent the coefficients of $b_1, b_2, \ldots, b_n$, respectively. Each row of the matrix thus represents one of the equations of (10.65).

If the first row of the matrix is divided by a_{11} and the result is multiplied in turn by $a_{21}, a_{31}, \ldots, a_{n1}$ and subtracted, respectively, from the second, third, $\ldots$, and nth rows, the result is an auxiliary matrix having zeros in place of the elements $a_{21}, a_{31}, \ldots, a_{n1}$. This corresponds exactly to the first step described in Sec. 10.3, namely, that of eliminating x_1 from all but the first equation. In fact, by following the steps of Sec. 10.3, we arrive at the set of equations represented by the augmented matrix

$$\begin{bmatrix} 1 & a'_{12} & a'_{13} & \cdots & a'_{1n} & c'_{11} & 0 & 0 & \cdots & 0 \\ 0 & 1 & a'_{23} & \cdots & a'_{2n} & c'_{21} & c'_{22} & 0 & \cdots & 0 \\ 0 & 0 & 1 & \cdots & a'_{3n} & c'_{31} & c'_{32} & c'_{33} & \cdots & 0 \\ \cdots & \cdots & \cdots & & \cdots & \cdots & \cdots & \cdots & & \cdots \\ 0 & 0 & 0 & \cdots & 1 & c'_{n1} & c'_{n2} & c'_{n3} & \cdots & c'_{nn} \end{bmatrix} \qquad (10.66)$$

By multiplying the last row of the above matrix by $a'_{1n}, a'_{2n}, \ldots,$ $a'_{n-1,n}$ and subtracting it from the first, second, $\ldots$, $(n-1)$th rows, respectively, one obtains zeros for all elements in the nth column which lie above the last element. Proceeding then to the $(n-1)$th column and going through a similar process, we obtain zeros everywhere in that column except for the unit element lying in the $(n-1)$th row. It is clear that by carrying through the procedure for each of the columns from the nth to the second, we obtain the augmented matrix

$$\begin{bmatrix} 1 & 0 & 0 & \cdots & 0 & c_{11} & c_{12} & c_{13} & \cdots & c_{1n} \\ 0 & 1 & 0 & \cdots & 0 & c_{21} & c_{22} & c_{23} & \cdots & c_{2n} \\ 0 & 0 & 1 & \cdots & 0 & c_{31} & c_{32} & c_{33} & \cdots & c_{3n} \\ \multicolumn{10}{c}{\cdots\cdots\cdots\cdots\cdots\cdots\cdots} \\ 0 & 0 & 0 & \cdots & 1 & c_{n1} & c_{n2} & c_{n3} & \cdots & c_{nn} \end{bmatrix} \qquad (10.67)$$

The equations corresponding to this matrix are

$$x_i = \sum_{k=1}^{n} c_{ik}b_k \qquad i = 1, 2, \ldots, n$$

therefore the square matrix formed from the c's alone is the inverse of A.

10.11. Inversion of a Matrix by the Crout Method. Since the Crout method is essentially the elimination method, the procedure is basically that of Sec. 10.10, where one goes from the given augmented matrix (10.64) to that of (10.66) and finally to that of (10.67). The calculations, however, are adapted to a modern desk-type calculator where a sum of products with or without a final division can be evaluated as a single, continuous machine operation. The rules given in Sec. 10.6 for finding an auxiliary matrix from the given matrix can be applied directly to obtain the *auxiliary matrix*

$$\begin{bmatrix} a'_{11} & a'_{12} & a'_{13} & \cdots & a'_{1n} & c'_{11} & 0 & 0 & \cdots & 0 \\ a'_{21} & a'_{22} & a'_{23} & \cdots & a'_{2n} & c'_{21} & c'_{22} & 0 & \cdots & 0 \\ a'_{31} & a'_{32} & a'_{33} & \cdots & a'_{3n} & c'_{31} & c'_{32} & c'_{33} & \cdots & 0 \\ \multicolumn{10}{c}{\cdots\cdots\cdots\cdots\cdots\cdots\cdots} \\ a'_{n1} & a'_{n2} & a'_{n3} & \cdots & a'_{nn} & c'_{n1} & c'_{n2} & c'_{n3} & \cdots & c'_{nn} \end{bmatrix} \qquad (10.68)$$

As before, the only meaningful elements of this matrix lie to the right of the diagonal elements $a'_{11}, a'_{22}, \ldots, a'_{nn}$, and these are identical to those of matrix (10.66).

The *final matrix* is the square matrix comprising the last n columns of (10.67) and is, therefore, the desired inverse of A. The method of determining the element c_{ij} of this matrix from the *auxiliary matrix* (10.68) is indicated by the following equations:

$$c_{nj} = c'_{nj}$$
$$c_{n-1,j} = c'_{n-1,j} - a'_{n-1,n}c_{nj}$$
$$c_{n-2,j} = c'_{n-2,j} - a'_{n-2,n}c_{nj} - a'_{n-2,n-1}c_{n-1,j}$$
$$\cdots \cdots \cdots \cdots \cdots \cdots \cdots \cdots \cdots \cdots \qquad (10.69)$$

$$c_{ij} = c'_{ij} - \sum_{k=i+1}^{n} a'_{ik}c_{kj}$$
$$\cdots \cdots \cdots \cdots \cdots \cdots$$

$$c_{1j} = c'_{1j} - \sum_{k=2}^{n} a'_{1k}c_{kj}$$

10.12. The Gram-Schmidt Orthogonalizing Process. The general matrix equation

$$AB = C \qquad (10.70)$$

written in terms of components becomes

$$\sum_{k=1}^{n} a_{ik}b_{kj} = c_{ij} \qquad (10.71)$$

where i and j may each take on any value from 1 to n inclusive. If the elements of the ith row of A are thought of as components of a vector a_i, and likewise if the elements of the jth column of B are taken to be the components of a vector b_j, then it is clear that Eq. (10.71) becomes

$$a_i \cdot b_j = c_{ij} \qquad (10.72)$$

where the dot indicates the *inner product* of the two vectors.† Therefore, Eq. (10.70) can be written

$$\begin{bmatrix} a_1 \\ a_2 \\ \cdot \\ a_n \end{bmatrix} (b_1, b_2, \cdots, b_n) = \begin{bmatrix} a_1 \cdot b_1 & a_1 \cdot b_2 & \cdots & a_1 \cdot b_n \\ a_2 \cdot b_1 & a_2 \cdot b_2 & \cdots & a_2 \cdot b_n \\ \cdots \cdots \cdots \cdots \cdots \cdots \cdots \\ a_n \cdot b_1 & a_n \cdot b_2 & \cdots & a_n \cdot b_n \end{bmatrix} \qquad (10.73)$$

If a matrix V satisfies the condition that its reciprocal V^{-1} is equal to its transpose V' (formed from V by interchanging rows and columns), the matrix is termed an *orthogonal matrix*. Therefore, for an orthogonal matrix

$$VV' = I \qquad (10.74)$$

where I is the unit matrix (see Sec. 10.9). Writing this equation in the vector notation above,

† The *inner*, or *dot*, *product* of two vectors $\mathbf{u}$ and $\mathbf{v}$ with components u_i and v_i, respectively, is given by the equation $\mathbf{u} \cdot \mathbf{v} = \sum_i u_i v_i$.

$$\begin{bmatrix} \mathbf{v}_1 \\ \mathbf{v}_2 \\ \mathbf{v}_3 \\ \cdot \\ \mathbf{v}_n \end{bmatrix} (\mathbf{v}_1, \mathbf{v}_2, \ldots, \mathbf{v}_n) = \begin{bmatrix} 1 & 0 & 0 & \cdots & 0 \\ 0 & 1 & 0 & \cdots & 0 \\ 0 & 0 & 1 & \cdots & 0 \\ \multicolumn{5}{c}{\cdots\cdots\cdots\cdots\cdots} \\ 0 & 0 & 0 & \cdots & 1 \end{bmatrix} \qquad (10.75)$$

Comparing this with Eq. (10.73), we see that the necessary and sufficient condition for a matrix V to be orthogonal is that the vectors formed from the individual rows form a set of orthogonal unit vectors, i.e.,

$$\mathbf{v}_i \cdot \mathbf{v}_j = \delta_{ij} = \begin{cases} 0 & \text{if } i \neq j \\ 1 & \text{if } i = j \end{cases} \qquad (10.76)$$

Given any square matrix A, which, as seen above, can be written in the form

$$\begin{bmatrix} \mathbf{a}_1 \\ \mathbf{a}_2 \\ \cdot \\ \mathbf{a}_n \end{bmatrix}$$

one may find its inverse A^{-1} by the following procedure. Construct from the vectors $\mathbf{a}_i$ a set of orthogonal vectors $\mathbf{w}_i$ defined by the equations

$$\mathbf{w}_1 = \mathbf{a}_1$$
$$\mathbf{w}_2 = \mathbf{a}_2 - \frac{\mathbf{a}_2 \cdot \mathbf{w}_1}{w_1^2} \mathbf{w}_1$$
$$\mathbf{w}_3 = \mathbf{a}_3 - \frac{\mathbf{a}_3 \cdot \mathbf{w}_1}{w_1^2} \mathbf{w}_1 - \frac{\mathbf{a}_3 \cdot \mathbf{w}_2}{w_2^2} \mathbf{w}_2$$
$$\cdots\cdots\cdots\cdots\cdots\cdots\cdots\cdots$$
$$\mathbf{w}_i = \mathbf{a}_i - \sum_{k=1}^{i-1} \frac{\mathbf{a}_i \cdot \mathbf{w}_k}{w_k^2} \mathbf{w}_k \qquad (10.77)$$
$$\cdots\cdots\cdots\cdots\cdots\cdots\cdots$$
$$\mathbf{w}_n = \mathbf{a}_n - \sum_{k=1}^{n-1} \frac{\mathbf{a}_n \cdot \mathbf{w}_k}{w_k^2} \mathbf{w}_k$$

where $w_k^2 = \mathbf{w}_k \cdot \mathbf{w}_k$. These equations can be written in the form

$$\begin{bmatrix} 1 & 0 & 0 & \cdots & 0 \\ b_{21} & 1 & 0 & \cdots & 0 \\ b_{31} & b_{32} & 1 & \cdots & 0 \\ \multicolumn{5}{c}{\cdots\cdots\cdots\cdots\cdots} \\ b_{n1} & b_{n2} & b_{n3} & \cdots & 1 \end{bmatrix} \begin{bmatrix} \mathbf{w}_1 \\ \mathbf{w}_2 \\ \mathbf{w}_3 \\ \cdot \\ \mathbf{w}_n \end{bmatrix} = \begin{bmatrix} \mathbf{a}_1 \\ \mathbf{a}_2 \\ \mathbf{a}_3 \\ \cdot \\ \mathbf{a}_n \end{bmatrix} \qquad (10.78)$$

where
$$b_{ik} = \frac{\mathbf{a}_i \cdot \mathbf{w}_k}{w_k^2} \qquad (10.79)$$

If the triangular matrix of (10.78) is designated by B and the matrix formed by the components of the set of vectors $\mathbf{w}_k$ is designated by W, Eq. (10.78) may be written

$$A = BW$$

The inverse of A is therefore seen to be

$$A^{-1} = W^{-1}B^{-1} \tag{10.80}$$

Now, since the vectors $\mathbf{w}_k$ are orthogonal,

$$\begin{pmatrix} \mathbf{w}_1 \\ \mathbf{w}_2 \\ \cdot \\ \mathbf{w}_n \end{pmatrix} \left(\frac{\mathbf{w}_1}{w_1^2}, \frac{\mathbf{w}_2}{w_2^2}, \cdots, \frac{\mathbf{w}_n}{w_n^2} \right) = \begin{pmatrix} 1 & 0 & \cdots & 0 \\ 0 & 1 & \cdots & 0 \\ \cdots\cdots\cdots\cdots \\ 0 & 0 & \cdots & 1 \end{pmatrix}$$

hence

$$W^{-1} = \left(\frac{\mathbf{w}_1}{w_1^2}, \frac{\mathbf{w}_2}{w_2^2}, \cdots, \frac{\mathbf{w}_n}{w_n^2} \right) = W'D \tag{10.81}$$

where W' is the transpose of W and

$$D = \begin{pmatrix} \dfrac{1}{w_1^2} & 0 & \cdots & 0 \\ 0 & \dfrac{1}{w_2^2} & \cdots & 0 \\ \cdots\cdots\cdots\cdots \\ 0 & 0 & \cdots & \dfrac{1}{w_n^2} \end{pmatrix} \tag{10.82}$$

Since B is a triangular matrix, its inverse can readily be found by one of the methods described in previous sections; this involves only about $n^3/6$ multiplications. Having obtained W', D, and B^{-1}, one finds A^{-1} from the equation [see Eqs. (10.80) and (10.81)]

$$A^{-1} = W'DB^{-1} \tag{10.83}$$

A careful analysis of this method reveals that the inversion of A^{-1} can be done with about $5n^3/3$ multiplications. This is somewhat larger than the n^3 needed for the elimination method.

10.13. Method of Iteration Applied to Simultaneous Linear Equations.† Consider the set of linear equations represented by the matrix equation

$$AX = B \tag{10.84}$$

Let

$$A = E + H \tag{10.85}$$

where E is a nonsingular matrix that is readily inverted. Now

† Frazer, Duncan, and Collar, *op. cit.*, p. 132:

$$X = A^{-1}B = (E + H)^{-1}B = [E(1 + E^{-1}H)]^{-1}B$$
$$= (1 + E^{-1}H)^{-1}E^{-1}B$$

therefore
$$X = (1 - F)^{-1}B' = (1 + F + F^2 + F^3 + \cdots)B' \qquad (10.86)$$
where
$$F = -E^{-1}H$$
$$B' = E^{-1}B \qquad (10.87)$$

provided the series converges.[†] It can be shown that the series converges provided the moduli of the latent roots of F are all less than unity.

The *latent roots*, or *characteristic roots*, of F are the roots of the polynomial equation in λ given by

$$\det |F - \lambda I| = 0 \qquad (10.88)$$

I being the unit matrix. This determinant, when written out, has the form

$$\begin{vmatrix} F_{11} - \lambda & F_{12} & F_{13} & \cdots & F_{1n} \\ F_{21} & F_{22} - \lambda & F_{23} & \cdots & F_{2n} \\ F_{31} & F_{32} & F_{33} - \lambda & \cdots & F_{3n} \\ \cdots\cdots\cdots\cdots\cdots\cdots\cdots\cdots\cdots\cdots\cdots\cdots \\ F_{n1} & F_{n2} & F_{n3} & \cdots & F_{nn} - \lambda \end{vmatrix} = 0 \qquad (10.89)$$

By multiplying Eq. (10.88) by the determinant of $-E$ (this does not affect the roots) we obtain an equivalent equation for the latent roots, namely,

$$\det |H + \lambda E| = 0 \qquad (10.90)$$

If the moduli of the roots of (10.88) or (10.90) are all less than unity, the series converges, and we can define as successive approximations to X, the required matrix, the matrices

$$X^{(0)} = B'$$
$$X^{(1)} = (I + F)B' = B' + FX^{(0)}$$
$$X^{(2)} = (I + F + F^2)B' = B' + FX^{(1)}$$
$$\cdots\cdots\cdots\cdots\cdots\cdots\cdots\cdots\cdots$$
$$X^{(i+1)} = (I + F + F^2 + \cdots + F^{i+1})B' = B' + FX^{(i)} \qquad (10.91)$$
$$\cdots\cdots\cdots\cdots\cdots\cdots\cdots\cdots\cdots\cdots\cdots$$

Thus, we have the matrix-iteration equations

$$X^{(i+1)} = B' + FX^{(i)} \qquad (10.92)$$

It is clear that as $i \to \infty$, $X^{(i)}$ contains more and more terms of the infinite series for X given in Eq. (10.86), and hence $X^{(i)} \to X$, provided, of course, the series is convergent.

If an approximate solution is already known, this can be used as a 0th approximation in place of B'.

[†] Convergence of an infinite series of matrices requires that the series for each individual element converge.

In the *tabular method of Morris*, E is chosen as a triangular matrix formed by inserting zeros for all the elements of A lying above the diagonal; all the elements on and below the diagonal are left unchanged. This means that H has zeros on and below the diagonal and the same elements as A above the diagonal. Since E is triangular, it is easily inverted. It can be shown for this case that if A is symmetrical and *positive semidefinite*, the series (10.86) converges, and hence the iteration of Eq. (10.92) converges to X, the required solution of Eq. (10.84). A is *positive semidefinite* if $X'AX \geq 0$ for any column matrix X.†

At the expense of $n^3 + n^2$ multiplications any matrix equation such as (10.84) can be converted into a matrix equation

$$\bar{A}X = \bar{B} \tag{10.93}$$

where $\bar{A}$ is symmetrical and positive semidefinite. This is done by multiplying Eq. (10.84) on both sides by A', the transpose of A; thus

$$\bar{A} = A'A \tag{10.94}$$

$\bar{A}$ is clearly symmetrical, since it is equal to its transpose. It is also positive semidefinite, since

$$X'\bar{A}X = X'A'AX = Y'Y \geq 0 \tag{10.95}$$

where $Y = AX$ is a column matrix. That $Y'Y$ is positive or zero becomes clear on writing out this product, since

$$Y'Y = (y_1, y_2, \ldots, y_n) \begin{Bmatrix} y_1 \\ y_2 \\ . \\ y_n \end{Bmatrix} = y_1^2 + y_2^2 + \cdots + y_n^2 \geq 0 \tag{10.96}$$

An alternative method to that of Morris is to choose E to be a diagonal matrix, or sometimes just the unit matrix I. The natural choice of a diagonal matrix is that having the diagonal elements of A.

Example 1. Solve the set of equations

$$\begin{align*} 25X_1 + 2X_2 + X_3 &= 69 \\ 2X_1 + 10X_2 + X_3 &= 63 \\ X_1 + X_2 + 4X_3 &= 43 \end{align*} \tag{10.97}$$

These can alternatively be written in the matrix form

$$\begin{bmatrix} 25 & 2 & 1 \\ 2 & 10 & 1 \\ 1 & 1 & 4 \end{bmatrix} \begin{bmatrix} X_1 \\ X_2 \\ X_3 \end{bmatrix} = \begin{bmatrix} 69 \\ 63 \\ 43 \end{bmatrix} \tag{10.98}$$

† Since AX is a single column matrix and X', the transpose of X, is a row matrix, the product $X'AX$ is a matrix having only one element and thus represents a single number. It is here assumed that the elements of all matrices are real numbers.

A 0th approximation to the solution of these equations can be obtained by neglecting all but the large diagonal terms. This amounts to solving the equations

$$
\begin{pmatrix} 25 & 0 & 0 \\ 0 & 10 & 0 \\ 0 & 0 & 4 \end{pmatrix} \begin{pmatrix} X_1^{(0)} \\ X_2^{(0)} \\ X_3^{(0)} \end{pmatrix} = \begin{pmatrix} 69 \\ 63 \\ 43 \end{pmatrix} \tag{10.99}
$$

This equation can be written

$$
EX^{(0)} = B \tag{10.100}
$$

where E is the diagonal matrix of (10.99) and the superscript (0) indicates the 0th approximation to the required column matrix X. The solution of Eqs. (10.99) and (10.100) is

$$
X^{(0)} = E^{-1}B \equiv B' = \begin{pmatrix} \frac{69}{25} \\ \frac{63}{10} \\ \frac{43}{4} \end{pmatrix} = \begin{pmatrix} 2.76 \\ 6.3 \\ 10.75 \end{pmatrix} \tag{10.101}
$$

A first approximation for the X's can be obtained by substituting in the terms omitted in the 0th approximation the approximations for the X's obtained above, namely, $X_1^{(0)} = 2.76$, $X_2^{(0)} = 6.3$, and $X_3^{(0)} = 10.75$. Equations (10.97) then become

$$
25X_1^{(1)} + 2X_2^{(0)} + X_3^{(0)} = 69
$$
$$
2X_1^{(0)} + 10X_2^{(1)} + X_3^{(0)} = 63
$$
$$
X_1^{(0)} + X_2^{(0)} + 4X_3^{(1)} = 43
$$

which can quickly be solved for the first approximations $X_1^{(1)}$, $X_2^{(1)}$, and $X_3^{(1)}$. The same procedure can be used to obtain the second, third, and higher approximations. In general,

$$
25X_1^{(i+1)} = 69 \qquad\quad - 2X_2^{(i)} - X_3^{(i)}
$$
$$
10X_2^{(i+1)} = 63 - 2X_1^{(i)} \qquad\quad - X_3^{(i)} \tag{10.102}
$$
$$
4X_3^{(i+1)} = 43 - X_1^{(i)} - X_2^{(i)}
$$

These equations serve as iteration equations to find better and better approximations to the required solutions X_1, X_2, X_3.

It is clear that Eqs. (10.102) can be written in term of matrices as

$$
EX^{(i+1)} = B - HX^{(i)} \tag{10.103}
$$

where E is the diagonal matrix given in Eq. (10.99) and

$$
H = \begin{pmatrix} 0 & 2 & 1 \\ 2 & 0 & 1 \\ 1 & 1 & 0 \end{pmatrix} \tag{10.104}
$$

By multiplying Eq. (10.103) by E^{-1} we obtain the equation

$$
X^{(i+1)} = B' + FX^{(i)} \tag{10.105}
$$

where B' and F are defined as before. This equation is, of course, identical with Eq. (10.92). Thus we see that the symbolic method used in obtaining Eq. (10.92) is equivalent to solving a simpler set of equations to obtain a 0th approximation and then making use of these values in the original set of equations to find a better approximation. The latter process is repeated as often as necessary.

For our particular example, B' is given in Eq. (10.101), and

$$F = -E^{-1}H = \begin{pmatrix} 0 & -0.08 & -0.04 \\ -0.2 & 0 & -0.1 \\ -0.25 & -0.25 & 0 \end{pmatrix} \qquad (10.106)$$

The iteration equations given by Eq. (10.105) are therefore

$$\begin{pmatrix} X_1^{(i+1)} \\ X_2^{(i+1)} \\ X_3^{(i+1)} \end{pmatrix} = \begin{pmatrix} 2.76 \\ 6.3 \\ 10.75 \end{pmatrix} + \begin{pmatrix} 0 & -0.08 & -0.04 \\ -0.2 & 0 & -0.1 \\ -0.25 & -0.25 & 0 \end{pmatrix} \begin{pmatrix} X_1^{(i)} \\ X_2^{(i)} \\ X_3^{(i)} \end{pmatrix}$$

The numerical values of the various iterates are given in the table below. The correct answers are exactly 2, 5, and 9.

TABLE 10.1. SUCCESSIVE APPROXIMATIONS TO THE SOLUTION OF EQS. (10.97)

	Number of the approximation						
	0th	1st	2d	3d	4th	5th	6th
X_1	2.76	1.826	2.0468	1.9881	2.0031	1.9992	2.0002
X_2	6.3	4.673	5.0863	4.9781	5.0057	4.9985	5.0004
X_3	10.75	8.485	9.1252	8.9667	9.0084	8.9978	9.0006

A more rapid convergence can be obtained as follows. In Eqs. (10.97) find $X_1^{(0)}$ as previously by neglecting in the first equation the terms in X_2 and X_3. However, instead of neglecting the terms in X_1 and X_3 in the second equation, neglect only the term in X_3, and substitute for X_1 the approximate value $X_1^{(0)}$. The second equation then yields $X_2^{(0)}$, an approximate value of X_2. Finally substitute $X_1^{(0)}$ and $X_2^{(0)}$ in the third equation, and determine $X_3^{(0)}$.

It is clear that the last paragraph is merely a description of the way one solves the equations

$$\begin{pmatrix} 25 & 0 & 0 \\ 2 & 10 & 0 \\ 1 & 1 & 4 \end{pmatrix} \begin{pmatrix} X_1^{(0)} \\ X_2^{(0)} \\ X_3^{(0)} \end{pmatrix} = \begin{pmatrix} 69 \\ 63 \\ 43 \end{pmatrix} \qquad (10.107)$$

Since these equations may be expressed by the matrix equation

$$EX^{(0)} = B$$

the solution may be written in the form

$$X^{(0)} = E^{-1}B = B' = \begin{pmatrix} 2.760 \\ 5.748 \\ 8.623 \end{pmatrix}$$

Moreover, the inverse of E is found to be

$$E^{-1} = \begin{pmatrix} 0.04 & 0 & 0 \\ -0.008 & 0.1 & 0 \\ -0.008 & -0.025 & 0.25 \end{pmatrix}$$

One can now substitute this 0th approximation in the terms originally neglected and find a first approximation in the same way, and then use this for a second approximation, etc. In this way one obtains a set of iteration equations that may be represented by Eqs. (10.92). For this choice of E, which differs from that described under the tabular method of Morris only in the interchange of rows and columns, the matrix F is given by

$$F = -E^{-1}H = \begin{pmatrix} 0 & -0.08 & -0.04 \\ 0 & 0.016 & -0.092 \\ 0 & 0.016 & 0.033 \end{pmatrix}$$

The successive approximations to X_1, X_2, X_3 are given in Table 10.2.

TABLE 10.2. SUCCESSIVE APPROXIMATIONS TO THE SOLUTION OF EQS. (10.97)

	0th	1st	2d	3d	4th	5th
X_1	2.760	......	......	......	......	2.0000
X_2	5.748	5.0467	5.0008	4.9999	5.0000	5.0000
X_3	8.623	8.9995	9.0007	9.0000	9.0000	9.0000

For our example the Morris method is much more rapidly convergent than the first method used. Notice, incidentally, that since F has zeros throughout the first column, it is not necessary to calculate the intermediate approximations to X_1.

10.14. Accuracy of Solutions. In the direct methods, such as the elimination method, for solving a system of linear equations or for inverting a matrix, round-off errors at each stage of the calculations are carried to the next stage. If the number of equations is quite large, these errors grow as the work progresses, and considerable care must be exercised to prevent their nullifying the final answer. Thus a great many extra figures may have to be carried along as guarding figures to keep the round-off error introduced at each step small. From this point of view, the iteration methods possess a certain advantage in that the round-off error of one iteration tends to be corrected in subsequent iterations.†

The number of extra figures theoretically needed, according to the analysis of von Neumann‡ gets to be almost prohibitive for the inversion of a matrix of the order of 100 or so. However, there is some indication that many of the matrices met in practice are better behaved than those admitted in this analysis. Thus Mitchell has successfully inverted a matrix of order 38 arising in economic studies without the loss of any significant figures.§

It would appear to be good practice, where possible, to carry through a few iterations of the type described in Sec. 10.13 using the answers obtained from a direct method. Thus any large errors due to round-off

† Bargmann, Montgomery, and von Neumann, *op. cit.*

‡ See J. von Neumann and H. H. Goldstine, Numerical Inverting of Matrices of High Order, *Bull. Am. Math. Soc.*, **53**: 1021 (1947).

§ H. F. Mitchell, Jr., Inversion of a Matrix of Order 38, *Math. Tables and Other Aids to Computation*, **3**: 161–166 (1948).

would be indicated, and their size would be reduced to within tolerable limits.

The method of conjugate gradients due to Hestenes and Stiefel† is particularly designed to eliminate the effect of round-off errors.

10.15. Nonlinear Sets of Equations. The solution of a set of nonlinear algebraic equations usually involves a great deal more work than that needed for linear systems. Since the latter already entails considerable computation time on even a large-scale automatic computer when n, the number of equations, is large, the solution of nonlinear systems may often be almost prohibitive. However, the method outlined below can be used where a solution is required.

Consider the set of real equations

$$f_i(x_k) = 0 \qquad i = 1, 2, \ldots, n \qquad (10.108)$$

where x_k stands for all the unknowns $x_1, x_2, \ldots, x_n$ entering this set of equations. One may reduce this set of equations to the problem of minimizing the function

$$M(x_k) = \sum_{i=1}^{n} f_i^2(x_k) \qquad (10.109)$$

It is at once evident that for real functions $f_i(x_k)$ the function $M(x_k)$ is greater than, or equal to, zero. Since it is zero if, and only if, Eqs. (10.108) are satisfied, the minimization of (10.109) leads uniquely to the desired solution of the set of equations (10.108).

It is likewise convenient to think of the x_k as components of a vector in an n-dimensional space and of $M(x_k)$ as a scalar function of position in this space. Then the solution sought corresponds to the point Q in the space at which $M(x_k)$ vanishes. At all other points it is positive. Consider the hypersurfaces of constant $M(x_k)$. These will encircle the point Q, at least for small values of $M(x_k)$.

The normal vector to these surfaces is the *gradient* of $M(x_k)$. It has the components

$$\frac{\partial M(x_k)}{\partial x_j} = 2 \sum_{i=1}^{n} f_i(x_k) \frac{\partial f_i(x_k)}{\partial x_j} \qquad (10.110)$$

Now, if one has a sufficiently accurate guess $x_j^{(0)}$ for the solution of Eqs. (10.108), one may represent such a guess as a point P_0 on some surface S_0 of constant $M(x_k)$ surrounding the point Q. Intuitively one

† M. R. Hestenes and E. Stiefel, Method of Conjugate Gradients for Solving Linear Systems, *Natl. Bur. Standards (U.S.) Rept.* 1659, 1952; *Natl. Bur. Standards J. Research* **49**: 409–436 (1952).

may decrease $M(x_k)$ by going inward from P_0 along the normal, i.e., in a direction opposite to the gradient, to some point P_1, whose components are given by

$$x_j^{(1)} = x_j^{(0)} - \lambda \frac{\partial M(x_k)}{\partial x_j} \tag{10.111}$$

The constant λ can then be chosen so as to minimize $M(x_k)$. In geometrical language, the point P_1 represents the point of tangency of the normal with the innermost surface of constant $M(x_k)$ pierced by this line.

It is as if one attempted to locate the core of an onion by driving a needle perpendicular to the outer surface at some point P_0 and removing each concentric shell of the onion until one found that shell just grazed by the needle. The point at which the needle touches this shell is the point P_1 analogous to the point P_1 above.

Assuming that one is able to find the $x_j^{(1)}$ corresponding to the point P_1 by the procedure above, one may reevaluate the gradient at this point and again proceed inward to some point P_2 giving the minimum value of $M(x_k)$. Proceeding in this way, it is clear, one will find a set of points P_i that converges to the minimum point Q of $M(x_k)$. The x_j of this point Q represents the solution to the set of equations (10.109). This method is referred to as the *method of steepest descent*.

It is usually tedious, however, to try to find the gradient at each point. Since it is not necessary for one to go in along the normal in order to improve the initial guess $x_k^{(0)}$, one prefers to take a more or less arbitrary line through P_0 and find the point P_1' on this line corresponding to a minimum value of $M(x_k)$. Having P_1', one moves along another arbitrary line to the next point P_2'. Thus one obtains a set of points P_i' each lying on a smaller and smaller hypersurface about Q. Clearly here, too, the series of points P_i' approaches Q, whose coordinates in turn represent the solution of Eqs. (10.108). A convenient choice of directions is obtained by varying one coordinate at a time in a fixed order. This corresponds to going from each point P_i' along a line parallel to one of the axes, say x_j.

To find the best point P_{i+1}' along this line, one sets $\partial M(x_k)/\partial x_j = 0$. This, by Eq. (10.110), requires that one find a solution of the nonlinear equation

$$\sum_{i=1}^{n} f_i(x_k) \frac{\partial f_i(x_k)}{\partial x_j} = 0 \tag{10.112}$$

treated as a function of x_j. Again it is not necessary to find the optimum point P_{i+1}'. Much less work is needed to locate it only approximately using a single application of the Newton-Raphson method (see Sec. 1.5).

PROBLEMS

1. Solve the set of equations

$$5x - y - 2z = 142$$
$$x - 3y - z = -30$$
$$2x - y - 3z = 5$$

using the method of elimination.

2. Apply the square-root method to the set of equations

$$3x - 2y + z \qquad = 4$$
$$-2x + 5y - z + w = 7$$
$$x - y - 11z - 2w = 2$$
$$y - 2z + 12w = 3$$

3. Use Chiò's method to evaluate the determinant

$$\begin{vmatrix} 15 & 1 & 2 & -3 \\ 5 & 6 & 4 & 4 \\ -10 & -3 & 2 & 1 \\ -5 & 3 & 4 & 0 \end{vmatrix}$$

4. Find the solution of the set of equations

$$a \qquad + 3c + d = 2$$
$$4a - b + c - 2d = 1$$
$$a + 3b \qquad + 5d = -1$$
$$3a - b + 2c + d = 3$$

using the method of postmultiplication.

5. Solve the set of equations in Prob. 3 using the Gram-Schmidt orthogonalizing process.

6. Use Crout's method to find the solution of the following set of equations:

$$3.12x - 2.31y + z \qquad = 4.04$$
$$2.82x + y - z + 1.42w = 11.88$$
$$x + y - z - 2.56w = 2$$
$$x + 3.66y \qquad - 3w = 3.94$$

7. Express x, y, z, and w in the following set of equations

$$x - 3y + z \qquad = a$$
$$2x + y \qquad - w = b$$
$$3x - 2y - z - 2w = c$$
$$4x - y \qquad + 3w = d$$

in terms of a, b, c, and d using the technique of Sec. 10.10.

8. Using the tabular method of Morris, set up an iterative scheme for solving the set of equations

$$6a - 2b + 3c = 5$$
$$a - 4b - 2c = -4$$
$$3a + b + 3c = 5$$

9. Set up some suitable iterative process using matrix notation to solve the set of equations

$$10x - y + 2z = 20$$
$$2x - y + 8z = 16$$
$$x - 20y - 2z = 20$$

and give the first three approximations obtained from it for the unknowns x, y, and z.

10. Obtain a solution of the following set of nonlinear equations:

$$x + y + z = 8.5$$
$$x^2 + y^2 + z^2 = 31.25$$
$$x^2 - z^3 = 21$$

11. Referring ahead to Chap. 13, devise an appropriate relaxation method for solving the following set of equations:†

$$19x - y + z - 2w = 1$$
$$x - 20y - z + w = 2$$
$$2x + y + 21z - w = 3$$
$$-x + 2y - 2z + 22w = 4$$

Carry out the solution to two decimal places.

12. Evaluate the determinant

$$D = \begin{vmatrix} 4 & 3 & -2 & 6 & 0 \\ 8 & -1 & 0 & 1 & 2 \\ 0 & 1 & 7 & 0 & 0 \\ 2 & 0 & 1 & 0 & 1 \\ 1 & 0 & 2 & -2 & 0 \end{vmatrix}$$

using Chiò's method.

† Such a method is illustrated in D. N. de G. Allen, "Relaxation Methods," McGraw-Hill Book Company, Inc., New York, 1954.

CHAPTER 11

INTERPOLATION IN TABLES OF TWO
OR MORE VARIABLES

11.1. Introduction. Frequently it is necessary to tabulate a function of two or more variables. Some of these variables may be in the nature of parameters. Thus we may wish to tabulate $F(x;\lambda)$ for a set of values of λ, say $\lambda_0, \lambda_1, \lambda_2, \ldots, \lambda_n$. If only these values of λ are ever to be used, the tabulation can be carried out as $n + 1$ univariate tables; but if one needs to be able to find $F(x;\lambda)$ for intermediate values of λ, then the tabulation should be looked upon as a bivariate table in the variables x and λ. The interval between the λ's must in the latter case be made small enough so that interpolation can be made with respect to λ, as well as with respect to x.

It is clear that, since the number of entries in a bivariate table is MN, where M and N are the number of points at which the separate variables need to be tabulated, one cannot usually afford the space for a tabulation in which linear interpolation is adequate. In fact, in order to cut down the number of entries, one usually is forced to rather high-order interpolation formulas. This means that the problem of interpolation is greatly increased for bivariate tables over that for univariate tables, merely because of the fact that the intervals between the values are usually larger.

Added to this basic difficulty is the fact that bivariate and multivariate interpolation have received only a very small amount of attention compared to that expended on univariate interpolation. Consequently one finds considerable arbitrariness, complexity of statement and notation, and even confusion in the various treatments of the subject. In the opinion of the writer, the most reliable discussion of the subject is given by Steffensen.†

---

† J. F. Steffensen, "Interpolation," 2d ed., pp. 203–223, Chelsea Publishing Company, New York, 1950. The treatment given is accurate but the notation is somewhat formidable. See also Karl Pearson, "On the Construction of Tables and on Interpolation, II, Bi-variate Interpolation, Tracts for Computers, III," Cambridge University Press, London, 1920.

Two basic methods of interpolation are available:
1. Interpolating separately for each of the variables.
2. Finding a function which can be simply evaluated in terms of elementary operations (addition, subtraction, multiplication, etc.) and which approximates the given function to the desired degree of accuracy over a small range of the variables.

The first method is simple in theory and permits one to use any of the interpolation formulas derived for univariate tables. Although the method requires, if mth-order interpolation is used throughout, $m + 2$ separate interpolations, it is somewhat shorter, in terms of the number of multiplications and divisions needed, than the second method.

The second method is required whenever one wishes an approximation to a function that holds for all values of independent variables over a small range, rather than at a specific point.   It will be shown, moreover, that the two methods are intimately related.   Thus, for example, one can obtain a more accurate error term for the first method by taking the point of view of the second method.   The main disadvantage of the second method is the complexity arising from the large number of approximating functions one might use.   It is hoped that the treatment given in this chapter will help remove some of this complexity by showing how the ideas and formulas of univariate interpolation can in a very natural way be generalized to bivariate and multivariate tables.   This makes possible the removal of most of the *ad hoc* assumptions which haunt_this subject.

Our discussion will be confined almost wholly to bivariate tables; however, the generalization to more than two variables is usually quite clear.   Moreover, since the number of formulas to be derived is quite large, we shall make use of the symbolic method of deriving formulas, as this method is by far the shortest available.

11.2. Bivariate Tables. If a function $F(x,y)$ is tabulated for equal intervals of the variables, we have the bivariate table shown in Table 11.1.   Here x_0 and y_0 designate any pair of values of the independent

TABLE 11.1. BIVARIATE TABLE

$\cdots$ $F(x_{-1},y_{-1})$	$F(x_{-1},y_0)$	$F(x_{-1},y_1)$	$F(x_{-1},y_2)$	$\cdots$ $F(x_{-1},y_n)$	$\cdots$
$\cdots$ $F(x_0,y_{-1})$	$F(x_0,y_0)$	$F(x_0,y_1)$	$F(x_0,y_2)$	$\cdots$ $F(x_0,y_n)$	$\cdots$
$\cdots$ $F(x_1,y_{-1})$	$F(x_1,y_0)$	$F(x_1,y_1)$	$F(x_1,y_2)$	$\cdots$ $F(x_1,y_n)$	$\cdots$
$\cdots$ $F(x_2,y_{-1})$	$F(x_2,y_0)$	$F(x_2,y_1)$	$F(x_2,y_2)$	$\cdots$ $F(x_2,y_n)$	$\cdots$
$\cdots$ $F(x_m,y_{-1})$	$F(x_m,y_0)$	$F(x_m,y_1)$	$F(x_m,y_2)$	$\cdots$ $F(x_m,y_n)$	$\cdots$

variables that one wishes to use as a reference point.   This table can be more conveniently treated by introducing the new variables u and v

defined by the equations,

$$u = \frac{x - x_0}{h}$$

$$v = \frac{y - y_0}{k}$$

(11.1)

where h and k are the intervals between the tabulated values of x and y, respectively. With these definitions,

$$F(x,y) = F(x_0 + uh,\ y_0 + vk) \equiv f(u,v)$$

(11.2)

and the tabulated values of $f(u,v)$ correspond to integral values of the variables u and v. To distinguish between tabulated values and the value of the function at intermediate points, we shall designate the tabulated values by subscripts; thus

$$f_{st} \equiv f(s,t)$$

(11.3)

Table 11.1 can then be replaced by Table 11.2.

TABLE 11.2

$\cdots$	$f_{-1,-1}$	f_{-10}	f_{-11}	f_{-12}	$\cdots$	f_{-1n}	$\cdots$
$\cdots$	$f_{0,-1}$	$\mathbf{f_{00}}$	f_{01}	f_{02}	$\cdots$	f_{0n}	$\cdots$
$\cdots$	$f_{1,-1}$	f_{10}	f_{11}	f_{12}	$\cdots$	f_{1n}	$\cdots$
$\cdots$	$f_{2,-1}$	f_{20}	f_{21}	f_{22}	$\cdots$	f_{2n}	$\cdots$
$\cdots$	$f_{m,-1}$	f_{m0}	f_{m1}	f_{m2}	$\cdots$	f_{mn}	$\cdots$

11.3. Interpolation Based upon Separate Treatment of the Variables. To find $f(u,v)$ in Table 11.2 when u and v are both nonintegers† one needs only to employ univariate formulas for interpolating first with respect to u and then with respect to v, or vice versa. Since f_{00} can be chosen at will, u and v can always be made to lie in the range zero to one if one so desires.

Thus $f(u,t)$ can first be found by means of any of the univariate-interpolation formulas using the tabulated values f_{st} lying in the tth column. Having obtained $f(u,t)$ in this way, for various values of t, one can find $f(u,v)$ by univariate interpolation from these computed values.

Suppose, in particular, that we use the Gregory-Newton formula for forward interpolation in all cases. Then treating t as just a parameter,‡

---

† If either u or v is an integer, we need only a single interpolation. If both are integers, $f(u,v)$ is one of the tabulated values.

‡ The subscript x on Δ_x merely indicates that the difference is with respect to the first variable and

$$u^{[k]} = u(u - 1)(u - 2) \cdots (u - k + 1)$$

$$f(u,t) \cong \left(1 + u\Delta_x + \frac{u^{[2]}}{2!}\Delta_x^2 + \frac{u^{[3]}}{3!}\Delta_x^3 + \cdots + \frac{u^{[m]}}{m!}\Delta_x^m\right)f_{0t} \quad (11.4)$$

[see Eq. (3.43)]. If t is given the values $0, 1, 2, \ldots, n$ in succession (this involves $n + 1$ applications of the formula), we can find $f(u,v)$ by interpolating from the set of values $f(u,0), f(u,1), f(u,2), \ldots, f(u,n)$. This can be done from the formula

$$f(u,v) \cong \left(1 + v\Delta_y + \frac{v^{[2]}}{2!}\Delta_y^2 + \frac{v^{[3]}}{3!}\Delta_y^3 + \cdots + \frac{v^{[n]}}{n!}\Delta_y^n\right)f(u,0) \quad (11.5)$$

since the differences involved are those one can construct from the above function values.

One can express this result symbolically by substituting for $f(u,0)$ in Eq. (11.5), using Eq. (11.4); thus one writes

$$f(u,v) \cong \left(1 + v\Delta_y + \frac{v^{[2]}}{2!}\Delta_y^2 + \cdots + \frac{v^{[n]}}{n!}\Delta_y^n\right)$$
$$\left(1 + u\Delta_x + \frac{u^{[2]}}{2!}\Delta_x^2 + \cdots + \frac{u^{[m]}}{m!}\Delta_x^m\right)f_{00} \quad (11.6)$$

The right-hand side of this equation represents a polynomial $P_{mn}(u,v)$ of degree m (or perhaps less) in u and degree n (or less) in v. In fact,

$$P_{mn}(u,v) = \sum_{s=0}^{m}\sum_{t=0}^{n}\frac{u^{[s]}}{s!}\frac{v^{[t]}}{t!}(\Delta_x^s\Delta_y^t f_{00}) \quad (11.7)$$

where the differences $\Delta_x^s\Delta_y^t f_{00}$ are constants that depend on the tabulated values of $f(u,v)$. Thus interpolating for each variable separately is equivalent to using $P_{mn}(u,v)$ as the approximating function.

The above differences are defined by the equations

$$\Delta_x^s\Delta_y^t f_{pq} = \Delta_x^{s-1}\Delta_y^t f_{p+1,q} - \Delta_x^{s-1}\Delta_y^t f_{pq}$$
$$\Delta_x^s\Delta_y^t f_{pq} = \Delta_x^s\Delta_y^{t-1} f_{p,q+1} - \Delta_x^s\Delta_y^{t-1} f_{pq} \quad (11.8)$$

For example,

$$\Delta_x\Delta_y f_{00} = \Delta_y f_{10} - \Delta_y f_{00} = f_{11} - f_{10} - f_{01} + f_{00}$$

and

$$\Delta_x^2\Delta_y f_{00} = \Delta_x\Delta_y f_{10} - \Delta_x\Delta_y f_{00} = (f_{21} - f_{20} - f_{11} + f_{10})$$
$$- (f_{11} - f_{10} - f_{01} + f_{00}) \quad (11.9)$$

If $u = s'$ is one of the integers $0, 1, 2, \ldots, m$ and $v = t'$ is one of the integers $0, 1, 2, \ldots, n$, this method of interpolation yields the tabulated value $f_{s't'}$. Therefore $P_{mn}(s',t') = f(s',t')$, and hence the plot of the surface of $P_{mn}(u,v)$ passes through the $(m + 1)(n + 1)$ ordinates $f(s',t')$

lying in the rectangle whose corners are at $(0,0)$, $(m,0)$, (m,n), and $(0,n)$. Here (s,t) stands for the point $u = s$, $v = t$. This rectangle will be referred to as the *rectangle of fit*.

Since one usually chooses to have u and v lie in the range zero to one, the interpolated point (u,v) lies in the upper left-hand corner of the above rectangle. Because of this fact one should ordinarily choose a central-difference formula as the interpolation formula, since the rectangle of fit would be nearly symmetrical about the origin, and hence the point (u,v) would lie near its center. Usually this results in greater accuracy. This advantage of the central-difference formulas was already evident in univariate interpolation, but here the advantage of such formulas over the Gregory-Newton type is even more striking.

Perhaps the easiest way to perform the interpolations is by the Lagrange interpolation formula for equal intervals [see Eqs. (5.25) and (5.26)] using tables of Lagrangian coefficients. This procedure dispenses with the difference tables that must be constructed in using the other interpolation formulas. The same points of coincidence between $P_{mn}(u,v)$ and $f(u,v)$ occur for Lagrange's formula as for the first central-difference formula of Gauss.

11.4. Interpolation by Means of an Approximating Function. The second method of interpolating is to approximate $F(x,y) = f(u,v)$ by a function $P(u,v)$ that can be evaluated in terms of the elementary operations (addition, subtraction, multiplication, and division);† i.e., one works directly with functions such as the $P_{mn}(u,v)$ discussed above.

Two basic problems have arisen in this approach to bivariate interpolation:

1. What form should $P(u,v)$ take? That is, in geometrical terms, what type of surface should be used to approximate $f(u,v)$? Should one, for instance, use a sphere, ellipsoid, hyperboloid, paraboloid, or perhaps some other surface, when a second-degree surface is called for?

2. At what values of u and v should $P(u,v)$ agree with the tabulated values, i.e., what should be the *region of fit* between $P(u,v)$ and $f(u,v)$? Or speaking geometrically, through which of the ordinates f_{st} should the surface corresponding to $P(u,v)$ pass?

A great many answers have been given to the above two questions, and this has resulted in considerable confusion. Our approach will be to seek an answer to these questions based not upon any *ad hoc* assumptions of what we consider to be satisfactory answers but rather upon a natural generalization of the principles of univariate interpolation.

By far the greatest attention in univariate interpolation has been given to parabolic interpolation, the approximation of a function by a polynomial. Perhaps the principal reason for this is that polynomials

† One might even include the taking of a square root as an elementary operation.

are so nicely expressed in terms of elementary operation and so very convenient to handle by differencing techniques. Experience has shown that for general use parabolic interpolation is by far the most useful type employed in univariate tables.

We shall therefore assume that the approximating function $P(u,v)$ is a polynomial in u and v and that our attention is thus to be confined to parabolic surfaces.

The particular points through which the parabolic surface is to pass will not be assumed in advance. Rather we shall generalize the formulas of univariate interpolation in a natural way and then investigate at which points (s,t) one has $P(s,t) = f(s,t)$.

11.5. The Triangular Gregory-Newton Interpolation Formula. As the generalization of the operators E and Δ used in univariate interpolation, let us define the four operators E_x, E_y, Δ_x, and Δ_y by the equations

$$E_x f_{st} = f_{s+1,t}$$
$$E_y f_{st} = f_{s,t+1}$$
$$\Delta_x f_{st} = f_{s+1,t} - f_{st} = (E_x - 1)f_{st} \qquad (11.10)$$
$$\Delta_y f_{st} = f_{s,t+1} - f_{st} = (E_y - 1)f_{st}$$

These yield the two symbolic equations

$$\Delta_x = E_x - 1 \quad \text{or} \quad E_x = 1 + \Delta_x$$

and
$$\Delta_y = E_y - 1 \quad \text{or} \quad E_y = 1 + \Delta_y \qquad (11.11)$$

It is clear from these definitions that

$$f(s,t) = f_{st} = E_x^s E_y^t f_{00} \qquad (11.12)$$

It is, therefore, natural to express $f(u,v)$ as

$$f(u,v) = E_x^u E_y^v f_{00} = (1 + \Delta_x)^u (1 + \Delta_y)^v f_{00} \qquad (11.13)$$

Expanding each operator by the binomial theorem, this equation becomes

$$f(u,v) = \left(1 + u\Delta_x + \frac{u^{[2]}}{2!}\Delta_x^2 + \frac{u^{[3]}}{3!}\Delta_x^3 + \cdots\right)$$
$$\left(1 + v\Delta_y + \frac{v^{[2]}}{2!}\Delta_y^2 + \frac{v^{[3]}}{3!}\Delta_y^3 + \cdots\right)f_{00} \qquad (11.14)$$

Multiplying this out, we have

$$f(u,v) = \left[1 + (u\Delta_x + v\Delta_y) + \frac{1}{2!}(u^{[2]}\Delta_x^2 + 2uv\Delta_x\Delta_y + v^{[2]}\Delta_y^2)\right.$$
$$\left. + \frac{1}{3!}(u^{[3]}\Delta_x^3 + 3u^{[2]}v\Delta_x^2\Delta_y + 3uv^{[2]}\Delta_x\Delta_y^2 + v^{[3]}\Delta_y^3) + \cdots\right]f_{00} \qquad (11.15)$$

By introducing the notation

$$(u\Delta_x + v\Delta_y)^{[p]} = u^{[p]}\Delta_x^p + pu^{[p-1]}v^{[1]}\Delta_x^{p-1}\Delta_y + \frac{p(p-1)}{2}\, u^{[p-2]}v^{[2]}\Delta_x^{p-2}\Delta_y^2$$
$$+ \cdots + v^{[p]}\Delta_y^p \quad (11.16)$$

Eq. (11.15) can be written

$$f(u,v) = \left[1 + (u\Delta_x + v\Delta_y)^{[1]} + \frac{1}{2!}(u\Delta_x + v\Delta_y)^{[2]} \right.$$
$$\left. + \frac{1}{3!}(u\Delta_x + v\Delta_y)^{[3]} + \cdots \right] f_{00} \quad (11.17)$$

If this series is terminated, we obtain a polynomial approximation to $f(u,v)$,

$$f(u,v) \cong P_m(u,v) = \left[1 + (u\Delta_x + v\Delta_y)^{[1]} + \frac{1}{2!}(u\Delta_x + v\Delta_y)^{[2]} + \cdots \right.$$
$$\left. + \frac{1}{m!}(u\Delta_x + v\Delta_y)^{[m]} \right] f_{00} \quad (11.18)$$

This by analogy to the univariate case is the *Gregory-Newton formula for forward interpolation*. The polynomial is of degree m in the variables u and v; i.e., the sum of the powers of u and v for any term is m, or less.†
Consider the special case $m = 3$:

$$P_3(u,v) = f_{00} + u(\Delta_x f_{00}) + v(\Delta_y f_{00})$$
$$+ \frac{1}{2!}[u^{[2]}(\Delta_x^2 f_{00}) + 2uv(\Delta_x\Delta_y f_{00}) + v^{[2]}(\Delta_y^2 f_{00})]$$
$$+ \frac{1}{3!}[u^{[3]}(\Delta_x^3 f_{00}) + 3u^{[2]}v(\Delta_x^2\Delta_y f_{00}) + 3uv^{[2]}(\Delta_x\Delta_y^2 f_{00}) + v^{[3]}(\Delta_y^3 f_{00})] \quad (11.19)$$

The differences in this equation are defined by Eqs. (11.8). They can also be expressed by means of the symbolic equations (11.11) in terms of the tabulated values f_{10}, f_{10}, f_{01}, etc.; thus

$$\begin{aligned}
\Delta_x f_{00} &= (E_x - 1)f_{00} = f_{10} - f_{00} \\
\Delta_y f_{00} &= (E_y - 1)f_{00} = f_{01} - f_{00} \\
\Delta_x^2 f_{00} &= (E_x^2 - 2E_x + 1)f_{00} = f_{20} - 2f_{10} + f_{00} \\
\Delta_y^2 f_{00} &= f_{02} - 2f_{01} - f_{00} \\
\Delta_x\Delta_y f_{00} &= (E_x - 1)(E_y - 1)f_{00} = f_{11} - f_{10} - f_{01} + f_{00} \quad (11.20) \\
\Delta_x^3 f_{00} &= f_{30} - 3f_{20} + 3f_{10} - f_{00} \\
\Delta_y^3 f_{00} &= f_{03} - 3f_{02} + 3f_{01} - f_{00} \\
\Delta_x^2\Delta_y f_{00} &= (E_x^2 - 2E_x + 1)(E_y - 1)f_{00} = f_{21} - 2f_{11} + f_{01} - f_{20} + 2f_{10} - f_{00} \\
\Delta_x\Delta_y^2 f_{00} &= f_{12} - 2f_{11} + f_{10} - f_{02} + 2f_{01} - f_{00}
\end{aligned}$$

† If the highest-order differences in Eq. (11.18), namely, those contained in the term $(1/m!)(u\Delta_x + v\Delta_y)^{[m]}f_{00}$, are all zero, the polynomial will actually be of lower degree than the mth.

Using these relations it is not difficult to show that

$$P_3(s,t) = f_{st} \qquad (11.21)$$

for the 10 pairs of values of s and t corresponding to the tabulated values below:

$$
\begin{array}{llll}
f_{00} & f_{01} & f_{02} & f_{03} \\
f_{10} & f_{11} & f_{12} & \\
f_{20} & f_{21} & & \\
f_{30} & & &
\end{array}
\qquad (11.22)
$$

These 10 points of fit are associated with the 10 arbitrary constants $a_1, a_2, \ldots, a_{10}$ in a third-degree polynomial

$$P_3(u,v) = a_1 + a_2 u + a_3 v + a_4 u^2 + a_5 uv + a_6 v^2 + a_7 u^3 + a_8 u^2 v$$
$$+ a_9 uv^2 + a_{10} v^3 \qquad (11.23)$$

In fact, it can be shown that the tabulated values given in (11.22) uniquely determine the coefficients a_1 to a_{10} in Eq. (11.23), and hence there is one, and only one, polynomial of the third degree which satisfies Eq. (11.21) at the 10 points.

Returning to the polynomial of degree m given by Eq. (11.18),

$$P_m(s,t) = f_{st}$$

for the $\frac{1}{2}(m + 1)(m + 2)$ tabulated values contained in the triangle of fit whose vertices are at f_{00}, f_{m0}, and f_{0m}. Again the polynomial is unique for these tabulated values.

The disadvantage of the Gregory-Newton formula is that u and v are generally between zero and one; hence, the interpolated point (u,v) lies near f_{00} and hence in the upper left-hand corner of the region of fit.

11.6. The Rectangular Gregory-Newton Interpolation Formula. If one terminates the series in Eq. (11.15) by assuming that $f(u,v)$ can be approximated by a polynomial $P_{mn}(u,v)$ of degree m (or less) in u and degree n (or less) in v, then the expansions of E_x^u and E_x^v in terms of differences can be terminated independently of one another, and one obtains Eq. (11.6). The polynomial $P_{mn}(u,v)$ is therefore given by Eq. (11.7). This is, consequently, just the polynomial obtained by interpolating for each variable separately and, as seen in Sec. 11.3, it has a rectangular region of fit.

Note again that if u and v are each between zero and one, the interpolated point lies at the upper left-hand corner of the region of fit, which in this case, however, is a rectangular one. This is undesirable since the closeness of fit can be expected to be poorer near the edges of the region of fit. One could, of course, overcome this by selecting the reference point $u = 0$, $v = 0$ some distance to the left and above the point (u,v)

for which one is interpolating. This, however, necessitates a u of the order of $m/2$ and a v of the order of $n/2$.

11.7. Descending-difference Tables. One might hope to organize the various differences that can be formed from a bivariate table in a three-dimensional scheme that could be looked upon as a natural generalization of the univariate difference table. This turns out to be impossible because of insufficient spaces to tabulate the higher-order differences. That this should occur is clear from the fact that the number of differences of a given order increases with the order. For each tabulated value there are 2 first differences, 3 second differences, 4 third differences, and $m + 1$ differences of the mth order; while for the univariate case there is only 1 difference of each order.

This difficulty may be overcome by thinking of the differences as a four-dimensional array of numbers. This array can be represented by tabulating with each entry in the two-dimensional table the two-dimensional array of differences associated with that entry. This scheme is shown in Table 11.3.†

TABLE 11.3. DESCENDING-DIFFERENCE TABLE

f_{00}	Δ_y	Δ_y^2	$\cdots$	f_{01}	Δ_y	Δ_y^2	$\cdots$	f_{02}	Δ_y	Δ_y^2	$\cdots$
Δ_x	$\Delta_x\Delta_y$	$\Delta_x\Delta_y^2$	$\cdots$	Δ_x	$\Delta_x\Delta_y$	$\Delta_x\Delta_y^2$	$\cdots$	Δ_x	$\Delta_x\Delta_y$	$\Delta_x\Delta_y^2$	$\cdots$
Δ_x^2	$\Delta_x^2\Delta_y$	$\Delta_x^2\Delta_y^2$	$\cdots$	Δ_x^2	$\Delta_x^2\Delta_y$	$\Delta_x^2\Delta_y^2$	$\cdots$	Δ_x^2	$\Delta_x^2\Delta_y$	$\Delta_x^2\Delta_y^2$	$\cdots$
$\cdots$				$\cdots$				$\cdots$			
f_{10}	Δ_y	Δ_y^2	$\cdots$	f_{11}	Δ_y	Δ_y^2	$\cdots$	f_{12}	Δ_y	Δ_y^2	$\cdots$
Δ_x	$\Delta_x\Delta_y$	$\Delta_x\Delta_y^2$	$\cdots$	Δ_x	$\Delta_x\Delta_y$	$\Delta_x\Delta_y^2$	$\cdots$	Δ_x	$\Delta_x\Delta_y$	$\Delta_x\Delta_y^2$	$\cdots$
Δ_x^2	$\Delta_x^2\Delta_y$	$\Delta_x^2\Delta_y^2$	$\cdots$	Δ_x^2	$\Delta_x^2\Delta_y$	$\Delta_x^2\Delta_y^2$	$\cdots$	Δ_x^2	$\Delta_x^2\Delta_y$	$\Delta_x^2\Delta_y^2$	$\cdots$
$\cdots$				$\cdots$				$\cdots$			
f_{20}	Δ_y	Δ_y^2	$\cdots$	f_{21}	Δ_y	Δ_y^2	$\cdots$	f_{22}	Δ_y	Δ_y^2	$\cdots$
Δ_x	$\Delta_x\Delta_y$	$\Delta_x\Delta_y^2$	$\cdots$	Δ_x	$\Delta_x\Delta_y$	$\Delta_x\Delta_y^2$	$\cdots$	Δ_x	$\Delta_x\Delta_y$	$\Delta_x\Delta_y^2$	$\cdots$
Δ_x^2	$\Delta_x^2\Delta_y$	$\Delta_x^2\Delta_y^2$	$\cdots$	Δ_x^2	$\Delta_x^2\Delta_y$	$\Delta_x^2\Delta_y^2$	$\cdots$	Δ_x^2	$\Delta_x^2\Delta_y$	$\Delta_x^2\Delta_y^2$	$\cdots$
$\cdots$				$\cdots$				$\cdots$			

In Table 11.3 the differences in each group are to be taken with respect to the tabulated value of that group. Thus, written out in full, the upper left-hand group should appear as in Table 11.4.

TABLE 11.4. LEADING DIFFERENCES OF A BIVARIATE DIFFERENCE TABLE

f_{00}	$\Delta_y f_{00}$	$\Delta_y^2 f_{00}$	$\Delta_y^3 f_{00}$	$\cdots$
$\Delta_x f_{00}$	$\Delta_x\Delta_y f_{00}$	$\Delta_x\Delta_y^2 f_{00}$	$\Delta_x\Delta_y^3 f_{00}$	$\cdots$
$\Delta_x^2 f_{00}$	$\Delta_x^2\Delta_y f_{00}$	$\Delta_x^2\Delta_y^2 f_{00}$	$\Delta_x^2\Delta_y^3 f_{00}$	$\cdots$
$\Delta_x^3 f_{00}$	$\Delta_x^3\Delta_y f_{00}$	$\Delta_x^3\Delta_y^2 f_{00}$	$\Delta_x^3\Delta_y^3 f_{00}$	$\cdots$
$\cdots$				

† This scheme was originally suggested to me by James E. Storer.

Table 11.3 is formed by repeated application of the following two rules:

1. Subtract any number in any given group from the number lying in the corresponding position in the group just below, and place the result just below the first number.

2. Subtract any number in any given group from the number lying in the corresponding position in the group just to the right, and place the result to the right of the given number.

11.8. Alternative Form for the Gregory-Newton Formula. If one writes down the array of functions shown in Table 11.5 and multiplies each of these functions by the corresponding difference in Table 11.4, one obtains the two-way series given by Eq. (11.15). If one neglects all differences above the mth order, one obtains the triangular Gregory-Newton formula for forward interpolation. Similarly, if one neglects all differences above the mth in u and the nth in v, one obtains the rectangular Gregory-Newton formula.

TABLE 11.5. FACTORIALS NEEDED IN THE GREGORY-NEWTON FORMULA

$$
\begin{array}{cccc}
1 & v & \dfrac{v^{[2]}}{2!} & \dfrac{v^{[3]}}{3!} \quad \cdots \\[2ex]
u & uv & u\dfrac{v^{[2]}}{2!} & u\dfrac{v^{[3]}}{3!} \quad \cdots \\[2ex]
\dfrac{u^{[2]}}{2!} & \dfrac{u^{[2]}}{2!}v & \dfrac{u^{[2]}}{2!}\dfrac{v^{[2]}}{2!} & \dfrac{u^{[2]}}{2!}\dfrac{v^{[3]}}{3!} \quad \cdots \\[2ex]
\dfrac{u^{[3]}}{3!} & \dfrac{u^{[3]}}{3!}v & \dfrac{u^{[3]}}{3!}\dfrac{v^{[2]}}{2!} & \dfrac{u^{[3]}}{3!}\dfrac{v^{[3]}}{3!} \quad \cdots
\end{array}
$$

. . . .

This is a convenient form for some purposes. Once the difference table is built up to contain the required differences with respect to f_{00}, one next constructs Table 11.5 by the following procedure. Given the numerical value of u, one multiplies it by $(u - 1)/2$, and that product by $(u - 2)/3$, etc., to obtain the elements of the first column. The same procedure is followed for v to determine the first row. All other elements are then formed by building up a multiplication table from these elements of the first row and of the first column.

11.9. Gregory-Newton Formulas Involving Backward Interpolation. The Gregory-Newton formula for forward interpolation [Eq. (11.18)] is obtained from

$$f(u,v) = E_x^u E_y^v f_{00}$$

by replacing the operators E_x^u and E_y^v by $(1 + \Delta_x)^u$ and $(1 + \Delta_y)^u$, in agreement with the operator equations (11.11), and then applying the binomial expansion. Since the ascending difference ∇ is defined by the operator equation $\nabla = 1 - E^{-1}$ [see Eq. (3.64)], one may use in place of E_x^u either $(1 + \Delta_x)^u$ or $(1 - \nabla_x)^{-u}$ and in place of E_y^v either $(1 + \Delta_y)^v$ or $(1 - \nabla_y)^{-v}$.

For example, one may express E_x^u in terms of ascending differences and E_y^v in terms of descending differences. In which case,

$$f(u,v) = (1 - \nabla_x)^{-u}(1 + \Delta_y)^v f_{00} = \left[1 + u\nabla_x + \frac{(u + 1)^{[2]}}{2!}\nabla_x^2 \right.$$
$$\left. + \frac{(u + 2)^{[3]}}{3!}\nabla_x^3 + \cdots \right]\left(1 + v\Delta_y + \frac{v^{[2]}}{2!}\Delta_y^2 + \frac{v^{[3]}}{3!}\Delta_y^3 + \cdots\right)f_{00}$$

Expanding this expression using the fact that $\nabla_x = \Delta_x E_x^{-1}$ (see Chap. 3) and assuming that $f(u,v)$ can be approximated satisfactorily by an mth-degree polynomial in u and v, one obtains the interpolation formula

$$f(u,v) = f_{00} + (u\,\Delta_x f_{-10} + v\,\Delta_y f_{00}) + \left[\frac{(u + 1)^{[2]}}{2!}\Delta_x^2 f_{-20} + uv\,\Delta_x\Delta_y f_{-10}\right.$$
$$+ \frac{v^{[2]}}{2!}\Delta_y^2 f_{00}\right] + \cdots + \left[\frac{(u + m - 1)^{[m]}}{m!}\Delta_x^m f_{-m0}\right.$$
$$+ \frac{(u + m - 2)^{[m-1]}}{(m - 1)!}v\,\Delta_x^{m-1}\Delta_y f_{1-m,0} + \cdots + \frac{v^{[m]}}{m!}\Delta_y^m f_{00}\right] \quad (11.24)$$

One might refer to this as the *backward-forward* (BF) Gregory-Newton formula—backward in x and forward in y—in which case, Eq. (11.18) would be designated as the (FF) formula. Clearly there are also two others, the (FB) and (BB) formulas.

The polynomial represented by the right-hand side of Eq. (11.24) agrees with $f(u,v)$ at all the tabulated points in the right triangle whose vertices are the points f_{00}, f_{-m0}, and f_{0m}. Thus one interpolates at the upper right-hand corner of the triangle of fit.

11.10. Linear Interpolation. Because of the importance of linear interpolation, we shall write down the four Gregory-Newton formulas, the (FF), (FB), (BF), and (BB), for this case:

$$
\begin{aligned}
(FF): &\quad f(u,v) = f_{00} + u\,\Delta_x f_{00} + v\,\Delta_y f_{00} + R_{FF} \\
(FB): &\quad f(u,v) = f_{00} + u\,\Delta_x f_{00} + v\,\Delta_y f_{0,-1} + R_{FB} \\
(BF): &\quad f(u,v) = f_{00} + u\,\Delta_x f_{-10} + v\,\Delta_y f_{00} + R_{BF} \\
(BB): &\quad f(u,v) = f_{00} + u\,\Delta_x f_{-10} + v\,\Delta_y f_{0,-1} + R_{BB}
\end{aligned}
\quad (11.25)
$$

where in terms of the given function $F(x,y) = f(u,v)$

$$R_{FF} = h^2\left[\frac{u(u - 1)}{2}\frac{\partial^2 F(\xi,y)}{\partial\xi^2} + uv\frac{\partial^2 F(\xi',\eta')}{\partial\xi'\,\partial\eta'} + \frac{v(v - 1)}{2}\frac{\partial^2 F(x,\eta)}{\partial\eta^2}\right]$$

$$R_{FB} = h^2\left[\frac{u(u - 1)}{2}\frac{\partial^2 F(\xi,y)}{\partial\xi^2} + uv\frac{\partial^2 F(\xi',\eta')}{\partial\xi'\,\partial\eta'} + \frac{v(v + 1)}{2}\frac{\partial^2 F(x,\eta)}{\partial\eta^2}\right]$$

$$R_{BF} = h^2\left[\frac{u(u + 1)}{2}\frac{\partial^2 F(\xi,y)}{\partial\xi^2} + uv\frac{\partial^2 F(\xi',\eta')}{\partial\xi'\,\partial\eta'} + \frac{v(v - 1)}{2}\frac{\partial^2 F(x,\eta)}{\partial\eta^2}\right] \quad (11.26)$$

$$R_{BB} = h^2\left[\frac{u(u + 1)}{2}\frac{\partial^2 F(\xi,y)}{\partial\xi^2} + uv\frac{\partial^2 F(\xi',\eta')}{\partial\xi'\,\partial\eta'} + \frac{v(v + 1)}{2}\frac{\partial^2 F(x,\eta)}{\partial\eta^2}\right]$$

In these equations ξ and ξ' lie between the smallest and largest of the values x, x_0, x_1, while η and η' are similarly related to y, y_0, y_1. These remainders are justified in Sec. 11.19.

Each of the above formulas (without the remainder term) approximates the surface defined by the function $f(u,v)$ by a plane, and is, in fact, determined by the value of f_{00} and two adjacent ordinates of the four neighboring ordinates f_{10}, f_{01}, f_{-10}, and $f_{0,-1}$. Thus the plane represented by the (FF) formula passes through the ordinates f_{00}, f_{01}, and f_{10}. The other three formulas represent the other three possible planes.

Note that linear interpolation separately with respect to u and v leads to a polynomial $Auv + Bu + Cv + D$, which, while linear in u and linear in v, is in fact of the second degree and therefore does not, ordinarily, represent a plane. This polynomial agrees with $f(u,v)$ at the four lattice points surrounding the point (u,v).

11.11. Interpolation Formulas of Gauss. Just as in the case of the Gregory-Newton formulas, since there is a forward (F) and a backward (B) formula of Gauss for univariate interpolation, there are now four formulas, (FF), (FB), (BF), and (BB), for bivariate interpolation. For trivariate interpolation there would be eight formulas and, in general, for interpolation in a table with n independent variables there would be 2^n formulas.

We shall derive here only the first of these four formulas. From the univariate-interpolation formula of Eq. (4.41) the following operator equation for E^u is established:

$$E^u = 1 + (u)_1\Delta + (u)_2\Delta^2E^{-1} + (u+1)_3\Delta^3E^{-1} + (u+1)_4\Delta^4E^{-2} + \cdots$$

$$= 1 + u\Delta + \frac{1}{2!}u(u-1)\Delta^2E^{-1} + \frac{1}{3!}u(u^2-1)\Delta^3E^{-1}$$

$$+ \frac{1}{4!}u(u^2-1)(u-2)\Delta^4E^{-2} + \cdots \quad (11.27)$$

therefore

$$f(u,v) = E_x^uE_y^vf_{00} = \left[1 + u\Delta_x + v\Delta_y + \frac{u(u-1)}{2!}\Delta_x^2E_x^{-1} \right.$$

$$+ uv\Delta_x\Delta_y + \frac{v(v-1)}{2!}\Delta_y^2E_y^{-1} + \frac{u(u^2-1)}{3!}\Delta_x^3E_x^{-1} + \frac{u(u-1)}{2!}v\Delta_x^2\Delta_yE_x^{-1}$$

$$+ u\frac{v(v-1)}{2!}\Delta_x\Delta_y^2E_y^{-1} + \frac{v(v^2-1)}{3!}\Delta_y^3E_y^{-1} + \cdots \right]f_{00} \quad (11.28)$$

To approximate $f(u,v)$ by a polynomial $P_m(u,v)$ of degree m all differences above the mth are set equal to zero. Thus

$$f(u,v) \cong P_m(u,v) \equiv f_{00} + u\,\Delta_x f_{00} + v\,\Delta_y f_{00} + \frac{u(u-1)}{2!}\,\Delta_x^2 f_{-10}$$

$$+ uv\,\Delta_x\Delta_y f_{00} + \frac{v(v-1)}{2!}\,\Delta_y^2 f_{0,-1} + \frac{u(u^2-1)}{3!}\,\Delta_x^3 f_{-10}$$

$$+ \frac{u(u-1)}{2!}\,v\,\Delta_x^2\Delta_y f_{-10} + u\,\frac{v(v-1)}{2!}\,\Delta_x\Delta_y^2 f_{0,-1} + \frac{v(v^2-1)}{3!}\,\Delta_y^3 f_{0,-1}$$

$$+ \cdots + \text{terms involving } m\text{th differences} \qquad (11.29)$$

where the terms involving mth differences vary in form depending on whether m is even or odd.

11.12. Rapid Method for Determining the Points of Fit. We saw in discussing the Gauss forward-interpolation formula [Gauss (I), see Chap. 4] that the points of fit between the given function $f(x)$ and the interpolation polynomial increased with the degree of the polynomial. The way in which new points were added is given below.

$$
\begin{array}{cccccccc}
x_{-3} & x_{-2} & x_{-1} & x_0 & x_1 & x_2 & x_3 & x_4 \\
\cdot & \cdot & \cdot & \downarrow & \cdot & \cdot & \cdot & \cdot \\
6 & 4 & 2 & 0 & 1 & 3 & 5 & 7
\end{array}
$$

The numbers give the order of the difference that must be present in the formula for the point to be a point of fit. Thus, if Gauss' formula is terminated at the fifth difference, the points of fit are those six points having numbers equal to, or less than, 5. It is convenient to omit the arguments x_s and the points and use only the numbers.

For bivariate interpolation one numbers the points along each axis (see Table 11.6) in the same way as above for univariate interpolation and then assigns to each of the other points the sum of numbers found at the base of perpendiculars drawn from that point to the two axes.

The interpretation is similar to that for univariate interpolation. Those points with numbers less than, or equal to, m are points of fit for the mth-degree polynomial of Eq. (11.29). Thus, for the fourth-degree polynomial obtained by terminating the interpolation formula of Gauss at the fourth differences the region of fit is that outlined by the dotted lines in Table 11.6.

Note that the region of fit is almost rectangular and that $u = 0$, $v = 0$, designated by 0, is very close to its center. This is true of all the central-difference formulas and accounts for their increased accuracy over the Gregory-Newton formulas. This advantage of interpolating near the center of the region of fit was noted in univariate interpolation, but one would expect this difference to be even more marked in bivariate tables.

It is significant that here, as in the Gregory-Newton formula, $m + 1$ points of fit are added in going from a polynomial of degree $m - 1$ to one of degree m.   This is in agreement with the number of coefficients of the mth degree in the latter polynomial.

11.13. Stirling's and Bessel's Formulas.   In univariate interpolation (see Chap. 4) Stirling's formula is just the average of the forward- and backward-interpolation formulas of Gauss.   Thus, if we designate the

TABLE 11.6. POINTS OF FIT FOR THE *(FF)* INTERPOLATION FORMULA OF GAUSS

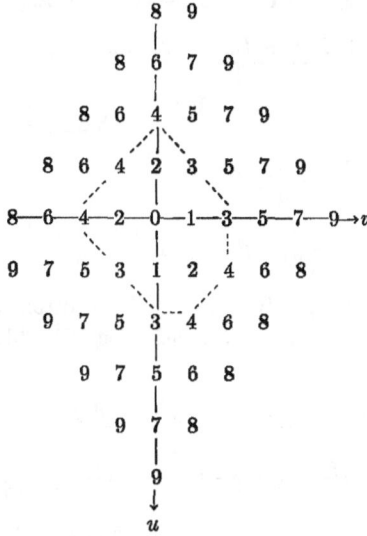

forward and backward operators with subscripts F and B, respectively, and Stirling's operator by the subscript S; then the latter formula, in operator form, may be written

$$f(u) = (E^u)_S f_0 = \left[\frac{(E^u)_F + (E^u)_B}{2} \right] f_0$$

Therefore, in bivariate interpolation Stirling's formula becomes

$$f(u,v) = (E_x^u)_S (E_y^v)_S f_{00} = \left[\frac{(E_x^u)_F + (E_x^u)_B}{2} \right] \left[\frac{(E_y^v)_F + (E_y^v)_B}{2} \right] f_{00}$$

$$= \tfrac{1}{4}[(E_x^u)_F (E_y^v)_F f_{00} + (E_x^u)_F (E_y^v)_B f_{00} + (E_x^u)_B (E_y^v)_F f_{00} + (E_x^u)_B (E_y^v)_B f_{00}] \quad (11.30)$$

This is just the average of the four interpolation formulas of Gauss, the *(FF)*, *(FB)*, *(BF)*, and *(BB)*.

Another way of proceeding is to obtain the operator E_S^u directly from Stirling's univariate formula or by averaging the forward and backward formulas of Gauss.   In either case [see Eq. (4.42)]

$$(E^u)_s = 1 + u\Delta \frac{1 + E^{-1}}{2} + \tfrac{1}{2}u^2\Delta^2 E^{-1} + \frac{1}{3!} u(u^2 - 1)\Delta^3 \frac{E^{-1} + E^{-2}}{2}$$

$$+ \frac{1}{4!} u^2(u^2 - 1)\Delta^4 E^{-2} + \cdots \quad (11.31)$$

therefore Stirling's formula may be written

$$f(u,v) = (E^u_x)_s(E^v_y)_s f_{00} = \left[1 + u\Delta_x \frac{1 + E^{-1}_x}{2} + v\Delta_y \frac{1 + E^{-1}_y}{2} + \tfrac{1}{2}u^2\Delta^2_x E^{-1}_x \right.$$

$$+ uv\Delta_x\Delta_y \frac{1 + E^{-1}_x + E^{-1}_y + E^{-1}_x E^{-1}_y}{4} + \tfrac{1}{2}v^2\Delta^2_y E^{-1}_y$$

$$+ \frac{1}{3!} u(u^2 - 1)\Delta^3_x \frac{E^{-1}_x + E^{-2}_x}{2} + \tfrac{1}{2}u^2 v\Delta^2_x\Delta_y \frac{E^{-1}_x + E^{-1}_x E^{-1}_y}{2}$$

$$+ \tfrac{1}{2}uv^2\Delta_x\Delta^2_y \frac{E^{-1}_y + E^{-1}_x E^{-1}_y}{2}$$

$$\left. + \frac{1}{3!} v(v^2 - 1)\Delta^3_y \frac{E^{-1}_y + E^{-2}_y}{2} + \cdots \right] f_{00} \quad (11.32)$$

Instead of using the infinite series, which often may not converge, one usually uses the truncated series [Eq. (11.33)] as a polynomial approximation:

$$f(u,v) \cong f_{00} + u \frac{\Delta_x f_{00} + \Delta_x f_{-10}}{2} + v \frac{\Delta_y f_{00} + \Delta_y f_{0,-1}}{2} + \tfrac{1}{2}u^2 \Delta^2_x f_{-10}$$

$$+ uv \frac{\Delta_x\Delta_y f_{00} + \Delta_x\Delta_y f_{-10} + \Delta_x\Delta_y f_{0,-1} + \Delta_x\Delta_y f_{-1,-1}}{4} + \tfrac{1}{2}v^2 \Delta^2_y f_{0,-1}$$

$$+ \frac{1}{3!} u(u^2 - 1) \frac{\Delta^3_x f_{-10} + \Delta^3_x f_{-20}}{2} + \tfrac{1}{2}u^2 v \frac{\Delta^2_x\Delta_y f_{-10} + \Delta^2_x\Delta_y f_{-1,-1}}{2}$$

$$+ \tfrac{1}{2}uv^2 \frac{\Delta_x\Delta^2_y f_{0,-1} + \Delta_x\Delta^2_y f_{-1,-1}}{2} + \frac{1}{3!} v(v^2 - 1) \frac{\Delta^3_y f_{0,-1} + \Delta^3_y f_{0,-2}}{2}$$

$$+ \cdots + \text{terms in the } m\text{th differences} \quad (11.33)$$

By a similar procedure, starting with Eq. (4.43), Bessel's formula can be shown to be

$$f(u,v) \cong P_m(U,V) \equiv \frac{f_{00} + f_{10} + f_{01} + f_{11}}{4} + U \frac{\Delta_x f_{00} + \Delta_x f_{01}}{2}$$

$$+ V \frac{\Delta_y f_{00} + \Delta_y f_{10}}{2} + \frac{1}{2!}\left(U^2 - \frac{1}{4}\right) \frac{\Delta^2_x f_{00} + \Delta^2_x f_{-10} + \Delta^2_x f_{01} + \Delta^2_x f_{-11}}{4}$$

$$+ UV \Delta_x\Delta_y f_{00} + \frac{1}{2!}\left(V^2 - \frac{1}{4}\right) \frac{\Delta^2_y f_{00} + \Delta^2_y f_{10} + \Delta^2_y f_{0,-1} + \Delta^2_y f_{1,-1}}{4}$$

$$+ \frac{1}{3!} U\left(U^2 - \frac{1}{4}\right) \frac{\Delta^3_x f_{-10} + \Delta^3_x f_{-11}}{2} + \frac{1}{2!}\left(U^2 - \frac{1}{4}\right) V \frac{\Delta^2_x\Delta_y f_{-10} + \Delta^2_x\Delta_y f_{00}}{2}$$

$$+ \frac{1}{2!} U\left(V^2 - \frac{1}{4}\right) \frac{\Delta_x\Delta^2_y f_{0,-1} + \Delta_x\Delta^2_y f_{00}}{2} + \frac{1}{3!} V\left(V^2 - \frac{1}{4}\right) \frac{\Delta^3_y f_{0,-1} + \Delta^2_y f_{1,-1}}{2}$$

$$+ \cdots + \text{terms involving } m\text{th differences} \quad (11.34)$$

where
$$U = u - \tfrac{1}{2}$$
$$V = v - \tfrac{1}{2}$$
(11.35)

It is evident from the formulas of this section that the complexity of notation needed for bivariate tables is very great. This illustrates well the relative simplicity of the procedure of interpolating first for one variable and then for the other (see Sec. 11.3), thus eliminating the need for any detailed notations.

11.14. Bivariate Interpolation Formulas in Lagrangian Form. Any of the interpolation formulas discussed so far can be written in Lagrangian form, i.e., in terms of the tabulated values by substituting for the differences their expressions in terms of the tabulated values.

Example 1. Express the (FF) interpolation formula of Gauss,

$$f(u,v) \cong P_2(u,v) \equiv f_{00} + u \, \Delta_x f_{00} + v \, \Delta_y f_{00} + \frac{u(u-1)}{2} \Delta_x^2 f_{-10} + uv \, \Delta_x \Delta_y f_{00}$$
$$+ \frac{v(v-1)}{2} \Delta_y^2 f_{0,-1} \quad (11.36)$$

in Lagrangian form.

From the fundamental definitions of differences given in Eqs. (11.8),

$$\Delta_x f_{00} = f_{10} - f_{00} \qquad \Delta_y f_{00} = f_{01} - f_{00}$$
$$\Delta_x^2 f_{-10} = f_{10} - 2f_{00} + f_{-10} \qquad \Delta_y^2 f_{0,-1} = f_{01} - 2f_{00} + f_{0,-1}$$
$$\Delta_x \Delta_y f_{00} = f_{11} - f_{10} - f_{01} + f_{00}$$

therefore

$$f(u,v) \cong P_2(u,v) = \frac{v(v-1)}{2} f_{0,-1} + \frac{u(u-1)}{2} f_{-10} + (1 + uv - u^2 - v^2)f_{00}$$
$$+ \frac{u(u-2v+1)}{2} f_{10} + \frac{v(v-2u+1)}{2} f_{01} + uv f_{11} \quad (11.37)$$

This equation can be written

$$P_2(u,v) = \sum_{i,j} A_{ij}(u,v) f_{ij} \quad (11.38)$$

where the summation is over the six pairs of values for u and v, $(0,-1)$, $(-1,0)$, $(0,0)$, $(1,0)$, $(0,1)$, and $(1,1)$, and the $A_{ij}(u,v)$ are the quadratic polynomials in u and v of Eq. (11.37). These A's are seen to have the following properties:

$$A_{ij}(i',j') = \delta_{ii'}\delta_{jj'} = \begin{cases} 1 & \text{if } i' = i \text{ and } j' = j \\ 0 & \text{otherwise} \end{cases} \quad (11.39)$$

By virtue of these properties, for (i',j') one of the six points above,

$$P_2(i',j') = \sum_{i,j} A_{ij}(i',j') f_{ij} = \sum_{i,j} \delta_{ii'}\delta_{jj'} f_{ij} = f_{i'j'}$$

and therefore $P_2(u,v)$ fits the given function $f(u,v)$ at each of the six points.

It can be shown that the $A_{ij}(u,v)$ of Example 1 are fully determined by requiring that they satisfy Eqs. (11.39) at the six points listed and that they be only second-degree polynomials. One might, therefore, seek to obtain interpolation polynomials in Lagrangian form for any arbitrary choice for points of fit by such a technique; however, one usually finds that there are either two few or two many conditions on the A's. Thus the expression of interpolating polynomials in difference form, as done in the previous sections, plays the important role of determining suitable regions of fit.

There are, however, two very important advantages of expressing interpolation polynomials in ordinate form:

1. One may interpolate in a table without first building up a difference table. This is done by adding together the tabulated values f_{ij} each multiplied by their proper weights $A_{ij}(u,v)$. Clearly one cannot afford to tabulate many of the differences in a bivariate (or multivariate) table.

2. One may often observe at a glance the region of fit between the polynomial and the given function.

We shall therefore develop general expressions for the $A_{ij}(u,v)$ for the two most important types of regions of fit, the rectangular and triangular regions of our previous formulas. For brevity, we shall refer to the Lagrangian interpolation formulas arising from these as the *rectangular* and *triangular* Lagrangian interpolation formulas, respectively.

11.15. The Rectangular Lagrangian Interpolation Formula. Suppose one desires to interpolate in a bivariate table by employing a polynomial $P_{mn}(u,v)$ that fits the tabulated function $f(u,v)$ at the $(m + 1)(n + 1)$ points lying in the rectangle whose vertices are f_{pq}, $f_{p,-q}$, f_{-pq}, and $f_{-p,-q}$. This choice requires $m = 2p$ and $n = 2q$ to be even numbers. In Lagrangian form

$$f(u,v) \cong P_{mn}(u,v) \equiv \sum_{i=-p}^{p} \sum_{j=-q}^{q} A_{ij}(u,v)f_{ij} \qquad (11.40)$$

and to determine the weighting function $A_{ij}(u,v)$ one may proceed in the following way.

Let $$\prod (u,v) = \prod_{i=-p}^{p} (u - i) \prod_{j=-q}^{q} (v - j) \qquad (11.41)$$

then $\Pi(u,v)$ is a polynomial of degree $m + 1$ in u and of $n + 1$ in v that vanishes at all the $(m + 1)(n + 1)$ points above; thus

$$\Pi(i,j) = 0 \qquad \text{for } -p \le i \le p \text{ and } -q \le j \le q \qquad (11.42)$$

where all letters take on only integral values. Moreover, let

$$\Pi_{ij}(u,v) = \frac{\Pi(u,v)}{(u - i)(u - j)} \qquad (11.43)$$

then $\Pi_{ij}(u,v)$ vanishes at all the points of the rectangle except the point $u = i,\ v = j$.

We can now write for the $A_{ij}(u,v)$ of Eq. (11.40)

$$A_{ij}(u,v) = \frac{\Pi_{ij}(u,v)}{\Pi_{ij}(i,j)} \tag{11.44}$$

since, by inspection, the functions so defined satisfy Eqs. (11.39) for our rectangular region of fit. That satisfaction of Eqs. (11.39) ensures a fit between $P_m(u,v)$ and $f(u,v)$ for all points $u = s,\ v = t$ in the rectangle can be seen directly from Eq. (11.40), since by Eq. (11.39)

$$P_{mn}(s,t) = \sum_{i=-p}^{p} \sum_{j=-q}^{q} A_{ij}(s,t)f_{ij} = \sum_{i=-p}^{p} \sum_{j=-q}^{q} \delta_{si}\delta_{tj}f_{ij} = f_{st} \equiv f(s,t)$$

An alternative method of deriving the rectangular Lagrangian interpolation formula, described below, brings out a very important property of the $A_{ij}(u,v)$.

The univariate Lagrangian interpolation formula for $m = p_1 + p_2$†

$$f(u) \cong P_m(u) \equiv \sum_{i=-p_1}^{p_2} A_i^m(u)f_i \tag{11.45}$$

can be written

$$E^u f_0 \cong P_m(u) = \Big[\sum_{i=-p_1}^{p_2} A_i^m(u)E^i \Big] f_0 \tag{11.46}$$

Thus the operator on the right-hand side is an approximation for E^u. The $A_i^m(u)$ are well known and have been tabulated by various people (see Chap. 5).

For bivariate interpolation one would expect that, in keeping with Eq. (11.46),

$$f(u,v) = E_x^u E_y^v f_{00} \cong \Big[\sum_{i=-p_1}^{p_2} A_i^m(u)E_x^i \Big] \Big[\sum_{j=-q_1}^{q_2} A_j^n(v)E_y^j \Big] f_{00}$$

$$= \sum_{i=-p_1}^{p_2} \sum_{j=-q_1}^{q_2} [A_i^m(u)A_j^n(v)]f_{ij} \tag{11.47}$$

where $m = p_1 + p_2$ and $n = q_1 + q_2$. Comparing this equation with Eq. (11.40), one deduces the relation

$$A_{ij}(u,v) = A_i^m(u)A_j^n(v) \tag{11.48}$$

which can be easily verified.

† The superscript m on the $A_i^m(u)$ designates the degree of the polynomial. While it was understood in the notation of Chap. 5, it is here employed to prevent confusion in later equations.

Equation (11.48) means that the weighting functions in bivariate interpolation are just the products of the weighting functions used for interpolating independently in the two directions.

This is not surprising in view of the fact that one can interpolate, using Lagrange's formula, first with respect to u and then with respect to v (see Sec. 11.3). In practice, moreover, it is usually easier to interpolate for each of the variables separately than to employ the bivariate weights obtained from Eq. (11.48).

11.16. The Triangular Lagrangian Interpolation Formula. If one chooses to fit a polynomial in u and v of degree m to the function $f(u,v)$ over a triangular region, one may again employ directly the Lagrangian form. The number of points involved will be $\frac{1}{2}(m+1)(m+2)$, in agreement with the number of coefficients in the polynomial, and the solution is again unique. If the triangular region of fit is oriented as in the Newton-Gregory (FF) formula, the formula will be seen to be

$$f(u,v) \cong P_m(u,v) \equiv \sum_{s=0}^{m} \sum_{t=0}^{m-s} A_{st}(u,v) f_{st} \tag{11.49}$$

where
$$A_{st}(u,v) = \frac{\displaystyle\prod_{i=0}^{s-1}(u-i)\prod_{j=0}^{t-1}(v-j)\prod_{k=s+t+1}^{m}(u+v-k)}{\displaystyle\prod_{i=0}^{s-1}(s-i)\prod_{j=0}^{t-1}(t-j)\prod_{k=s+t+1}^{m}(s+t-k)} \tag{11.50}$$

It is necessary here to adopt the convention that a product with limits such as $\Pi_{k=m+1}^{m}(\cdots)$ is always 1; so that, for example,

$$A_{s,m-s}(u,v) = \frac{\displaystyle\prod_{i=0}^{s-1}(u-i)\prod_{j=0}^{m-s-1}(v-j)}{\displaystyle\prod_{i=0}^{s-1}(s-i)\prod_{j=0}^{m-s-1}(t-j)} \tag{11.51}$$

Since $A_{st}(u,v)$ has m linear factors, it is of degree m in u and v. Moreover, it is clear by inspection that

$$A_{st}(i,j) = \delta_{si}\delta_{tj} \tag{11.52}$$

This set of conditions is seen, by reference to Eq. (11.49), to be sufficient to ensure agreement between the polynomial and the given function at all tabulated values within the triangular region.

For regions of fit other than the rectangular and triangular regions treated above one has difficulty in expressing the interpolation polynomial in Lagrangian form. Thus in Example 1 (Sec. 11.14) the functions

$A_{ij}(u,v)$ are not all factorable into linear factors. It is not clear, therefore, how one would write the interpolation polynomial for Gauss' formula directly in Lagrangian form.

11.17. Divided Differences for Functions of Two Variables. The concept of divided differences for unequally spaced values can be generalized to a function of two variables $f(x,y)$ by means of the defining equations

$$f(x_0, x_1, \ldots, x_m; y_0, y_1, \ldots, y_n)$$

$$= \frac{1}{x_0 - x_m} [f(x_0, \ldots, x_{m-1}; y_0, \ldots, y_n) - f(x_1, \ldots, x_m; y_0, \ldots, y_n)]$$

$$= \frac{1}{y_0 - y_n} [f(x_0, \ldots, x_m; y_0, \ldots, y_{n-1}) - f(x_0, \ldots, x_m; y_1, \ldots, y_n)]$$

$$\text{(11.53)}$$

Here, as for the univariate case, the labeling of the x's and y's can be done in any arbitrary way independent of their magnitudes. If $x_0 = x_m$ or $y_0 = y_n$, the expressions in Eq. (11.53) must be replaced by their limiting values. As particular examples of divided differences

$$f(x_0, x_1; y_0) = \frac{1}{x_0 - x_1} [f(x_0, y_0) - f(x_1, y_0)] = \frac{f(x_0, y_0)}{x_0 - x_1} + \frac{f(x_1, y_0)}{x_1 - x_0}$$

$$f(x_0, x_1; y_0, y_1) = \frac{1}{y_0 - y_1} [f(x_0, x_1; y_0) - f(x_0, x_1; y_1)] = \frac{f(x_0, y_0)}{(x_0 - x_1)(y_0 - y_1)}$$

$$+ \frac{f(x_0, y_1)}{(x_0 - x_1)(y_1 - y_0)} + \frac{f(x_1, y_0)}{(x_1 - x_0)(y_0 - y_1)} + \frac{f(x_1, y_1)}{(x_1 - x_0)(y_1 - y_0)}$$

and in general

$$f(x_0, \ldots, x_m; y_0, \ldots, y_n) = \sum_{i=0}^{m} \sum_{j=0}^{n} \frac{f(x_i, y_j)}{\displaystyle\prod_{\substack{s=0 \\ s \neq i}}^{m} (x_i - x_s) \prod_{\substack{t=0 \\ t \neq j}}^{n} (y_j - y_t)} \quad \text{(11.54)}$$

This general expression [Eq. (11.54)] for a divided difference of two variables is seen to hold for $f(x_0, x_1; y_0)$, $f(x_0; y_0, y_1)$, and $f(x_0, x_1; y_0, y_1)$. It can then be proved by mathematical induction to hold generally as long as the x's and y's are all distinct.

It is clear from the form of Eq. (11.54) that the use of a divided difference $f(x_0, \ldots, x_m; y_0, \ldots, y_n)$ implies that the function $f(x,y)$ is tabulated for all (x_i, y_j), for which $0 \le i \le m$ and $0 \le j \le n$.

By the same argument, based on Rolle's theorem, as was used to prove Eq. (5.59),

$$f(a_0, \ldots, a_p; b_0, \ldots, b_q) = \frac{1}{p!} \frac{\partial^p}{\partial \xi^p} f(\xi; b_0, \ldots, b_q)$$

where ξ lies between the least and greatest of the numbers $a_0, a_1, \ldots, a_p$. Applying, in turn, this same reasoning to $f(\xi;b_0, \ldots, b_q)$, one obtains the useful relation

$$f(a_0, \ldots, a_p; b_0, \ldots, b_q) = \frac{1}{p!} \frac{1}{q!} \frac{\partial^{p+q}}{\partial \xi^p \, \partial \eta^q} f(\xi, \eta) \qquad (11.55)$$

where ξ and η lie between the least and greatest of the numbers $a_0, a_1, \ldots, a_p$ and $b_0, b_1, \ldots, b_q$, respectively.

11.18. Newton's Divided-difference Interpolation Formula for a Rectangular Region. By exactly the same technique as was used to develop Newton's divided-difference interpolation formula for one variable [see the development leading to Eqs. (5.56) and (5.57)] one has†

$$f(x,y) = f(x_0,y) + (x - x_0)f(x_0,x_1;y) + (x - x_0)(x - x_1)f(x_0,x_1,x_2;y)$$
$$+ \cdots + (x - x_0)(x - x_1) \cdots (x - x_{m-1})f(x_0, \ldots, x_m;y)$$
$$+ (x - x_0) \cdots (x - x_m)f(x,x_0, \ldots, x_m;y)$$
$$= \sum_{i=0}^{m} \left[\prod_{s=0}^{i-1} (x - x_s)f(x_0, \ldots, x_i;y) \right] + \prod_{s=0}^{m} (x - x_s)f(x,x_0, \ldots, x_m;y)$$

The latter part of the above expression requires the adoption of the convention, employed earlier, that for any value of k

$$\prod_{s=k}^{k-1} (x - x_s) = 1 \qquad (11.56)$$

Likewise

$$f(x_0, \ldots, x_i;y) = \sum_{j=0}^{n} \left[\prod_{t=0}^{j-1} (y - y_t)f(x_0, \ldots, x_i;y_0, \ldots, y_j) \right]$$
$$+ \prod_{t=0}^{n} (y - y_t)f(x_0, \ldots, x_i;y,y_0, \ldots, y_n)$$

and therefore

$$f(x,y) = \sum_{i=0}^{m} \sum_{j=0}^{n} \left[\prod_{s=0}^{i-1} (x - x_s) \prod_{t=0}^{j-1} (y - y_t)f(x_0, \ldots, x_i;y_0, \ldots, y_j) \right]$$
$$+ R_{mn} \qquad (11.57)$$

† Here and elsewhere the convention has been adopted that only those quantities involving the product index are to be included in the terms of a product; thus

$$\prod_{s=0}^{m} (x - x_s)f(x,x_0, \ldots, x_m;y) = \left[\prod_{s=0}^{m} (x - x_s) \right] f(x,x_0, \ldots, x_m;y)$$

where the remainder is

$$R_{mn} = \prod_{t=0}^{n} (y - y_t) \sum_{i=0}^{m} \left[\prod_{s=0}^{i-1} (x - x_s) f(x_0, \ldots, x_i; y, y_0, \ldots, y_n) \right]$$

$$+ \prod_{s=0}^{m} (x - x_s) f(x, x_0, \ldots, x_m; y) \quad (11.58)$$

The right-hand side of Eq. (11.57) with R_{mn} removed represents an interpolation polynomial $P_{mn}(x,y)$ of mth degree in x and nth degree in y. Since R_{mn} vanishes at all points (x_i, y_j), for which $0 \le i \le m$ and $0 \le j \le n$, in some rectangular region Γ, the polynomial fits the given function $f(x,y)$ at all these $(m + 1)(n + 1)$ points in the region.

The remainder term can be simplified by making use of the equation

$$f(x; y, y_0, \ldots, y_n) = \sum_{i=0}^{m} \left[\prod_{s=0}^{i-1} (x - x_s) f(x_0, \ldots, x_i; y, y_0, \ldots, y_n) \right]$$

$$+ \prod_{s=0}^{m} (x - x_s) f(x, x_0, \ldots, x_m; y, y_0, \ldots, y_n)$$

Eliminating the terms under the summation sign between this equation and Eq. (11.58), one obtains the simpler expression

$$R_{mn} = \prod_{s=0}^{m} (x - x_s) f(x, x_0, \ldots, x_m; y) + \prod_{t=0}^{n} (y - y_t) f(x; y, y_0, \ldots, y_n)$$

$$- \prod_{s=0}^{m} (x - x_s) \prod_{t=0}^{n} (y - y_t) f(x, x_0, \ldots, x_m; y, y_0, \ldots, y_n) \quad (11.59)$$

for the remainder term.

Thus, using Eq. (11.55), the remainder term can be written

$$R_{mn} = \frac{1}{m!} \prod_{s=0}^{m} (x - x_s) \frac{\partial^m}{\partial \xi^m} f(\xi, y) + \frac{1}{n!} \prod_{t=0}^{n} (y - y_t) \frac{\partial^n}{\partial \eta^n} f(x, \eta)$$

$$- \frac{1}{m! n!} \prod_{s=0}^{m} (x - x_s) \prod_{t=0}^{n} (y - y_t) \frac{\partial^{m+n}}{\partial \xi'^m \partial \eta'^n} f(\xi', \eta') \quad (11.60)$$

The variables ξ' and η', like ξ and η, lie between the least and greatest of the numbers x, x_0, . . . , x_m and y, y_0, . . . , y_n, respectively.

Note that the interpolation formula (11.57) is quite general, since the series of numbers x_0, x_1, . . . , x_m and the series y_0, y_1, . . . , y_n may be given any arbitrary magnitudes and signs. In particular, all the polynomials $P_{mn}(u,v)$ obtained previously are obtainable from this formula. Moreover, any number of the x's or y's may be identical. One can, for example, choose $x_i = x_0$ and $y_j = y_0$, for $0 \le i \le m$ and $0 \le j \le n$. Equation (11.57) will then yield a variation of Taylor's series for a function of two independent variables.†

11.19. Obtaining Interpolation Polynomials of the Form $P_m(x,y)$. By an obvious modification of the procedure used to obtain Eq. (11.57) one may derive the equation [note convention given by Eq. (11.56)]

$$f(x,y) = \sum_{i=0}^{m} \sum_{j=0}^{m-i} \left[\prod_{s=0}^{i-1} (x - x_s) \prod_{t=0}^{j-1} (y - y_t) f(x_0, \ldots ,x_i;y_0, \ldots ,y_j) \right] + R_m \quad (11.61)$$

where

$$R_m = \sum_{i=0}^{m} \left[\prod_{s=0}^{i-1} (x - x_s) \prod_{t=0}^{m-i} (y - y_t) f(x_0, \ldots ,x_i;y,y_0, \ldots ,y_{m-i}) \right] + \prod_{s=0}^{m} (x - x_s) f(x,x_0, \ldots ,x_m;y) \quad (11.62)$$

Written in terms of derivatives by the application of Eq. (11.55),

$$R_m = \frac{1}{(m + 1)!} \prod_{t=0}^{m} (y - y_t) \frac{\partial^{m+1} f(x_0,\eta_0)}{\partial \eta_0^{m+1}}$$

$$+ \frac{1}{(m + 1)!} \prod_{s=0}^{m} (x - x_s) \frac{\partial^{m+1} f(\xi_{m+1},y)}{\partial \xi_{m+1}^{m+1}}$$

$$+ \sum_{i=1}^{m} \frac{1}{i!(m - i + 1)!} \prod_{s=0}^{i-1} (x - x_s) \prod_{t=0}^{m-i} (y - y_t) \frac{\partial^{m+1} f(\xi_i,\eta_i)}{\partial \xi_i^i \partial \eta_i^{m-i+1}} \quad (11.63)$$

where for $1 \le i \le m$, ξ_i and η_i lie between the least and greatest of the numbers $x_0, x_1, \ldots , x_i$ and $y, y_0, \ldots , y_{m-i}$, respectively, while ξ_{m+1} lies between the numbers $x, x_0, \ldots , x_m$, and η_0 between $y, y_0, \ldots , y_m$. Because of its complexity this remainder is of little use.

Note that the remainder term above is not symmetrical in x and y. It would be desirable to have a remainder of no greater complexity which preserves this symmetry. For linear interpolation ($m = 1$) this can be done, since by a slight alteration of procedure one can obtain the symmetrical expression for the remainder

$$R = (x - x_0)(x - x_1) f(x,x_0,x_1;y_0) + (x - x_0)(y - y_0) f(x,x_0;y,y_0) + (y - y_0)(y - y_1) f(x_0;y,y_0,y_1) \quad (11.64)$$

† See Steffensen, *op. cit.*, p. 207.

In terms of derivatives, therefore,

$$R = \tfrac{1}{2}\left[(x - x_0)(x - x_1)\frac{\partial^2 f(\xi, y_0)}{\partial \xi^2} + 2(x - x_0)(y - y_0)\frac{\partial^2 f(\xi', \eta')}{\partial \xi' \, \partial \eta'} \right.$$
$$\left. + (y - y_0)(y - y_1)\frac{\partial^2 f(x_0, \eta)}{\partial \eta^2} \right] \quad (11.65)$$

The polynomial approximation to $f(x,y)$ given by Eq. (11.61) can, of course, be applied to a function that is tabulated at equal intervals of x and of y. Despite the fact that the spacing is even, one may choose the x_i (and y_j) to be an arbitrary set of integral multiples of the spacing h and k, respectively. This permits one to write any of the formulas of Newton or of Gauss by a proper choice of the x_i and y_j. For example, if $x_i = ih$ and $y_j = -jk$, Eq. (11.61) becomes the Gregory-Newton (FB) interpolation formula. While if $x_{2i-1} = i$, $x_{2i} = -i$ and $y_{2j-1} = j$, $y_{2j} = -j$, it becomes the (FF) formula of Gauss.

Clearly, therefore, one may specialize the remainder term of Eq. (11.63) to these cases and obtain thereby the proper remainder terms for the Gregory-Newton and Gauss formulas.

11.20. Cubature Formulas. Cubature formulas are formulas for approximating the value of a double integral

$$I = \iint_S F(x,y) \, dx \, dy \quad (11.66)$$

where S is the region over which the integral is desired. They bear the same relation to bivariate-interpolation theory as quadrature formulas bear to univariate interpolation.

One first breaks down the region S into regions S_i, usually rectangular; thus $I = \Sigma_{i=1}^N I_i$, where

$$I_i = \iint_{S_i} F(x,y) \, dx \, dy \quad (11.67)$$

Suppose that the function $F(x,y)$ is tabulated at equally spaced intervals, the interval in x being h, and the interval in y being k; then it is convenient to write

$$F(x,y) = F(x_0 + uh, \ y_0 + vk) \equiv f(u,v) \quad (11.68)$$

In terms of u and v

$$I_i = hk \iint_{S_i} f(u,v) \, du \, dv \quad (11.69)$$

where S_i is now understood to be the range expressed in terms of u and v.

If one now approximates the function $f(u,v)$ by any of the polynomials obtained from the bivariate-interpolation formulas of this chapter, one may perform the integration and obtain I_i in terms of the parameters occurring in the interpolation formula. Either these may be the tabulated values $f_{st} = f(s,t)$, where s and t are integers, or they may be the various differences (see Sec. 11.3) formed from them.

11.21. Newton-Cotes Cubature Formulas. Let the function $f(u,v)$ in Eq. (11.69) be approximated by a polynomial $P(u,v)$ whose region of fit with $f(u,v)$ is a rectangle. Such a polynomial expressed in Lagrangian form is given by Eq. (11.47). Suppose the region S_i over which the integration is performed is also rectangular, $a \le u \le b$, and $c \le v \le d$; then Eq. (11.69) becomes

$$I_i = hk \int_a^b du \int_c^d dv\, f(u,v)$$

$$\cong hk \int_a^b du \int_c^d dv \sum_{i=-p_1}^{p_2} \sum_{j=-q_1}^{q_2} A_i^m(u) A_j^n(v) f_{ij} \qquad (11.70)$$

or
$$I_i = hk \sum_{i,j} w_{ij} f_{ij} \qquad (11.71)$$

where
$$w_{ij} = \int_a^b A_i^m(u)\, du \int_c^d A_j^n(v)\, dv = w_i^m w_j^n \qquad (11.72)$$

Here w_i^m are the weights for the quadrature formula (see Chap. 7)

$$\int_a^b f(u)\, du = \sum_{i=-p_1}^{p_2} w_i^m f_i \qquad (11.73)$$

for a function of one variable [where $f_i = f(i)$], and w_j^n are the weights of the quadrature formula

$$\int_c^d f(v)\, dv = \sum_{j=-q_1}^{q_2} w_j^n f_j \qquad (11.74)$$

Thus any two quadrature formulas, such as those given by Eqs. (11.73) and (11.74), can be combined as in Eqs. (11.71) and (11.72) to give a cubature formula.

Often the same quadrature formula is used for integrating with respect to v as is used for u. Thus if one employs the trapezoidal rule

$$\int_0^h F(x)\, dx = h \int_0^1 f(u)\, du = \frac{h}{2}(f_0 + f_1)$$

the corresponding cubature formula is

$$\int_0^h \int_0^k F(x,y)\, dx\, dy = \frac{hk}{4}(f_{00} + f_{10} + f_{01} + f_{11}) \qquad (11.75)$$

Since the tabulated values f_{00}, f_{10}, f_{01}, and f_{11} are any four adjacent ones bearing the following spacial relationship to one another in the xy plane:

$$f_{00} \quad f_{01}$$
$$f_{10} \quad f_{11}$$

it is convenient to represent the cubature formula (11.75) in abbreviated form by a pattern of weights. Thus Eq. (11.75) may be written

$$\int_0^h \int_0^k F(x,y) \, dx \, dy = \frac{hk}{4} \begin{Bmatrix} 1 & 1 \\ 1 & 1 \end{Bmatrix} \tag{11.76}$$

If one uses Simpson's rule in place of the trapezoidal rule, one obtains the corresponding cubature formula

$$\int_{-h}^h \int_{-k}^k F(x,y) \, dx \, dy = \frac{hk}{9} \begin{Bmatrix} 1 & 4 & 1 \\ 4 & 16 & 4 \\ 1 & 4 & 1 \end{Bmatrix} \tag{11.77}$$

These are probably the two most generally used cubature formulas.

PROBLEMS

1. Prove by mathematical induction that the definitions of Eqs. (11.8) lead to the following expressions for differences:

$$\Delta_x^s \Delta_y^t f_{pq} = \sum_{i=0}^s (-1)^{s-i} \binom{s}{i} \Delta_y^t f_{p+i,q}$$

$$= \sum_{i=0}^s \sum_{j=0}^t (-1)^{s+t-i-j} \binom{s}{i} \binom{t}{j} f_{p+i,q+j}$$

2. Show that the expressions for the differences in Prob. 1 follow formally, very readily, from Eqs. (11.10).
3. Using the following property of the binomial coefficients

$$\sum_{s=i}^p (-1)^{s-i} \binom{p}{s} \binom{s}{i} = \binom{p}{i} \sum_{k=0}^{p-i} (-1)^k \binom{p-i}{k} = \begin{cases} 1 & \text{if } i = p \\ 0 & \text{if } i < p \end{cases}$$

show that Eq. (11.7) gives $P_{mn}(p,q) = f_{pq}$ for $0 \le p \le m$ and $0 \le q \le n$.
4. Show that the polynomial $P_3(u,v)$ of Eq. (11.19) does satisfy Eq. (11.21) for the required 10 pairs of values of s and t.
5. Prove that the coefficients of the third-degree polynomial $P_3(u,v)$ in Eq. (11.23) are uniquely determined by the tabulated values shown in Eq. (11.22).
6. Construct a descending-difference table for the polynomial

$$P_2(u,v) = 3u^2 + 2uv - 4v^2 + u - v + 7$$

using intervals of $h = k = 0.2$. Restrict the tabulation to the region $0 \le u \le 1$ and $0 \le v \le 1$.

7. A polynomial in u and v usually vanishes along some locus in the uv plane. Determine the locus for each of the coefficient polynomials of the f_{ij} in Eq. (11.37). Note that each one passes through all but one of the six points in the region of fit.

8. Use the Lagrangian coefficients to interpolate for $f(4.2,3.5)$ if for integral values of m and n

$$f(m,n) = \frac{m!}{n!(m - n)!}$$

Use four-point formulas for both variables.

9. Given the bivariate table for $f(x,y)$

y \ x	0.800	0.805	0.810	0.815
0.30	0.85 837	0.85 502	0.85 150	0.84 781
0.31	0.84 940	0.84 585	0.84 214	0.83 824
0.32	0.84 022	0.83 648	0.83 257	0.82 846
0.33	0.83 085	0.82 692	0.82 280	0.81 848

find by the use of the Gregory-Newton bivariate-interpolation formula (triangular form) the value of $f(0.304,0.802)$.

10. Determine the coefficients in the following cubature formula:

$$\int_{x_0}^{x_0+h} \int_{y_0}^{y_0+k} f(x,y)\, dx\, dy = hk[A_{00}f(x_0,y_0) + A_{10}f(x_0 + h, y_0) + A_{01}f(x_0, y_0 + k)$$
$$+ A_{20}f(x_0 + 2h, y_0) + A_{11}f(x_0 + h, y_0 + k) + A_{02}f(x_0, y_0 + 2k)]$$

11. Find the volume under the surface $f(x,y)$ over the region for which $x_0 < x < x_0 + h$ and $y_0 < y < y_0 + k$ in terms of the nearest 16 ordinates $f_{st} = f(x_0 + sh, y_0 + tk)$.

CHAPTER 12

EXPRESSION OF A PARTIAL DIFFERENTIAL EQUATION AS A PARTIAL DIFFERENCE EQUATION

One of the most powerful methods for solving a partial differential equation numerically is the method of finite differences, in which one first approximates the differential equation by a difference equation and then solves this resulting difference equation. It is the purpose of this chapter to explain this procedure in general terms, leaving for subsequent chapters the actual application of the method to the important partial differential equations of applied mathematics.

12.1. Classification of the Partial Differential Equations of the Second Order. Given a partial differential equation of the second order, with two independent variables x and y, of the form

$$a(x,y)\phi_{xx} + b(x,y)\phi_{xy} + c(x,y)\phi_{yy} + f(x,y,\phi_x,\phi_y) = 0 \qquad (12.1)$$

i.e., one that is linear in the second-order partial derivatives,

$$\phi_{xx} = \frac{\partial^2\phi}{\partial x^2} \qquad \phi_{xy} = \frac{\partial^2\phi}{\partial x\,\partial y} \qquad \text{and} \qquad \phi_{yy} = \frac{\partial^2\phi}{\partial y^2} \qquad (12.2)$$

it is possible to classify it on the basis of the expression $b^2 - 4ac$ formed from the functions a, b, and c. Furthermore, this expression determines the nature of the characteristic quadratic form

$$Q(v,w) = av^2 + bvw + cw^2 \qquad (12.3)$$

The classification is as follows:

1. *Elliptic Type*† (if $b^2 - 4ac < 0$ at all points of the region). This together with the assumption that $a > 0$ ensures that the characteristic quadratic form $Q(v,w)$ is positive definite, i.e., positive for all real values of v and w except when $v = w = 0$. Such a differential equation can always be reduced to the form‡

$$\phi_{\xi\xi} + \phi_{\eta\eta} = G(\xi,\eta,\phi,\phi_\xi,\phi_\eta) \qquad (12.4)$$

---

† The name applied to each type of differential equation is that of the associated conic, namely, the locus of points in the vw plane for which $Q(v,w) = Av + Bw + C$, where A, B, and C are constants.

‡ See J. D. Tamarkin and W. Feller, "Partial Differential Equations," pp. 1–5 (mimeographed notes, Brown University), 1941.

by means of transformations of the type

$$\xi = \xi(x,y) \quad \text{and} \quad \eta = \eta(x,y) \tag{12.5}$$

The boundary conditions are generally given by prescribing ϕ, its normal derivative, or a linear combination of these on the boundary of the region in which ϕ is to be determined, i.e., boundary conditions of the *Dirichlet* type.

Probably the most fundamental equation of this type is *Laplace's* equation

$$\nabla^2 \phi = \frac{\partial^2 \phi}{\partial x^2} + \frac{\partial^2 \phi}{\partial y^2} = 0 \tag{12.6}$$

2. *Parabolic Type* (if $b^2 - 4ac = 0$ at all points of the region considered). Such a differential equation by means of suitable equations of the form of Eqs. (12.5) can be put in the form

$$\phi_{\eta\eta} = F(\xi, \eta, \phi, \phi_\xi, \phi_\eta) \tag{12.7}$$

The one-dimensional heat-flow equation

$$\frac{\partial^2 T}{\partial x^2} - \frac{1}{a} \frac{\partial T}{\partial t} = 0 \tag{12.8}$$

is of this type. The initial value of T at some time t_0 (and its value, or that of its derivative, on the boundaries $x = c_1$ and $x = c_2$) usually constitutes the boundary conditions.

3. *Hyperbolic Type* (if $b^2 - 4ac > 0$ at all points of the region). Any differential equation of this type can be reduced to the form

$$\phi_{\xi\eta} = H(\xi, \eta, \phi, \phi_\xi, \phi_\eta) \tag{12.9}$$

by again employing a suitable transformation of the form of Eqs. (12.5).

One usually prescribes ϕ and its first derivatives† on some line. The determination of ϕ over the whole area from the differential equation and these boundary values is termed the *Cauchy* problem.

A basic equation of this type is the one-dimensional wave equation, satisfied, for instance, by the vibrating string, namely,

$$\frac{\partial^2 \phi}{\partial x^2} - \frac{1}{v^2} \frac{\partial^2 \phi}{\partial t^2} = 0 \tag{12.10}$$

where $v = v(x)$ is the phase velocity of the propagation at each point.

As a further classification we shall divide all differential equations into the two groups:

1. *Linear* equations, in which ϕ and all its various derivatives enter linearly. For such equations any sum, or linear combination, of solu-

---

† A directional derivative normal to the line suffices, since the derivative along the line is determined by specifying ϕ.

tions of the differential equations is again a solution. This makes it possible to express a solution that satisfies the particular boundary conditions assigned in terms of a set of fundamental solutions.

Most of the classical theory of applied physics is summarized by a relatively few linear partial differential equations.

2. *Nonlinear* equations. For these equations solutions cannot be combined linearly to form other solutions. For instance, if ϕ_1 and ϕ_2 are solutions of the differential equations, $\phi_1 + \phi_2$ and $\phi_1 - \phi_2$ need not be solutions.

These equations have received increasing attention in recent years, and involve very serious difficulties from the analytical point of view. The advent of large-scale digital computing machines has, however, opened up this field to numerical methods of solution.

12.2. Two-dimensional Lattices and Nets. Consider the set of values obtained from a function $f(x,y)$ by inserting only certain discrete values $x = x_s$ and $y = y_t$, where s and t may take on various integral values. If the lattice of points (x_s,y_t) is close enough together, this set of values, $f(x_s,y_t)$, can be used in numerical calculations in place of the function $f(x,y)$. The reasonableness of the approach

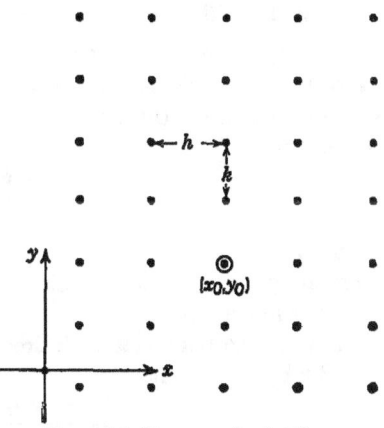

FIG. 12.1. Rectangular lattice.

lies in our ability to infer by interpolation the value of $f(x,y)$ for nonlattice points. Of course, this procedure involves an approximation, but the accuracy can be improved by increasing the number of lattice points.

As an example of this, consider the rectangular lattice formed when $x_s = x_0 + sh$ and $y_t = y_0 + tk$. This lattice is shown in Fig. 12.1, and the problem of interpolating from it has been covered in Chap. 11.

There are, strangely enough, only three basic two-dimensional lattices that can be formed from regular polygons, namely, (1) square, (2) triangular, and (3) hexagonal. Other suitable lattices, however, are obtainable from these by conformal mapping. If one draws in the sides of the elementary squares, triangles, or hexagons, the resulting lines form a *net*. That there are no other regular polygons which can be used as units of a net can be seen as follows. Let θ be the interior angle formed by any two adjacent sides of such a polygon. In order for this polygon to fit at its vertices with its neighbors to form a net, one must set $N\theta = 360°$, where N is an integer. Clearly $N > 2$, and also, since the equilateral triangle has the smallest angle of any regular polygon, $\theta \geq 60°$. Thus the only possibilities are $N = 3, 4, 5,$ and 6. Of these, $N = 5$ is impossible, and the other three yield the above basic nets.†

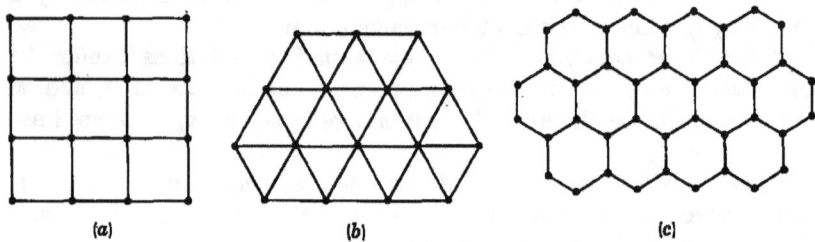

(a)   (b)   (c)

Fig. 12.2. The three basic nets: (a) square; (b) triangular; (c) hexagonal.

12.3. Approximation for the Two-dimensional Laplacian in a Square Lattice. In problems involving a variation of ϕ in only two of the three space coordinates, the Laplacian of ϕ expressed in cartesian coordinates reduces to

$$\frac{\partial^2 \phi}{\partial x^2} + \frac{\partial^2 \phi}{\partial y^2} \tag{12.11}$$

Therefore, in a square lattice this two-dimensional Laplacian can be approximated by replacing the second derivatives by their expression in terms of finite differences.

Suppose the Laplacian is desired at (x_0, y_0) in Fig. 12.1. We can define u and v by the equations

$$u = \frac{x - x_0}{h} \qquad v = \frac{y - y_0}{k} \tag{12.12}$$

and let

$$\phi_{st} \equiv \phi(x_s, y_t) \tag{12.13}$$

In which case, by the use of Stirling's interpolation formula [see Eq. (4.21)],

$$\phi(x, y_0) = \phi_{00} + u\,\mu\delta_x\phi_{00} + \frac{1}{2!}\,u^2\,\delta_x^2\phi_{00} + \frac{1}{3!}\,u(u^2 - 1)\,\mu\delta_x^3\phi_{00}$$

$$+ \frac{1}{4!}\,u^2(u^2 - 1)\,\delta^4\phi_{00} + \cdots \tag{12.14}$$

and hence

$$\frac{\partial^2}{\partial x^2}\,\phi(x, y_0)\,\bigg|_{x = x_0} = \frac{1}{h^2}\,\frac{\partial^2}{\partial u^2}\,\phi(x, y_0)\,\bigg|_{u = 0}$$

$$= \frac{1}{h^2}\,(\delta_x^2\phi_{00} - \tfrac{1}{12}\delta_x^4\phi_{00} + \cdots) \tag{12.15}$$

† A net formed of nonregular polygons may be useful in special cases; thus a net of parallelograms would be natural if the region in which a solution is desired is a parallelogram.

where the subscript x on the δ_x indicates that the differences are formed with respect to x. In particular (see Sec. 4.1),

$$\delta_x^2 \phi_{00} = \phi_{-10} - 2\phi_{00} + \phi_{10}$$

and $\quad\quad \delta_x^4 \phi_{00} = \phi_{-20} - 4\phi_{-10} + 6\phi_{00} - 4\phi_{10} + \phi_{20}$ \hfill (12.16)

Since the second derivative with respect to y can be handled similarly, we see that

$$\left(\frac{\partial^2 \phi}{\partial x^2} + \frac{\partial^2 \phi}{\partial y^2}\right)_{x_0, y_0} = \frac{1}{h^2}\left(\delta_x^2 \phi_{00} - \tfrac{1}{12}\delta_x^4 \phi_{00} + \cdots\right)$$

$$+ \frac{1}{k^2}\left(\delta_y^2 \phi_{00} - \tfrac{1}{12}\delta_y^4 \phi_{00} + \cdots\right) \quad (12.17)$$

where $\quad\quad \delta_y^2 \phi_{00} = \phi_{0,-1} - 2\phi_{00} + \phi_{01}$

and $\quad\quad \delta_y^4 \phi_{00} = \phi_{0,-2} - 4\phi_{0,-1} + 6\phi_{00} - 4\phi_{01} + \phi_{02}$ \hfill (12.18)

It is customary to use only the first approximation to the Laplacian obtained by neglecting the fourth and all higher differences in Eq. (12.17). Moreover, for a square net, $h = k$; therefore

$$\left(\frac{\partial^2 \phi}{\partial x^2} + \frac{\partial^2 \phi}{\partial y^2}\right)_{x_0, y_0} \cong \nabla_k^2 \phi_{00} \quad (12.19)$$

where

$$\nabla_k^2 \phi_{00} \equiv \frac{1}{h^2}\left(\delta_x^2 \phi_{00} + \delta_y^2 \phi_{00}\right)$$

$$= \frac{1}{h^2}\left(\phi_{10} + \phi_{01} + \phi_{-10} + \phi_{0,-1} - 4\phi_{00}\right) \quad (12.20)$$

If one uses a single subscript to designate the relative position of the ϕ's employed in Eq. (12.20), the expression looks somewhat simpler. Let $\phi_0 = \phi_{00}$, $\phi_1 = \phi_{10}$, $\phi_2 = \phi_{01}$, $\phi_3 = \phi_{-10}$, and $\phi_4 = \phi_{0,-1}$; then

$$\nabla_k^2 \phi_0 \equiv \frac{1}{h^2}\left(\phi_1 + \phi_2 + \phi_3 + \phi_4 - 4\phi_0\right) = \frac{1}{h^2}\left\{\begin{array}{ccc} & \phi_2 & \\ \phi_3 & -4\phi_0 & \phi_1 \\ & \phi_4 & \end{array}\right\} \quad (12.21)$$

Thus, purely symbolically

$$\nabla_k^2 \equiv \frac{1}{h^2} \quad$$ $$\quad (12.22)$$

For sake of completeness it is desirable to obtain the second-order approximation to the Laplacian by taking into account the fourth differences. Thus with neglect of only sixth and higher differences

$$\left(\frac{\partial^2\phi}{\partial x^2} + \frac{\partial^2\phi}{\partial y^2}\right)_{x_0,y_0} \cong \nabla_h'^2\phi_{00}$$

where

$$\nabla_h'^2\phi_{00} = \frac{1}{h^2}\left(\delta_x^2\phi_{00} - \tfrac{1}{12}\delta_x^4\phi_{00} + \delta_y^2\phi_{00} - \tfrac{1}{12}\delta_y^4\phi_{00}\right)$$

$$= \frac{1}{12h^2}\left[-(\phi_{20} + \phi_{02} + \phi_{-20} + \phi_{0,-2})\right.$$

$$\left. + 16(\phi_{10} + \phi_{01} + \phi_{-10} + \phi_{0,-1}) - 60\phi_{00}\right] \quad (12.23)$$

or symbolically,

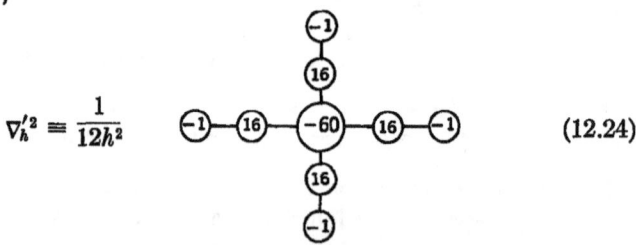

$$\nabla_h'^2 \equiv \frac{1}{12h^2} \qquad\qquad (12.24)$$

Although the approximation of the Laplacian by $\nabla_h'^2$ instead of ∇_h^2 should yield better results for a given net spacing h, because of the added complication of the former, it has not often been used in the literature. This may be due, in part, to the need for special weights, other than those above, at points adjacent to the boundary.

12.4. Approximation for the Laplacian in a Triangular Lattice. The second derivative of a function $\phi(x,y)$ in any desired direction can be expressed in terms of three second derivatives, $\partial^2\phi/\partial x^2$, $\partial^2\phi/\partial y^2$, and $\partial^2\phi/\partial x\,\partial y$. To derive this relation, let x' and y' be defined by the equations,

$$x = x'\cos\theta - y'\sin\theta$$
$$y = x'\sin\theta + y'\cos\theta \qquad\qquad (12.25)$$

This means geometrically that the prime coordinate system is rotated counterclockwise by the angle θ with respect to the unprimed axes. Therefore, the second derivative of $\phi(x,y)$ in a direction θ is nothing more than

$$\frac{\partial^2\phi}{\partial x'^2} = \frac{\partial}{\partial x'}\left(\frac{\partial\phi}{\partial x}\frac{\partial x}{\partial x'} + \frac{\partial\phi}{\partial y}\frac{\partial y}{\partial x'}\right) = \frac{\partial}{\partial x'}\left(\cos\theta\frac{\partial\phi}{\partial x} + \sin\theta\frac{\partial\phi}{\partial y}\right)$$

$$= \left(\cos\theta\frac{\partial}{\partial x} + \sin\theta\frac{\partial}{\partial y}\right)^2\phi$$

$$= \cos^2\theta\frac{\partial^2\phi}{\partial x^2} + \sin^2\theta\frac{\partial^2\phi}{\partial y^2} + 2\sin\theta\cos\theta\frac{\partial^2\phi}{\partial x\,\partial y} \qquad (12.26)$$

We are now in position to express the Laplacian in the triangular net of Fig. 12.3 in terms of the second partial derivatives in the three basic directions, a, b, and c. The values for the angle θ (the angle measured with respect to the positive x axis) for these three directions

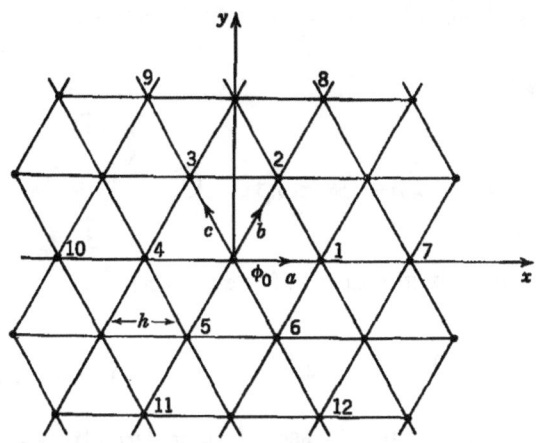

FIG. 12.3. Triangular net.

are, respectively, 0, 60, and 120°; therefore

$$\frac{\partial^2 \phi}{\partial a^2} = \frac{\partial^2 \phi}{\partial x^2}$$

$$\frac{\partial^2 \phi}{\partial b^2} = \frac{1}{4}\frac{\partial^2 \phi}{\partial x^2} + \frac{3}{4}\frac{\partial^2 \phi}{\partial y^2} + \frac{\sqrt{3}}{2}\frac{\partial^2 \phi}{\partial x\, \partial y}$$ (12.27)

$$\frac{\partial^2 \phi}{\partial c^2} = \frac{1}{4}\frac{\partial^2 \phi}{\partial x^2} + \frac{3}{4}\frac{\partial^2 \phi}{\partial y^2} - \frac{\sqrt{3}}{2}\frac{\partial^2 \phi}{\partial x\, \partial y}$$

Adding these equations, one has

$$\frac{\partial^2 \phi}{\partial a^2} + \frac{\partial^2 \phi}{\partial b^2} + \frac{\partial^2 \phi}{\partial c^2} = \frac{3}{2}\left(\frac{\partial^2 \phi}{\partial x^2} + \frac{\partial^2 \phi}{\partial y^2}\right)$$ (12.28)

If one now expresses the second derivatives with respect to a, b, and c in terms of differences, as indicated in Eq. (12.16), one obtains from Eq. (12.28) an expression for Laplacian in terms of the even-order central differences in these three directions. In particular, if the fourth and all higher differences are neglected,

$$\left(\frac{\partial^2 \phi}{\partial a^2}\right)_0 \cong \frac{1}{h^2}\left(\phi_4 - 2\phi_0 + \phi_1\right)$$

$$\left(\frac{\partial^2 \phi}{\partial b^2}\right)_0 \cong \frac{1}{h^2}\left(\phi_5 - 2\phi_0 + \phi_2\right)$$ (12.29)

$$\left(\frac{\partial^2 \phi}{\partial c^2}\right)_0 \cong \frac{1}{h^2}\left(\phi_6 - 2\phi_0 + \phi_3\right)$$

and hence the Laplacian at the point 0 is

$$\left(\frac{\partial^2 \phi}{\partial x^2} + \frac{\partial^2 \phi}{\partial y^2}\right)_0 \cong \frac{2}{3h^2}(\phi_1 + \phi_2 + \phi_3 + \phi_4 + \phi_5 + \phi_6 - 6\phi_0)$$

$$\equiv \nabla_h^2 \phi_0 \quad (12.30)$$

or symbolically†

$$\frac{\partial^2}{\partial x^2} + \frac{\partial^2}{\partial y^2} \cong \nabla_h^2 \equiv \frac{2}{3h^2}$$

$$(12.31)$$

Again, by including fourth differences one gets the second-order approximation

$$\frac{\partial^2 \phi}{\partial a^2} = \frac{1}{12h^2}(-\phi_{10} + 16\phi_4 - 30\phi_0 + 16\phi_1 - \phi_7)$$

and similar expressions for the second-order derivatives with respect to b and c. The second approximation to the Laplacian, which we shall again distinguish from the first approximation by a prime, is therefore

$$\nabla_h'^2 \phi_0 \equiv \frac{1}{18h^2}[-(\phi_7 + \phi_8 + \phi_9 + \phi_{10} + \phi_{11} + \phi_{12})$$

$$+ 16(\phi_1 + \phi_2 + \phi_3 + \phi_4 + \phi_5 + \phi_6) - 90\phi_0] \quad (12.32)$$

or symbolically

$$\frac{\partial^2}{\partial x^2} + \frac{\partial^2}{\partial y^2} \cong \nabla_h'^2 \equiv \frac{1}{18h^2}$$

$$(12.33)$$

Again, however, only ∇_h^2 [given by Eq. (12.31)] is commonly found in the literature.

12.5. Approximation for the Laplacian in a Hexagonal Lattice. To obtain an approximation for the Laplacian for the hexagonal net of Fig. 12.4 we shall assume that a polynomial

$$P_3(x,y) = a + bx + cy + dx^2 + ey^2 + fxy + gx^3 + jx^2y + kxy^2$$
$$+ my^3 \quad (12.34)$$

† This is obviously not the same $\nabla_h^2 \phi$ defined in Eq. (12.22), but since it is the analogous expression for a triangular net, and since no confusion is likely to arise from its use, we do not need to employ a new symbol.

is made to agree with ϕ at the points 0, 1, 2, 3, . . . , 9, by proper choice of theıcoefficients a, b, c, . . . , k, m. Then the Laplacian at $x = 0$ and $y = 0$,

$$\nabla^2\phi_0 \cong \nabla^2 P_3(x,y) = 2(d + e)$$
$$(12.35)$$

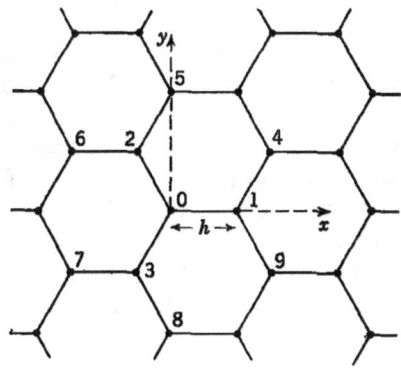

may be expressed in terms of the values of $P_3(x,y)$ at the above points by finding a suitable expression for $d + e$.

Suppose one adds the ϕ's at a set of these points that are symmetrically placed with respect to x and y axes. This amounts to adding $P_3(x,y)$ at these same points, and hence the only terms in the polyno-

FIG. 12.4. Hexagonal net.

mial that will contribute to the sum will be those symmetrical in x and y. Referring to Eq. (12.34), we see that only the terms a, dx^2, and ey^2 will contribute. In particular [as seen by substituting the coordinates of the points in Eq. (12.34)]

$$\phi_4 + \phi_5 + \phi_6 + \phi_7 + \phi_8 + \phi_9 = 6a + 9h^2(d + e)$$

and since $\phi_0 = a$, we have from Eq. (12.35) that

$$\nabla^2\phi_0 \cong 2(d + e) = \nabla_h^2\phi_0$$
$$\equiv \frac{2}{9h^2}(\phi_4 + \phi_5 + \phi_6 + \phi_7 + \phi_8 + \phi_9 - 6\phi_0) \quad (12.36)$$

This is not the usual expression obtained for the Laplacian, since ordinarily one makes use of only the four points 0, 1, 2, and 3. This latter expression can be obtained by adding ϕ at the points 1, 2, and 3. Since this set of points is symmetrical only with respect to y, the terms that are odd in x may not drop out. In particular, since the coordinates of these three points are, respectively,

$$(h,0) \qquad \left(-\frac{h}{2}, \frac{\sqrt{3}}{2}h\right) \quad \text{and} \quad \left(-\frac{h}{2}, \frac{-\sqrt{3}}{2}h\right)$$

from Eq. (12.34)

$$\phi_1 + \phi_2 + \phi_3 = 3a + \frac{3h^2}{2}(d + e) + \frac{3h^3}{4}(g - k)$$

Since $\phi_0 = a$, one has

$$\nabla^2\phi_0 \cong 2(d + e) = \frac{4}{3h^2}(\phi_1 + \phi_2 + \phi_3 - 3\phi_0) + h(g - k)$$

and one makes an error of the order h in setting

$$\nabla^2 \phi_0 \cong \bar{\nabla}_k^2 \phi_0 \equiv \frac{4}{3h^2} (\phi_1 + \phi_2 + \phi_3 - 3\phi_0) \tag{12.37}$$

The approximation for the Laplacian given by Eq. (12.37), although generally advanced as adequate for the solution of partial differential equations (whenever a hexagonal net is to be employed), is seen to be very poor, except for an extremely fine net. Southwell, who has proposed the hexagonal net as a possible type, has not found them practical. Probably the poorness of the approximation given by Eq. (12.37) accounts for this situation. The $\nabla_k^2 \phi$ given by Eq. (12.36) approximates $\nabla^2 \phi$ to the order of h^2, the same order of approximation as given by $\nabla_k^2 \phi$ for the rectangular and triangular nets. In fact the points $\phi_4, \phi_5, \ldots, \phi_9$ lie in a hexagon about the point 0, and therefore the scheme given in Eq. (12.36) amounts essentially to the scheme given in Eqs. (12.30) and (12.31) but with the spacing h replaced by $\sqrt{3}\, h$.

12.6. Approximations for the Time Derivatives. In forming a lattice it is customary to take the same space points at each instant of time rather than following the more general procedure of applying any regular lattice to the space-time region being investigated. The simplification thus maintained permits one, in deriving the equivalent difference expressions for the time derivatives, to consider only that one space point at which the derivative is desired. This point may belong to a square, triangular, or hexagonal lattice.

Let the value of a function ϕ at some point designated by s be $\phi_s(t)$, and consider the values of $\phi_s(t)$ at equally spaced times t_i (i = integer), so that $t_{i+1} - t_i = p$ = constant. These values may be designated as ϕ_s^i, i.e.,

$$\phi_s^i \equiv \phi_s(t_i) \tag{12.38}$$

Since the only variable here considered is t, the problem of expressing the time derivatives in terms of differences is not essentially different from that given in Chap. 7.

If the Gregory-Newton formula is employed, by analogy with Eq. (7.4),

$$\frac{\partial \phi_s(t)}{\partial t}\bigg|_{t=t_0} = \frac{1}{p} (\Delta \phi_s^0 - \tfrac{1}{2}\Delta^2 \phi_s^0 + \tfrac{1}{3}\Delta^3 \phi_s^0 - \cdots)$$

As an approximation it is customary to use only the first term of this series; therefore,

$$\frac{\partial \phi_s(t)}{\partial t}\bigg|_{t=t_0} \cong \frac{1}{p} \Delta \phi_s^0 = \frac{\phi_s^1 - \phi_s^0}{p} \tag{12.39}$$

For later reference, the next higher approximation is

$$\frac{\partial \phi_s(t)}{\partial t}\bigg|_{t=t_0} \cong \frac{1}{p} (\Delta \phi_s^0 - \tfrac{1}{2}\Delta^2 \phi_s^0) = \frac{-\phi_s^2 + 4\phi_s^1 - 3\phi_s^0}{2p} \tag{12.40}$$

If one differentiates Stirling's formula instead of the Gregory-Newton interpolation formula, one observes that, analogous to Eq. (7.7),

$$\frac{\partial \phi_s(t)}{\partial t}\bigg|_{t=t_0} = \frac{1}{p}\left(\frac{\Delta\phi_s^{-1} + \Delta\phi_s^0}{2} - \frac{1}{6}\frac{\Delta^3\phi_s^{-2} + \Delta^3\phi_s^{-1}}{2} + \cdots\right) \quad (12.41)$$

hence one obtains as a first approximation

$$\frac{\partial \phi_s(t)}{\partial t}\bigg|_{t=t_0} \cong \frac{1}{2p}(\Delta\phi_s^{-1} + \Delta\phi_s^0) = \frac{\phi_s^1 - \phi_s^{-1}}{2p} \quad (12.42)$$

Although this expression for the derivative is more accurate than that given by Eq. (12.39), it involves a knowledge of ϕ at the prior time t_{-1}. Especially in the case of differential equations of the hyperbolic type this fact makes the expression given in Eq. (12.39) more to one's liking. The next higher approximation is then given by

$$\frac{\partial \phi_s(t)}{\partial t}\bigg|_{t=t_0} \cong \frac{1}{12p}(-\phi_s^2 + 8\phi_s^1 - 8\phi_s^{-1} + \phi_s^{-2}) \quad (12.43)$$

An approximation for the second derivative is obtained in a completely analogous way by twice differentiating the Gregory-Newton interpolation formula; thus†

$$\frac{\partial^2\phi_s(t)}{\partial t^2}\bigg|_{t=t_0} = \frac{1}{p^2}(\Delta^2\phi_s^0 - \Delta^3\phi_s^0 + \tfrac{11}{12}\Delta^4\phi_s^0 - \cdots) \quad (12.44)$$

Hence the first approximation is

$$\frac{\partial^2\phi_s(t)}{\partial t^2}\bigg|_{t=t_0} \cong \frac{1}{p^2}\Delta^2\phi_s^0 = \frac{\phi_s^2 - 2\phi_s^1 + \phi_s^0}{p^2} \quad (12.45)$$

while the second is

$$\frac{\partial^2\phi_s(t)}{\partial t^2}\bigg|_{t=t_0} \cong \frac{1}{p^2}(\Delta^2\phi_s^0 - \Delta^3\phi_s^0)$$
$$\cong \frac{-\phi_s^3 + 4\phi_s^2 - 5\phi_s^1 + 2\phi_s^0}{p^2} \quad (12.46)$$

It is generally agreed that one does better to use the approximations obtained from Stirling's formula. In place of Eq. (12.44) one then has

$$\frac{\partial^2\phi_s(t)}{\partial t^2}\bigg|_{t=t_0} = \frac{1}{p^2}(\Delta^2\phi_s^{-1} - \tfrac{1}{12}\Delta^4\phi_s^{-2} + \cdots) \quad (12.47)$$

therefore, as a first approximation

$$\frac{\partial^2\phi_s(t)}{\partial t^2}\bigg|_{t=t_0} \cong \frac{1}{p^2}\Delta^2\phi_s^{-1} = \frac{\phi_s^1 - 2\phi_s^0 + \phi_s^{-1}}{p^2} \quad (12.48)$$

† This formula may be obtained at once from the lozenge diagram of Fig. 7.2 by going down the diagonal from Y_0 and applying the rules in Sec. 4.8 for the use of lozenge diagrams.

This is the approximation for the second derivative that is used almost universally.

Example 1. Approximate the two-dimensional heat-flow equation by a finite-difference equation expressed in terms of a square lattice.
The differential equation for heat flow in two dimensions is

$$\frac{\partial^2 \phi}{\partial x^2} + \frac{\partial^2 \phi}{\partial y^2} - \frac{1}{a}\frac{\partial \phi}{\partial t} = 0 \tag{12.49}$$

where $a = \kappa/c\rho$ is the *thermal diffusivity* and κ, c, ρ are the *thermal conductivity, specific heat*, and *density* of the medium. If we require that this equation be satisfied at all lattice points at the time $t_j = t_0 + pj$, where j is any positive or negative integer, and use the approximations for the derivatives given in Eqs. (12.21) and (12.39), we obtain the difference equation

$$\frac{1}{h^2}(\phi_1^j + \phi_2^j + \phi_3^j + \phi_4^j - 4\phi_0^j) - \frac{1}{ap}(\phi_0^{j+1} - \phi_0^j) = 0$$

and therefore

$$\phi_0^{j+1} = \phi_0^j + \frac{ap}{h^2}(\phi_1^j + \phi_2^j + \phi_3^j + \phi_4^j - 4\phi_0^j) \tag{12.50}$$

Here ϕ_0^j is the value of ϕ at any of the points in the lattice at the time $t = t_j$, and ϕ_1^j, ϕ_2^j, ϕ_3^j, ϕ_4^j are the values of ϕ at the four neighboring points.
If one lets

$$m = \frac{ap}{h^2} \tag{12.51}$$

then

$$\phi_0^{j+1} = (1 - 4m)\phi_0^j + m(\phi_1^j + \phi_2^j + \phi_3^j + \phi_4^j) \tag{12.52}$$

This equation expresses the temperature ϕ of any lattice point at t_{j+1} in terms of its temperature at t_j and the average temperature of the four surrounding points at this prior time. Since the initial values of the temperature at $t = t_0$ are usually specified in the boundary conditions, Eq. (12.52), when applied to each interior point of a lattice, gives the temperature at $t_1 = t_0 + p$. A second application of this procedure gives the values of ϕ at t_2. Thus, by repetition of the process one obtains the approximate temperature behavior for all subsequent times.

The use of the approximation for the derivative of Eq. (12.39) instead of the more accurate expression of Eq. (12.42) is clearly required if a simple determination of temperature at $t = t_1$ is to be obtained. One can, however, by applying special consideration to the beginning values, introduce the more accurate expression for the first time derivative given by Eq. (12.42), but it turns out that the resulting difference equation is completely unstable with respect to the accumulation of round-off error. Therefore this difference equation is generally discarded in favor of the less accurate but more stable equation of (12.52).

PROBLEMS

1. (a) Show that a partial differential equation of the elliptic type linear in the second derivatives can always be reduced to the form given in Eq. (12.4). (b) Show that the characteristic quadratic form of Eq. (12.3) is positive definite for an elliptic-type equation.

2. Consider a very general homogeneous linear partial differential equation

$$a\phi_{xx} + b\phi_{xy} + c\phi_{yy} + d\phi_x + e\phi_y = 0$$

of the second degree, where the coefficients $a, b, \ldots, e$ may be functions of x and y. Prove that, if ϕ_1 and ϕ_2 are solutions of this equation, it will also have as a solution any linear combination $\phi = c_1\phi_1 + c_2\phi_2$ of these solutions.

3. If ϕ_1 and ϕ_2 are two solutions of the nonlinear differential equation $\phi_{xx} + \phi_y^2 = 0$, show that $\phi = \phi_1 + \phi_2$ cannot be a solution unless ϕ_1 or ϕ_2 is a function of x alone.

4. Look up the expression for the Laplacian in the following coordinate systems: (a) cylindrical, (b) spherical, (c) general curvilinear.

5. Show that, if n is a unit vector, $f(n \cdot r - vt)$ is a particular solution of Eq. (12.10) and represents a wave moving in the direction of n with a velocity v. What is the nature of the surfaces of constant f?

6. (a) Show that the differential equation of a rectangular drumhead vibrating at a single frequency is given by

$$\frac{\partial^2\phi}{\partial x^2} + \frac{\partial^2\phi}{\partial y^2} + \lambda^2\phi = 0$$

where ϕ is the displacement and $\lambda = \omega\sqrt{M/T}$, ω being the angular frequency, T the tension, and M the mass per unit area. (b) Show that the characteristic values of λ are

$$\lambda_{mn} = \pi\left(\frac{m^2}{a^2} + \frac{n^2}{b^2}\right)^{\frac{1}{2}}$$

where m and n are arbitrary positive integers.

7. Derive the finite-difference approximation to the Laplacian for a rhombic net. Note that such a net can also be thought of as made up of isosceles triangles.

8. Obtain in terms of a modulus, similar to the m used in Eq. (12.52), a finite-difference approximation to the wave equation (12.10).

9. (a) Set up a finite-difference approximation to the differential equation of Prob. 6, and apply it to a square drumhead represented by four internal points of a 16-point square net, the displacement function ϕ being zero at all boundary points. (b) Find those values of λ which permit a nontrivial solution, and compare your answer with the first few λ's of Prob. 6, which were obtained directly from the solution of the differential equation.

10. Determine the proper finite-difference equation for heat flow in three dimensions when cylindrical symmetry is present. Note that only two space dimensions are necessary but that the weights to be used are functions of r, the radial distance from the axis.

CHAPTER 13

SOLUTION OF PARTIAL DIFFERENTIAL EQUATIONS
OF THE ELLIPTIC TYPE

As stated in Chap. 12, the boundary conditions for partial differential equations of the elliptic type are usually specified by giving the value of the function sought, its normal derivative, or a linear combination of both on some closed boundary. As an example, one might be interested in determining the electrostatic potential between a charged sphere and a cubical box enclosing this sphere. The potentials of the sphere and of the box fully determine the potential at each point of the space included between the outside surface of the sphere and the inside surface of the box. One needs only to solve Laplace's equation with these boundary conditions.

We shall consider first the two most important special cases of an elliptic partial differential equation, namely, Laplace's and Poisson's equations, and then later some typical examples of other elliptic equations. Our discussion will be confined almost entirely to two-dimensional problems and square lattices.

LAPLACE'S EQUATION

13.1. Difference Equations Are Reducible to Sets of Linear Algebraic Equations. It is convenient to combine our discussion of the basic methods with a simple example. Suppose that the potential ϕ satisfying Laplace's equation is required for the region contained within a given square (see Fig. 13.1). The potential ϕ is zero on three sides of the square. On the fourth, or top, side $\phi = 1$ at the middle point and drops off linearly as one approaches the corners, where $\phi = 0$.

Let a square net of 25 lattice points be used to approximate the behavior of ϕ. If one designates by ϕ_s the potential at the interior lattice points s, where $s = 1, 2, 3, \ldots, 9$, and by ϕ_t, $t = 10, 11, 12, \ldots, 25$, the potential at the boundary points, then the numerical solution involves determining the values of the ϕ_s in terms of the ϕ_t.

As shown in Chap. 12, Laplace's equation

$$\nabla^2\phi = \frac{\partial^2\phi}{\partial x^2} + \frac{\partial^2\phi}{\partial y^2} = 0 \tag{13.1}$$

can be approximated by the partial difference equation indicated symbolically by

$$\nabla_h^2 \phi_s = \frac{1}{h^2} \left\{ \begin{matrix} & 1 & \\ 1 & -4 & 1 \\ & 1 & \end{matrix} \right\} \phi_s = 0 \qquad (13.2)$$

where h is the net spacing. The symbol enclosed in braces merely indicates that one adds together the potential at the four closest neighbors of ϕ_s and then subtracts $4\phi_s$ from this sum. For example, if $s = 2$,

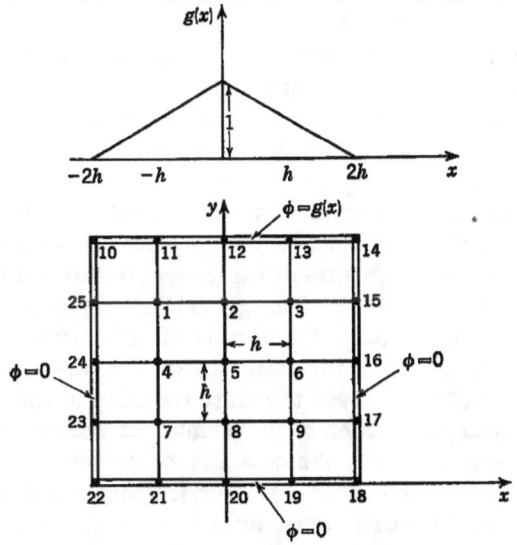

FIG. 13.1. Use of a square net to determine the potential within a square region.

the difference equation (13.2) is written

$$\frac{1}{h^2} (\phi_1 + \phi_{12} + \phi_3 + \phi_5 - 4\phi_2) = 0 \qquad (13.3)$$

One can, of course, multiply through by h^2.

This Laplace difference equation applied to each of the nine interior points ϕ_s gives rise to the system of nine simultaneous equations

$$
\begin{aligned}
-4\phi_1 + \phi_2 + \phi_4 &= -(\phi_{11} + \phi_{25}) \\
\phi_1 - 4\phi_2 + \phi_3 + \phi_5 &= -\phi_{12} \\
\phi_2 - 4\phi_3 + \phi_6 &= -(\phi_{13} + \phi_{15}) \\
\phi_1 - 4\phi_4 + \phi_5 + \phi_7 &= -\phi_{24} \\
\phi_2 + \phi_4 - 4\phi_5 + \phi_6 + \phi_8 &= 0 \\
\phi_3 + \phi_5 - 4\phi_6 + \phi_9 &= -\phi_{16} \\
\phi_4 - 4\phi_7 + \phi_8 &= -(\phi_{21} + \phi_{23}) \\
\phi_5 + \phi_7 - 4\phi_8 + \phi_9 &= -\phi_{20} \\
\phi_6 + \phi_8 - 4\phi_9 &= -(\phi_{17} + \phi_{19})
\end{aligned}
\qquad (13.4)
$$

The right-hand sides of these equations are known, since they represent the boundary points.† It is not too difficult to solve this set of equations by the standard analytical methods. If, however, a large number of interior points were used, one would need to resort to one of the methods used for solving high-order sets of linear equations given in Chap. 10.

The above is a particular example of a general property of the finite-difference method of solving linear partial differential equations, namely, the difference equation that replaces the differential equation can always be reduced to a set of n simultaneous linear equations, where n is the number of interior points of the lattice.

Since the solution of high-order linear equations has been successfully carried out using automatic computing machines,‡ a practical method that may be employed for problems involving less than about a hundred interior points is the reduction of the difference equation to a set of linear algebraic equations.

13.2. Making Use of Symmetries. If one solves the square lattice of Fig. 13.1 using the given boundary conditions, one finds that $\phi_1 = \phi_3$, $\phi_4 = \phi_6$, and $\phi_7 = \phi_9$. This result can clearly be inferred directly from the symmetry about the y axis of the given boundary conditions. If one makes use of this in advance, the number of unknowns can be reduced to six. Later examples will illustrate further how the presence of symmetry in a problem can be used to reduce the labor of computation.

13.3. Preliminary Reduction of the Number of Unknowns. One may also reduce the number of unknowns and hence the number of simultaneous equations to about half by observing that every other point in each row can be expressed as the average of its neighbors; hence it need not be included as one of the unknowns. For example, in our problem we need only determine the three unknowns ϕ_2, $\phi_4 = \phi_6$, and ϕ_8, since in terms of these

$$\phi_3 = \phi_1 = \tfrac{1}{4}(\phi_2 + \phi_4 + \phi_{11} + \phi_{25}) = \tfrac{1}{4}(\phi_2 + \phi_4 + \tfrac{1}{2})$$
$$\phi_5 = \tfrac{1}{4}(\phi_2 + \phi_4 + \phi_6 + \phi_8) = \tfrac{1}{4}(\phi_2 + 2\phi_4 + \phi_8) \qquad (13.5)$$
$$\phi_9 = \phi_7 = \tfrac{1}{4}(\phi_4 + \phi_8 + \phi_{21} + \phi_{23}) = \tfrac{1}{4}(\phi_4 + \phi_8)$$

The three simultaneous equations that must be solved for ϕ_2, ϕ_4, and ϕ_8 are

$$13\phi_2 - 4\phi_4 - \phi_8 = 5$$
$$-4\phi_2 + 24\phi_4 - 4\phi_8 = 1 \qquad (13.6)$$
$$-\phi_2 - 4\phi_4 + 13\phi_8 = 0$$

These are obtained by substitution from Eqs. (13.5) into Eqs. (13.4).

† Note that the potentials of the corner boundary points ϕ_{10}, ϕ_{14}, ϕ_{18}, and ϕ_{22} do not influence the potentials at the interior points.

‡ The parallel but somewhat more difficult problem of inversion of a high-order matrix is described by H. F. Mitchell; see *Math. Tables and Other Aids to Computation*, **3:** 161–166 (1948).

If we had retained the boundary values ϕ_t in the equations, the first equation of Eqs. (13.6) would read

$$\phi_{11} + 4\phi_{12} + \phi_{13} + \phi_{15} + 2\phi_6 + \phi_8 + 2\phi_4 + \phi_{26} - 13\phi_2 = 0$$

which, using the symbolic notation of Chap. 12, may be written

$$\frac{1}{h^2} \left\{ \begin{array}{ccccc} & & 0 & & \\ & 1 & 4 & 1 & \\ 1 & 0 & -13 & 0 & 1 \\ & 2 & 0 & 2 & \\ & & 1 & & \end{array} \right\} \phi_2 = 0 \qquad (13.7a)$$

For the second equation we need to rotate this difference equation and write

$$\frac{1}{h^2} \left\{ \begin{array}{ccccc} & & 1 & & \\ & 1 & 0 & 2 & \\ 0 & 4 & -13 & 0 & 1 \\ & 1 & 0 & 2 & \\ & & 1 & & \end{array} \right\} \phi_4 = 0 \qquad (13.7b)$$

Similarly the same difference equation rotated applies to the third equation. These composite equations can be obtained directly from the difference equation (13.2), as will be shown below. If a larger number of points had been employed, one could use for points not adjacent to the sides the difference equation

$$\frac{1}{h^2} \left\{ \begin{array}{ccccc} & & 1 & & \\ & 2 & 0 & 2 & \\ 1 & 0 & -12 & 0 & 1 \\ & 2 & 0 & 2 & \\ & & 1 & & \end{array} \right\} \phi_s = 0 \qquad (13.8)$$

Equations (13.7) and (13.8) can be obtained as linear combinations of the difference equation (13.2) applied to neighboring points; thus in the case of Eq. (13.8)

$$\left\{ \begin{array}{ccccc} & & 1 & & \\ & 2 & 0 & 2 & \\ 1 & 0 & -12 & 0 & 1 \\ & 2 & 0 & 2 & \\ & & 1 & & \end{array} \right\} = 4 \left\{ \begin{array}{ccccc} & & 0 & & \\ & 0 & 1 & 0 & \\ 0 & 1 & -4 & 1 & 0 \\ & 0 & 1 & 0 & \\ & & 0 & & \end{array} \right\}$$

$$+ \left\{ \begin{array}{ccccc} & & 1 & & \\ & 1 & -4 & 1 & \\ 0 & 0 & 1 & 0 & 0 \\ & 0 & 0 & 0 & \\ & & 0 & & \end{array} \right\}$$

$$+ \text{ 3 other orientations of the latter} \qquad (13.9)$$

The numerical values of ϕ for the nine interior points of the problem pictured in Fig. 13.1 are obtained by solving Eqs. (13.6) for ϕ_2, $\phi_4 = \phi_6$, and ϕ_8 and then applying the averaging process, expressed by Eqs. (13.5), to find ϕ at the other five points. The results are given in Table 13.1, together with the correct values as obtained by solving the problem analytically. A glance at the percentage of error establishes the fact that the net mesh is too coarse for this problem, i.e., the number of interior points is much too small. This is due primarily to the fact that the boundary values of ϕ have a discontinuous derivative at the mid-point of the upper boundary. One should ordinarily reduce the net size around such a point. The net makes use of only three points on the upper boundary and would be the same for any change in the potential applied to the upper boundary that keeps these three points fixed.

TABLE 13.1. POTENTIALS AT THE NINE INTERIOR POINTS OF THE SQUARE REGION IN FIG. 13.1

Point	Exact computed value multiplied by 224	Approximate decimal equivalent	Analytical solution	Error, %
$\phi_1 = \phi_3$	59	0.2634	0.2530	4.1
ϕ_2	96	0.4286	0.3762	13.9
$\phi_4 = \phi_6$	28	0.1250	0.1136	10.0
ϕ_5	42	0.1875	0.1623	15.5
$\phi_7 = \phi_9$	11	0.0491	0.0431	13.9
ϕ_8	16	0.0714	0.0610	17.0

Courant[†] has suggested a method of reducing the number of unknowns to N where N^2 is the number of interior points of a square. This does not include any reduction due to symmetries. One takes as the unknowns the values of ϕ in one of the rows or columns adjacent to one of the boundaries and expresses the ϕ at all the other points in terms of these N values. How this may be done is seen by again taking the problem of Fig. 13.1.

Let these unknowns for our problem be $\phi_1 = \phi_3$ and ϕ_2; then from the difference equation (13.2)

$$-4\phi_1 + \phi_2 + \phi_4 + \phi_{25} + \phi_{11} = 0$$

or
$$\phi_4 = 4\phi_1 - \phi_2 - \phi_{25} - \phi_{11} = 4\phi_1 - \phi_2 - \tfrac{1}{2} = \phi_6 \quad (13.10)$$

Likewise

$$\phi_5 = 4\phi_2 - \phi_1 - \phi_3 - \phi_{12} = -2\phi_1 + 4\phi_2 - 1$$
$$\phi_7 = 4\phi_4 - \phi_1 - \phi_5 - \phi_{24} = 17\phi_1 - 8\phi_2 - 1 = \phi_9 \quad (13.11)$$
$$\phi_8 = 4\phi_5 - \phi_2 - \phi_4 - \phi_6 = -16\phi_1 + 17\phi_2 - 3$$

† R. Courant, "Advanced Methods in Applied Mathematics," pp. 66–76 (mimeographed notes New York University Lectures), 1941.

The next step, which brings in the known values at the bottom boundary, results in two simultaneous equations in ϕ_1 and ϕ_2; thus

$$\phi_{21} = 4\phi_7 - \phi_4 - \phi_8 - \phi_{23}$$
$$\phi_{20} = 4\phi_8 - \phi_6 - \phi_7 - \phi_9$$

which by Eqs. (13.11) reduces to

$$160\phi_1 - 96\phi_2 = 1$$
$$-96\phi_1 + 80\phi_2 = 9$$

Solving for ϕ_1 and ϕ_2, one has

$$\phi_1 = \tfrac{59}{224} \quad \text{and} \quad \phi_2 = \tfrac{96}{224}$$

These are, of course, the same values as obtained previously. The other ϕ's are already expressed in terms of these two.

Suppose, however, that one expresses ϕ_1 and ϕ_2 as three-place decimals,

$$\phi_1 = 0.263 \quad \text{and} \quad \phi_2 = 0.429$$

then by Eqs. (13.11)

$$\phi_7 = 0.039 = \phi_9$$
$$\phi_8 = 0.085$$

By Table 13.1 these should have come out as

$$\phi_7 = 0.049 = \phi_9$$
$$\phi_8 = 0.071$$

This large discrepancy is due to the amplification of the round-off error that occurs when one obtains successive rows (or columns) from a knowledge of the two previous rows (or columns). This amplification, by reference to Eqs. (13.10) and (13.11), could be as much as 6 per row or about 6^N (N being the number of rows) for the last row filled in. One speaks of this method of solution as being *unstable* with respect to round-off errors. Thus, Courant's method is practical only if the solutions are expressed in exact form, which necessitates the use of fractions or the availability of a large number of extra digits.

It is clear that the restriction of the method to a square region is in no way essential but was used merely to simplify the discussion.

13.4. Boundary Conditions of the Cauchy Type.† As has been stated in Chap. 12, the natural boundary conditions for equations of the elliptic type, such as Laplace's equation, are those of the Dirichlet and Neumann types, where the function, its normal derivative, or a linear combination of both is specified on a closed boundary. Suppose, how-

---

† For a more rigorous discussion see R. Courant, K. Friedrichs, and H. Lewy, Über die partiellen Differenzengleichungen der mathematischen Physik, *Math. Ann.*, **100**: 32-74 (1928).

ever, that we specify the boundary conditions for Laplace's equation by giving the value of ϕ and its normal derivative on some line. Suppose for convenience that this is a row (or column) of net points of a lattice. In terms of the Laplace difference equation, specifying the derivative is equivalent to specifying ϕ on the next row (or column) of the lattice. Thus, in finite-difference terms the *Cauchy condition* specifies two adjacent rows (or columns) of our lattice.

We saw in Sec. 13.3 that the knowledge of two adjacent rows fully determines the value of ϕ in the next row and, using these values, those in the next row, and so on. However, the errors in the initial specifications are amplified by possibly as much as 6^N (by the time the last row is reached), where N is the number of rows. The solution is thus unstable for large N. Since the differential equation is obtained by letting the number of rows N in the lattice increase indefinitely, it is clear that the solution of Laplace's differential equation is completely unstable with respect to a specification of boundary values of the Cauchy type. Any attempt to determine these boundary conditions experimentally is therefore useless. This illustrates the complete artificiality of a Cauchy-type boundary condition for Laplace's equation. It turns out that only where the functions that specify ϕ and its normal derivative are analytical mathematical functions can any meaning be given to the Cauchy conditions for elliptic differential equations.

13.5. Symbolic Methods. One may combine the basic difference equation of (13.2) for neighboring points in such a way as to obtain the composite-difference equation

$$\left\{\begin{array}{ccccc} & 1 & 2 & 1 & \\ 1 & 0 & 0 & 0 & 1 \\ 2 & 0 & -16 & 0 & 2 \\ 1 & 0 & 0 & 0 & 1 \\ & 1 & 2 & 1 & \end{array}\right\}\phi_s = 0 \qquad (13.12)$$

where the boldfaced number indicates the weight given ϕ_s itself and the remaining numbers indicate the weights of neighboring ϕ. This is done in a way analogous to that shown in Eq. (13.9).

Applying the difference equation (13.12) to the example shown in Fig. 13.1, letting $\phi_s = \phi_5$, one obtains immediately an expression for the potential of the center point in terms of the boundary points, namely,

$$\phi_5 = \tfrac{1}{16}[\phi_{11} + \phi_{13} + \phi_{15} + \phi_{17} + \phi_{19} + \phi_{21} + \phi_{23} \\ + \phi_{25} + 2(\phi_{12} + \phi_{16} + \phi_{20} + \phi_{24})] \quad (13.13)$$

Thus, the composite-difference equation (13.12) contains implicitly the expression of the center point of a nine-point square in terms of the boundary points.

If the number of interior points for a square is N^2, then the expression of the center point in terms of the boundary is shown in Table 13.2 for $N = 1, 3, 5$, and 7. Since the composite-difference equation has the same numbers on each side of the square, only the first row of numbers and the number at the center are given.

Table 13.2 also gives the value of ϕ computed for the center of the square of Fig. 13.1 for these various values of N (of course, the net shown in Fig. 13.1 no longer applies). The error between these values, obtained from nets with $N^2 = 1, 9, 25$, and 49 interior points, and the true value 0.1623, obtained analytically, is seen to be roughly inversely proportional to the number of interior points.

TABLE 13.2. COMPOSITE-DIFFERENCE EQUATIONS GIVING THE POTENTIAL AT THE CENTER OF THE SQUARE OF FIG. 13.1 FOR A NET OF N^2 INTERIOR POINTS

N	Number at the center	Numbers in the first row									ϕ at the center of square	Error
1	-4	0	1	0							0.2500	0.0877
3	-16	0	1	2	1	0					0.1875	0.0252
5	-104	0	3	6	8	6	3	0			0.1731	0.0108
7	-544	0	9	18	26	30	26	18	9	0	0.1682	0.0059

If one adds to the composite-difference equation (13.12) the two others

$$\left\{\begin{array}{ccccc} & 11 & 37 & 11 & \\ 11 & 0 & -112 & 0 & 11 \\ 7 & 0 & 0 & 0 & 7 \\ 3 & 0 & 0 & 0 & 3 \\ & 3 & 5 & 3 & \end{array}\right\}\phi_s = 0$$

$$\left\{\begin{array}{ccccc} & 67 & 22 & 7 & \\ 67 & -224 & 0 & 0 & 7 \\ 22 & 0 & 0 & 0 & 6 \\ 7 & 0 & 0 & 0 & 3 \\ & 7 & 6 & 3 & \end{array}\right\}\phi_s = 0 \quad (13.14)$$

the nine-point solution of the potential within a square for any given values of ϕ on the boundary is fully provided for. All the other difference equations needed can be obtained by simply rotating the last two, those of Eqs. (13.14).

Since the potential at any net point in the square is some linear function of the known boundary values, these composite-difference equations can be looked upon simply as a method of tabulating the coefficients of these linear expressions; for example, compare Eqs. (13.12) with (13.13). Moskovitz[†] has developed a method of obtaining these coefficients, which he expresses in decimal form, by solving the difference equation symbolically.

13.6. Liebmann's Method. The matrix of the coefficients of the set of linear equations arising from the Laplace difference equation has -4 for the elements of its principal diagonal [see, for example, Eqs. (13.4)]. Since the other elements are either 0 or 1, these diagonal terms are dominant. One is led, therefore, by the discussion of Chap. 10, to expect that such simultaneous equations can be solved by some method of successive approximation. Such a method is that introduced by Liebmann.[‡]

The symbolic equation (13.2) may be multiplied through by h^2 and written

$$\left\{ \begin{matrix} & 1 & \\ 1 & 0 & 1 \\ & 1 & \end{matrix} \right\} \phi_s - 4\phi_s = 0 \qquad (13.15)$$

from which one has

$$\phi_s = \tfrac{1}{4} \left\{ \begin{matrix} & 1 & \\ 1 & 0 & 1 \\ & 1 & \end{matrix} \right\} \phi_s \qquad (13.16)$$

This equation states that ϕ_s *is just the average of the values of ϕ at the four neighboring points.* Liebmann's method consists in improving the values initially guessed for ϕ by repeated application of Eq. (13.16). One passes from point to point in the lattice replacing the previous value of ϕ at each point by the average of the ϕ's for the four neighboring points. The itinerary may be quite arbitrary but must include all interior points of the lattice.

As this process is repeated, the successive values of ϕ at each lattice point tend to approach as a limit the corresponding solution of the linear equations arising from the Laplace difference equation. That this Liebmann process does converge, although rather slowly, can be inferred from the fact that each application of Eq. (13.16) averages the errors in the four neighboring points. Thus, if these errors are not all of the same sign, they will tend to average out, and ϕ_s is likely to be improved. In no case can the error in ϕ_s be greater than that of its least correct

[†] D. Moskovitz, The Numerical Solution of Laplace's and Poisson's Equations, *Quart. Appl. Math.*, **2**: 148–163 (1944).

[‡] H. Liebmann, Die angenäherte Ermittlung harmonischer Funktionen und konformer Abbildungen (nach Ideen von Boltzmann und Jacobi), *Sitzber. Math. Physik. Kl. bayer. Akad. Wiss. München*, **47**: 385–416 (1918).

neighbor. Thus there is no possible mechanism for the range of error for the lattice as a whole to increase. Besides the averaging out of the errors that occurs for points away from the boundary there is the added fact that the boundary values are not in error, and hence the application of Eq. (13.16) to points adjacent to the boundary actually tends to reduce the error, as well as to average it out. More information on the speed of convergence of the Liebmann process will be given when the Shortley-Weller modification of the Liebmann process is discussed (Sec. 13.11).

The Liebmann process is clearly admirably suited to a large-scale calculating machine in so far as the mathematical routine is very simple; however, since the ϕ's for all points in the lattice must be made available over and over again, the storage and transfer problems place severe restrictions on the number of lattice points used. Larger storage capacities in some machines help in this direction. There remains, however, the fundamental objection to this method that the rate of convergence is discouragingly slow.

13.7. Treatment of Curved Boundaries. In general, the boundary of the region investigated will fall between the points of the lattice. This

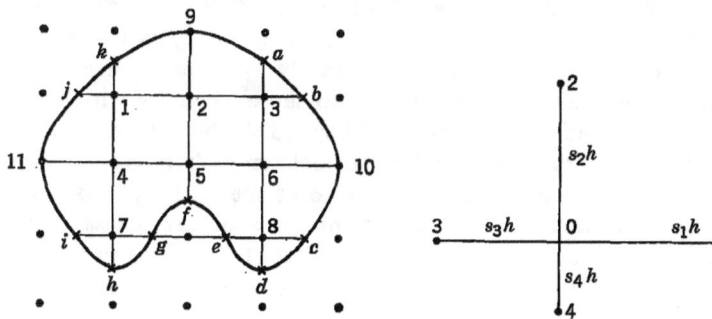

FIG. 13.2. Interlattice boundary points.   FIG. 13.3. Unequal-arm star.

gives rise to a class of points that have one or more of their immediate neighbors lying outside the boundary; note, for example, the points 1, 3, 5, 7, and 8 in Fig. 13.2. One may treat these points in several ways:

1. One may introduce interlattice boundary points, such as the points a to k in Fig. 13.2, and make use of second divided differences (see Chap. 5) as approximations for the second derivatives occurring in the Laplacian. For the general unequal-arm figure (referred to as a *star* by Southwell) shown in Fig. 13.3 Laplace's difference equation becomes

$$\frac{1}{h^2}\left[\frac{2}{s_1(s_1+s_3)}\phi_1 + \frac{2}{s_2(s_2+s_4)}\phi_2 + \frac{2}{s_3(s_1+s_3)}\phi_3 \right.$$
$$\left. + \frac{2}{s_4(s_2+s_4)}\phi_4 - \left(\frac{2}{s_1 s_3}+\frac{2}{s_2 s_4}\right)\phi_0\right] = 0 \quad (13.17)$$

often referred to as the *four-point influence* formula. Usually, however, only two arms are shortened, and the difference equation simplifies. Letting $s_1 = s_2 = 1$; $s_3 = s$, $s_4 = t$; and $\phi_3 = \phi_s$, $\phi_4 = \phi_t$, Eq. (13.17) becomes

$$\frac{1}{h^2}\left[\frac{2}{1+s}\,\phi_1 + \frac{2}{1+t}\,\phi_2 + \frac{2}{s(1+s)}\,\phi_s + \frac{2}{t(1+t)}\,\phi_t - \frac{2(s+t)}{st}\,\phi_0\right] = 0$$

$$(13.18)$$

Since ϕ_s and ϕ_t are boundary values and hence constant, Eq. (13.18) can be put in the form

$$\phi_0 = A\phi_1 + B\phi_2 + C \qquad (13.19)$$

where A, B, and C are constants that are calculated once and for all for that particular star. This is the Liebmann improvement formula for points near the boundary that have only two legitimate neighbors.

2. A second method is to freeze the values of ϕ at the lattice points nearest the boundary during the Liebmann iterative process and then from time to time to change these values to make them consistent with the given boundary conditions. This is done by interpolation, which in most cases can be done sufficiently accurately by graphical means. One can restrict these semifixed points to those which lie within the area, or one can include points lying just beyond the boundary.

13.8. Obtaining Initial Guesses for the Liebmann Process. By and large the intuitive feeling for the values of ϕ at the lattice points is by far the most important tool used in obtaining a good first guess. The closer this guess, of course, the more quickly does the Liebmann process yield a solution. However, there are certain aids that often prove helpful.

One such aid is the knowledge that all the maxima and minima of ϕ must occur on the boundary. One should, therefore, always assign values of ϕ that lie between the greatest and smallest values of ϕ given on the boundary. Thus, for the example discussed earlier (see Fig. 13.1), one would assume for initial values of ϕ at the nine interior points numbers lying between zero and one.

To obtain still better values one may employ a very rough net that can be easily solved. Graphical interpolation can then be used to yield the necessary values. For problems involving a large number of interior points it usually pays to go one step farther and obtain a solution of the problem for one or two coarser nets before going to the final net. How this may best be done is indicated in Sec. 13.9.

A very crude guess can be made of the potential at each point by employing the four-point influence formula (13.17), where the points

1, 2, 3, and 4 are taken on the boundary. Frocht and Levens† suggest, however, that one take the values of ϕ predicted for a given point by this formula for various choices of the four boundary points and plot them as a function of the angle (the angle one arm of the star of Fig. 13.3 makes with the x axis). The average value of this ϕ, considered as a function of the angle, is then assumed to be a good approximation to the actual potential at that lattice point.

13.9. Change of Net. As indicated in Sec. 13.8, it is usually better to start with one or more coarser nets before going to the final net. This is due primarily to the fact that the convergence of the Liebmann process is much more rapid for such nets, in consequence of the smaller number of interior points.

One may obtain a good set of lattice values for the finer net, once the coarser net has been solved, by the following procedure. One inserts a new point at the center of each square of the original lattice and assigns to these lattice points a value of ϕ that is the average of ϕ at the four corners of the square (see Fig. 13.4). This gives rise to a square lattice that runs diagonally across

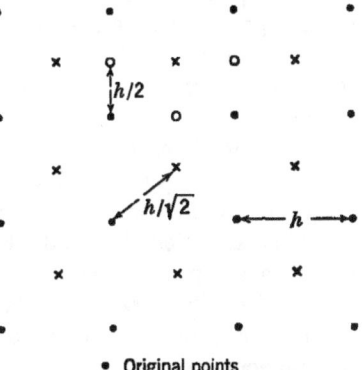

- Original points
- First inserted points
- Second inserted points

FIG. 13.4. Change to a finer net.

the original lattice and has a spacing of $h/\sqrt{2}$ (h being the spacing of the original lattice). One may then proceed to improve the values assigned to each lattice point, by the Liebmann process or any of the other iterative processes treated later. Note that a repetition of this procedure yields a still finer net of spacing $h/2$, which lies parallel to the original net. Of course, more boundary points must be brought in as the net spacing is thus reduced.

This change of net is often made only over small portions of the original net. The portions chosen are those at which ϕ is expected to change rapidly. For instance, in the example of Fig. 13.1 a smaller net is called for near the middle of the top boundary. If one desires to reduce the spacing over a portion of the net to one-half the original spacing, one must actually make the two reductions of spacing shown in Fig. 13.4. Therefore, there must be a small transition region in which the diagonal lattice (spacing $h/\sqrt{2}$) is used. This interlocking of the

† See M. M. Frocht and M. M. Levens, A Rational Approach to the Numerical Solution of Laplace's Equation, *J. Appl. Phys.*, **12**: 596–604 (1941).

various-sized nets must be borne in mind when one improves the values of the over-all lattice by Liebmann's process.

13.10. Minor Modifications of Liebmann's Method. In the Liebmann process the improved values at any point, obtained by averaging the potential at the four neighboring points, are used in all subsequent improvements of these neighboring points. One may, however, use the old values throughout until improved values have been obtained for all interior points of the lattice. This is the scheme used by Kormes† in adapting the Liebmann process to punch-card machines. Since it makes use of the new values only after the entire lattice has been scanned, the Kormes process converges even more slowly than the Liebmann. Its sole advantage is that it is much better suited to punch-card machines. This advantage, however, is unimportant for automatic electronic computers.‡

Another modification of the Liebmann process is to scan the lattice, improving every other point—these points form a lattice of spacing $\sqrt{2}\,h$—and then use these improved values to improve those skipped in the first scanning. Actually this process can be looked upon as merely the Liebmann process for a special itinerary.

13.11. The Shortley-Weller Method.§ In the Shortley-Weller method Liebmann's process is modified to make possible the improvement simultaneously of a block of points in place of a single point. Perhaps the most advantageous number of points to be included in such a group is 9, although one might wish to go to blocks of 25 points. Where one cannot fit in blocks of 9 or 25 points, one may use 4 points (see Fig. 13.5). The improvement formula for each of these three blocks of points will be considered below.

For a block of 9 points one first improves the center point by the direct application of the composite-difference equation (13.12). Thus in Fig. 13.5, for the block of 9 near the center of the figure

$$p = \tfrac{1}{16}\left(\sum_{i=1}^{8} t_i + 2\sum_{i=1}^{4} s_i\right) \tag{13.20}$$

† See Mark Kormes, Numerical Solution of the Boundary Value Problem for the Potential Equation by Means of Punched Cards, *Rev. Sci. Instr.*, **14**: 248–250 (1943).

‡ L. F. Richardson as early as 1910 devised a method that resembles that of Kormes in that new values are not used as they are obtained to improve succeeding points. His method, however, converges much more rapidly than that of Kormes if certain parameters in his improvement formulas are chosen judiciously to fit the individual problem; however, his method does not have the general utility of Liebmann's process. See L. F. Richardson, *Trans. Roy. Soc. (London)*, **210A**: 307–357(1910), and L. F. Richardson, *Math. Gaz.*, **12**: 415–421 (1925).

§ See G. H. Shortley and R. Weller, The Numerical Solution of Laplace's Equation, *J. Appl. Phys.*, **9**: 334–348 (1938).

For the purpose of such improvement formulas the values of the points adjacent to the block are treated as if they were correct. If they are not boundary points, they themselves must be corrected, either as a member of another block of points, which would be the case for s_1 and t_1

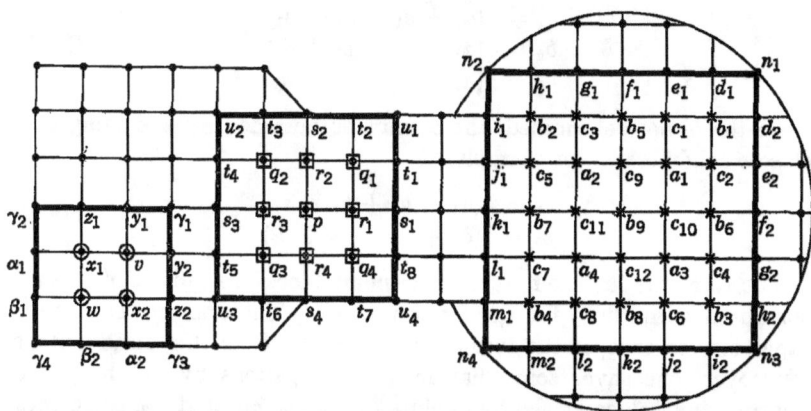

FIG. 13.5. Division of a lattice into blocks of 4, 9, and 25 points to permit use of the improvement formulas of Shortley and Weller.

in Fig. 13.5, or as individual points, which would be true of t_3 and t_6. In the latter case one would, of course, use the Liebmann formula (13.16).

Having found p, one may determine q_1, q_2, q_3, q_4 by averaging their four diagonal neighbors. Thus, for example,

$$q_1 = \tfrac{1}{4}(p + s_1 + s_2 + u_1) \quad (13.21)$$

Having the q_i, one then may determine the r_i by the Liebmann process, i.e., by averaging their four closest neighbors. For example,

$$r_1 = \tfrac{1}{4}(p + q_1 + q_4 + s_1) \quad (13.22)$$

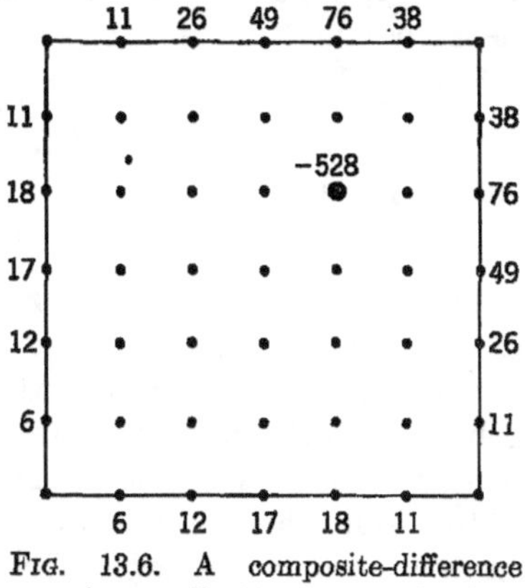

In dealing with the 25-point block, one determines first the a_i. This is done by employing the composite-difference equation shown in Fig. 13.6. Applying it to the block of 25 points of Fig. 13.5, one obtains as an expression for a_1, in terms of the points bordering the block,

FIG. 13.6. A composite-difference equation applicable to a block of 25 points.

$$\begin{aligned}
a_1 = \tfrac{1}{528}[&38(d_1 + d_2) + 76(e_1 + e_2) + 49(f_1 + f_2) + 26(g_1 + g_2)\\
&+ 11(h_1 + h_2 + i_1 + i_2) + 18(j_1 + j_2) + 17(k_1 + k_2)\\
&+ 12(l_1 + l_2) + 6(m_1 + m_2)] \quad (13.23)
\end{aligned}$$

The other three a_i are found from completely analogous expressions obtainable by rotating the composite-difference equation of Fig. 13.6.

When these four values a_1, a_2, a_3, a_4 have been found, the b_i in Fig. 13.5 may be found by averaging their diagonal neighbors. For example,

$$b_1 = \tfrac{1}{4}(a_1 + e_1 + e_2 + n_1)$$
$$b_5 = \tfrac{1}{4}(a_1 + a_2 + e_1 + g_1) \qquad (13.24)$$
$$b_9 = \tfrac{1}{4}(a_1 + a_2 + a_3 + a_4)$$

Once the b_i are determined, the c_i may be obtained by averaging their closest neighbors. For example,

$$c_1 = \tfrac{1}{4}(a_1 + b_1 + b_5 + e_1)$$
$$c_9 = \tfrac{1}{4}(a_1 + a_2 + b_5 + b_9) \qquad (13.25)$$

The formulas for the a_i for the 25-point block are quite cumbersome compared with those for the 9-point block. There is, however, more rapid convergence of the former that partly offsets this disadvantage. Moreover, one saves somewhat in the calculations by not having to improve the values at the five points b_9, c_9, c_{10}, c_{11}, and c_{12}, since these are not needed in the improvement of any other points.

The 4-point block (see Fig. 13.5) is improved by first finding v from the formula

$$v = \tfrac{1}{24}[7(y_1 + y_2) + 2(z_1 + z_2 + \alpha_1 + \alpha_2) + (\beta_1 + \beta_2)] \quad (13.26)$$

Having v, we determine the value at w by averaging its four diagonal neighbors, i.e.,

$$w = \tfrac{1}{4}(v + \alpha_1 + \alpha_2 + \gamma_4) \qquad (13.27)$$

Finally the x_i are obtained by averaging their four closest neighbors. For example,

$$x_1 = \tfrac{1}{4}(v + w + z_1 + \alpha_1) \qquad (13.28)$$

Note that in all these block-improvement formulas one relates some of the points to their closest neighbors and others to their diagonal neighbors. This means that the process does not converge exactly to a solution of the difference equation of (13.2), but instead it converges to a hybrid set of difference conditions, some of which are identical with (13.2), and others of which are similar in form but are rotated 45° and have a spacing of $\sqrt{2}\,h$. This error is unimportant in relation to the initial error arising from replacing a continuous function satisfying Laplace's differential equation by a function having values only at the lattice points and satisfying the difference equation (13.2). In fact, it would appear reasonable on this basis to simplify the coefficients of Eq. (13.23); e.g., the coefficient 49 might be taken as 50, and 11 as 10, etc., and the normalizing factor $\frac{1}{528}$ changed appropriately.

13.12. Improvement Matrix for the Liebmann Process. Let us first investigate the rate of convergence of the Liebmann process. Choose some definite itinerary for the improvement of interior values ϕ_s of the given lattice, the improvement consisting of replacing the value at a point by the average of its four neighboring points. The mathematical expression of this improvement process is given by Eq. (13.16), and it indicates that the new value at each point is a linear function of the four surrounding points.

We shall now seek to show that the new values ϕ'_s at each interior point obtained after a complete scanning of the interior points of the lattice are some linear function of the old values ϕ_s and the boundary values ϕ_t.† The first point improved, say the point a, is assigned the average of the values of the ϕ for its four neighbors. It is, therefore, clearly a linear function of the old values, which may include one or more of the boundary values ϕ_t. The new value ϕ'_b assigned to the second point b is likewise a linear function of values of its neighbors, some of which may be boundary points. However, one of its neighbors might be the first improved point a, and if so, one must substitute the expression for ϕ'_a in the linear expression for ϕ'_b. Since both functions involved are linear functions, the resulting expression for ϕ'_b in terms of the unimproved values ϕ_s is linear. Similarly for a third point c, the initial expression for ϕ'_c may involve one, or both, of the improved values ϕ'_a and ϕ'_b; however, since both of these are linear functions of the old values, ϕ'_c will again be expressible as a linear function of the unimproved values. Clearly the argument is applicable to all the remaining points of the lattice.

Stated mathematically, each scanning of the interior points of the lattice using the prescribed itinerary leads to a set of new values ϕ'_s given by the set of linear equations

$$\phi'_s = \sum_{i=1}^{n} T_{si}\phi_i + \sum_{j=n+1}^{m} U_{sj}\phi_j \qquad (13.29)$$

The constants T_{si} and U_{sj} are independent of the values chosen for ϕ_i and ϕ_j. Since the boundary values are not changed,

$$\phi'_j = \phi_j \qquad j = n+1, n+2, \ldots, m \qquad (13.30)$$

Let the true solution of the finite-difference equation be a set of values ω_s. Since the ω_s satisfy the difference equation (13.2), the value of each ω_s is exactly the average of its neighbors, and hence the Liebmann process applied to such a set would not change any of the ω_s. In terms

† As stated previously, one may designate the interior points by the subscript s, where $s = 1, 2, 3, \ldots, n$, and the boundary points by t, where $t = n + 1, n + 2, \ldots, m$.

of Eq. (13.29), therefore,

$$\omega_s' = \sum_{i=1}^{n} T_{si}\omega_i + \sum_{j=n+1}^{m} U_{sj}\omega_j = \omega_s \qquad (13.31)$$

It is convenient to let

$$\phi_s = \omega_s + e_s \qquad (13.32)$$

where the e_s can be interpreted as the errors in the potentials ϕ_s with reference to the true solution ω_s of the Laplace difference equation. Since the values on the boundary are given,

$$\phi_j = \omega_j \quad \text{and} \quad e_j = 0 \quad j = n + 1, \ldots, m \qquad (13.33)$$

On the other hand, by subtracting Eq. (13.31) from Eq. (13.29), using Eq. (13.32), one sees that

$$e_s' = \sum_{i=1}^{n} T_{si}e_i \qquad (13.34)$$

or in matrix notation

$$e' = Te \qquad (13.35)$$

Here e' and e are thought of as column matrices, or vectors (see Chap. 10). The n-by-n matrix T is dependent on the lattice and also to some extent on the scanning itinerary but not at all on the error vector e. T is termed the *improvement matrix*, since repeated applications of T, corresponding to repeated scannings of the lattice, tend to reduce the errors e_s at each point to zero.

Let the vector $\phi^{(k)}$ with components $\phi_s^{(k)}$ represent the potentials at the inside lattice points after the kth improvement, i.e., the kth scanning of the lattice, and let

$$\phi^{(k)} = \omega + e^{(k)} \qquad (13.36)$$

If $e^{(0)}$ is the initial error in $\phi^{(0)}$ on assigning values $\phi_s^0 = \phi_s$ to the interior points, then, as seen by repeated application of Eq. (13.35),

$$e^{(k)} = T^k e^{(0)} \qquad (13.37)$$

Convergence clearly occurs if, and only if,

$$\lim_{k \to \infty} e^{(k)} = \lim_{k \to \infty} T^k e^{(0)} = 0$$

and this is true for an arbitrary $e^{(0)}$ if, and only if, the absolute magnitude of all the characteristic roots of T [see Eq. (10.88)] is less than unity, as it is for the Liebmann process.

13.13. Rate of Improvement Determined by Characteristic Values of T. It is a well-known property of matrices that any n-by-n matrix T may be expressed in the form

$$T = S^{-1}T_0 S \qquad (13.38)$$

where S^{-1} is the reciprocal of the square matrix S and T_0 is of the form[†]

$$T_0 = \begin{pmatrix} \lambda_1 & k_1 & 0 & 0 & \cdots & 0 & 0 \\ 0 & \lambda_2 & k_2 & 0 & \cdots & 0 & 0 \\ 0 & 0 & \lambda_3 & k_3 & \cdots & 0 & 0 \\ 0 & 0 & 0 & \lambda_4 & \cdots & 0 & 0 \\ \cdots & \cdots & \cdots & \cdots & \cdots & \cdots & \cdots \\ 0 & 0 & 0 & 0 & \cdots & \lambda_{n-1} & k_{n-1} \\ 0 & 0 & 0 & 0 & \cdots & 0 & \lambda_n \end{pmatrix} \qquad (13.39)$$

In the latter the k_i are either zero or one, and the λ_i are the *characteristic values* of the matrix, which may be ordered such that $\lambda_i \leq \lambda_j$ for $i > j$. Moreover, we may require that

$$\lambda_{s+1} = \lambda_s \quad \text{if } k_s = 1 \qquad (13.40)$$

Thus, if all the λ_i are different, T_0 is a diagonal matrix.

Consider the base vectors ϵ_i, $i = 1, 2, 3, \ldots, n$, defined by the requirement that the jth component of ϵ_i is δ_{ij}.[‡] In terms of these base vectors one may define a set of fundamental vectors

$$\psi_i = S^{-1}\epsilon_i \quad i = 1, 2, \ldots, n \qquad (13.41)$$

Assuming that the initial error vector $e^{(0)}$ can be expanded in terms of these ψ_i, one has

$$e^{(0)} = \sum_{i=1}^{n} a_i \psi_i \qquad (13.42)$$

and by Eq. (13.37)

$$e^{(k)} = \sum_{i=1}^{n} a_i T^k \psi_i \qquad (13.43)$$

Thus, a knowledge of the effect of T on the ψ_i is sufficient.

From Eqs. (13.38), (13.39), and (13.41)

$$T\psi_i = S^{-1}T_0\epsilon_i = S^{-1}(\lambda_i\epsilon_i + k_{i-1}\epsilon_{i-1}) = \lambda_i\psi_i + k_{i-1}\psi_{i-1} \qquad (13.44)$$

and if it happens that $k_{i-1} = 0$,

$$T^k\psi_i = \lambda_i^k\psi_i \qquad (13.45)$$

Hence the error distribution given by such a ψ_i would approach zero as $k \to \infty$ if $|\lambda_i| < 1$. In particular, this error will be decreased by a factor

[†] See R. A. Frazer, W. J. Duncan, and A. R. Collar, "Elementary Matrices and Some Applications to Dynamics and Differential Equations," pp. 93–95, Cambridge University Press, London, 1952.

[‡] δ_{ij} is Kronecker's delta. It has the value of 1 if $i = j$ and the value of 0 if $i \neq j$.

λ_i each time the net is scanned, i.e., each time an improvement of all the values of the lattice has been made.

If, on the other hand,

$$k_{i-1} = k_{i-2} = \cdots = k_{i-m} = 1 \tag{13.46}$$
$$k_{i-m-1} = 0$$

by Eq. (13.40)

$$\lambda_i = \lambda_{i-1} = \lambda_{i-2} = \cdots = \lambda_{i-m} = \lambda \tag{13.47}$$

and hence Eq. (13.44) becomes

$$T\psi_i = \lambda\psi_i + \psi_{i-1} \tag{13.48}$$

Thus
$$T^2\psi_i = \lambda(\lambda\psi_i + \psi_{i-1}) + (\lambda\psi_{i-1} + \psi_{i-2})$$
$$= \lambda^2\psi_i + 2\lambda\psi_{i-1} + \psi_{i-2}$$

and
$$T^3\psi_i = \lambda^3\psi_i + 3\lambda^2\psi_{i-1} + 3\lambda\psi_{i-2} + \psi_{i-3}$$

Proceeding in this way, we can readily establish that

$$\psi_i^{(k)} = T^k\psi_i = \sum_{r=0}^{p} \binom{k}{r}\lambda^{k-r}\psi_{i-r} \tag{13.49}$$

where $\binom{k}{r}$ are the binomial coefficients and

$$p = \begin{cases} k & \text{if } k \leq m \\ m & \text{if } k > m \end{cases} \tag{13.50}$$

If the k in Eq. (13.49) is large, the effect of applying T to $\psi_i^{(k)}$ is very nearly that of multiplying by λ. This is seen as follows. For large values k

$$\binom{k+1}{r} = \frac{(k+1)!}{(k+1-r)!r!} = \binom{k}{r}\frac{k+1}{k+1-r} \cong \binom{k}{r} \tag{13.51}$$

hence, by Eq. (13.49)

$$\psi_i^{(k+1)} = \sum_{r=0}^{m} \binom{k+1}{r}\lambda^{k+1-r}\psi_{i-r} \cong \lambda\psi_i^{(k)} \tag{13.52}$$

Therefore, whether we start with a ψ_i satisfying Eq. (13.45) or one satisfying Eq. (13.48), the effect, after a while, of applying the matrix T is to multiply the distribution of error expressed by that ψ_i by the characteristic value λ_i associated with it.

Since the initial error $e^{(0)}$ can be expanded in terms of the ψ_i, as in Eq. (13.43), it is clear that repeated application of T reduces to zero the coefficients of those ψ_i associated with the smaller λ's faster than those with larger λ's. In fact, after sufficient applications of T, one is left with an error distribution that corresponds to that for the largest λ_i. By the convention adopted earlier, this is λ_1. From this point on, each application of T is equivalent to multiplying this error distribution by λ_1. One may, therefore, speak of the rate of convergence of the Liebmann process in terms of the magnitude of λ_1.

Shortley and Weller showed that in square regions having $(p - 1)^2 = n$ interior points, where p is large,

$$- \ln \lambda_1 \cong 1 - \lambda_1 = \frac{\pi^2}{p^2} \qquad (13.53)$$

The number of traverses which must be made in order to reduce the error to a definite fraction ρ of its initial value is

$$\frac{- \ln \rho}{- \ln \lambda_1}$$

thus the value of $|\ln \lambda_1| = - \ln \lambda_1$ is a measure of the rapidity of convergence of the process.

Note that doubling the number of points in a region not only doubles the amount of computation for each scanning but also reduces the rate of convergence per scanning (measured by $|\ln \lambda_1|$) to about one-half. This fact makes it economical to start with one or more coarser nets before going to the final net.

13.14. Rate of Convergence for the Shortley-Weller Method. It is clear that in the block-improvement methods of Shortley and Weller the value of ϕ at each point is expressed as a linear function of its value at certain other points. To be sure, these values at the latter points may, as in the Liebmann process, be expressed as linear functions of those at still other points. However, by the same argument as used for the Liebmann process new values of the ϕ's at each point, after a complete scanning of the lattice, are a linear function of the old values, and one again has an improvement matrix T.

Shortley and Weller determined the asymptotic value of the convergence index $|\ln \lambda_1|$ for a square region as p becomes large

$$[(p - 1)^2 = n = \text{number of interior points}]$$

The results are given in Table 13.3. It is assumed in each case that the

TABLE 13.3. COMPARISON OF THE RAPIDITY OF CONVERGENCE (MEASURED BY $|\ln \lambda_1|$) FOR THE VARIOUS BLOCK-IMPROVEMENT FORMULAS OF SHORTLEY AND WELLER

Number of points in a block	$p^2 \|\ln \lambda_1\|$
1	π^2
4	$2\pi^2$†
9	$\frac{7}{2}\pi^2$
25	$\frac{58}{11}\pi^2$

† This value varies slightly depending on which of the four points is improved first in each of the various blocks.

number of interior points in the square is such that it is possible to use a single size of block throughout. The 1-point-block process is, of course, the Liebmann process.

13.15. Use of the Difference Function. For hand computation the labor of computation can be reduced by applying the improvement formulas to the difference between the original guess ϕ and the first improved value

$$\phi^{(1)} = T\phi$$

Let δ represent this difference, which is clearly

$$\delta = T\phi - \phi \tag{13.54}$$

Since ϕ and $\phi^{(1)}$ have the same values on the boundary, δ is zero on the boundary. Successive application of the improvement matrix to δ, as given by Eq. (13.54), yields the set of equations

$$T\delta = T^2\phi - T\phi$$
$$T^2\delta = T^3\phi - T^2\phi \tag{13.55}$$
$$\cdots\cdots\cdots$$
$$T^{k-1}\delta = T^k\phi - T^{k-1}\phi$$

Adding together Eqs. (13.54) and (13.55), one has

$$\sum_{p=0}^{k-1} T^p\delta = T^k\phi - \phi = \phi^{(k)} - \phi \tag{13.56}$$

Therefore, since the true solution ω is the limit of $\phi^{(k)}$,

$$\omega = \lim_{k\to\infty} \phi^{(k)} = \phi + \sum_{p=0}^{\infty} T^p\delta \tag{13.57}$$

13.16. Extrapolation. Since δ is zero on the boundary, it can be expanded in terms of the modes ψ_i used to expand the error vector $e^{(0)}$. Therefore, in exact analogy to the argument of Sec. 13.13, for a large enough k,

$$T(T^k\delta) \cong \lambda_1(T^k\delta) \tag{13.58}$$

where λ_1, the largest characteristic value, is associated with ψ_1.

This being the case,

$$\sum_{p=k}^{\infty} T^p\delta = \sum_{r=0}^{\infty} T^{k+r}\delta \cong \sum_{r=0}^{\infty} \lambda_1^r T^k\delta = \frac{T^k\delta}{1 - \lambda_1}$$

therefore, by Eq. (13.57)

$$\omega \cong \phi + \sum_{p=0}^{k-1} T^p\delta + \frac{T^k\delta}{1 - \lambda_1}$$

If one designates by $\delta^{(p)}$ the value obtained by p applications of the improvement matrix T to δ, this equation becomes

$$\omega \cong \phi + \sum_{p=0}^{k-1} \delta^{(p)} + \frac{\delta^{(k)}}{1 - \lambda_1} \tag{13.59}$$

In order for $\delta^{(k)}$ to satisfy Eq. (13.58), i.e.,

$$T\delta^{(k)} \cong \lambda_1 \delta^{(k)} \tag{13.60}$$

the coefficient b_1 in the expansion

$$\delta^{(k)} = \sum_{r=1}^{n} b_r \psi_r$$

must be very large compared with the rest of the b's; or, in other words, $\delta^{(k)}$ must have very nearly the same distribution as ψ_1.

Now ψ_1 describes the lowest characteristic function of T, and this function is, in general, pillow-shaped, having the same sign for all points of the region. The higher modes ψ_2, ψ_3, etc., have nodal lines through the area.

The amount of time saved by extrapolation, using Eq. (13.59), depends on how large k must be taken before Eq. (13.60) holds. Assuming that ψ_1 and ψ_2 are true characteristic vectors of T, i.e., $k_1 = k_2 = 0$ in Eq. (13.39),

$$T\psi_1 = \lambda_1 \psi_1$$
$$T\psi_2 = \lambda_2 \psi_2$$

which in practice appears nearly always to be the case, and letting

$$\delta = b_1 \psi_1 + b_2 \psi_2 + \sum_{p=3}^{n} b_p \psi_p$$

we have

$$\delta^{(k)} = T^k \delta = b_1 \lambda_1^k \psi_1 + b_2 \lambda_2^k \psi_2 + \text{even smaller terms}$$

Now k must be taken large enough so that the second term is small compared with the first, i.e.,

$$b_2 \lambda_2^k \ll b_1 \lambda_1^k$$

Therefore, how large k must be for extrapolation to be useful depends on the relative magnitudes of λ_1 and λ_2.

13.17. Relaxation Method of Southwell. One may define the *residue* R_s at each interior point s of a lattice for Laplace's difference equation by the equation

$$R_s \equiv \left\{ \begin{array}{ccc} & 1 & \\ 1 & -4 & 1 \\ & 1 & \end{array} \right\} \phi_s \tag{13.61}$$

where the same symbolic notation is employed on the right as in Eq. (13.2). For the actual solution ω of the Laplace difference equation (13.2) the residues at all interior points must be zero. Southwell's relaxation method consists in reducing the residues that are present for the initial values ϕ_s to zero by suitably changing these ϕ_s. When this reduction is carried to such a point that all the residues differ from zero by only 1 or 2 in the last significant figure retained, one considers the ϕ_s that are then assigned as constituting a sufficiently accurate solution of the difference equation.

The simple relaxation procedure consists in changing the value of ϕ_s at the point s having the largest residue (in absolute value) in such a

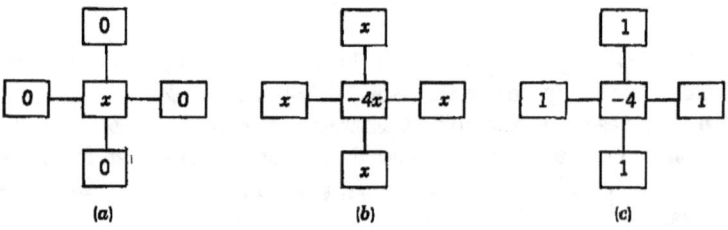

FIG. 13.7. Effect on the residues when one changes the dependent variable ϕ at only one point (the center one in each diagram): (a) change in the ϕ's; (b) change in the residues; (c) so-called relaxation pattern.

way that that residue is reduced to zero, or practically so. This can easily be done by adding to ϕ_s a quantity x which is equal approximately to $\frac{1}{4}R_s$. Since Eq. (13.61) may be written

$$R_s \equiv \phi_{s_1} + \phi_{s_2} + \phi_{s_3} + \phi_{s_4} - 4\phi_s$$

where ϕ_{s_1}, ϕ_{s_2}, ϕ_{s_3}, and ϕ_{s_4} are the values of ϕ at the four neighboring points, the new residue R'_s is

$$R'_s \equiv \phi_{s_1} + \phi_{s_2} + \phi_{s_3} + \phi_{s_4} - 4(\phi_s + x) = R_s - 4x \quad (13.62)$$

and therefore differs from zero only in so far as x differs from $\frac{1}{4}R_s$. It will become apparent that there is no particular advantage in making x exactly $\frac{1}{4}R_s$.

The adding of x to ϕ_s not only subtracts $4x$ from the residue R_s, but it will add x to each of the residues R_{s_1}, R_{s_2}, R_{s_3}, R_{s_4} of the four neighboring points. This follows from the fact that ϕ_s enters into the formula for these residues in the same way as, for example, ϕ_{s_1} enters into Eq. (13.62).

Figure 13.7 summarizes the results of this discussion. Note that the relaxation pattern in (c) is a mnemonic aid for assigning the right changes to the residues, as given in (b).

Example 1.   Use the simple relaxation procedure to determine to about two decimal places the solution of the Laplace difference equation for the lattice shown below.

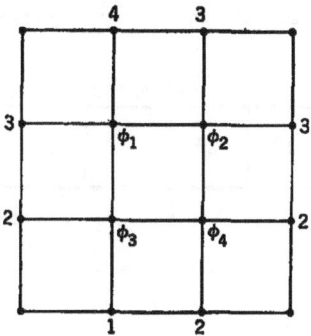

The values at the corner of the square need not be specified, since they do not enter the difference equations.   As a first guess let $\phi_1 = \phi_2 = 3$ and $\phi_3 = \phi_4 = 2$; then the residues at these four interior points, by Eq. (13.61), are

$$R_1 = \phi_2 + \phi_3 + 3 + 4 - 4\phi_1 = 0 \qquad R_2 = -1 \qquad R_3 = 0 \qquad R_4 = 1$$

In Work Sheet I these initial ϕ's are tabulated just to the left of their respective points, the residues being tabulated to the right.   To eliminate decimals we shall multiply through by 100 and designate the new values by primes.   As each ϕ'_s is changed,

WORK SHEET I

the amount of change x is written above the original value of ϕ_s.   The resulting changes in the residues are taken care of by crossing out the old values and listing the new values just above them.

Table 13.4 is given here to make clear the order in which the changes shown in the work sheet were made.   Note that one always works with the largest residue remaining, reducing it to any convenient small value.   The relaxation here illustrates its great advantage for hand computation, since it is clear that one can, for the most part, change the ϕ's by round numbers such as 20, 25, etc. In no case will it be necessary to use nonintegral numbers if one has previously multiplied through by that power of 10 which makes the final answer, expressed to the desired number of decimals, an integer.

TABLE 13.4

Initial values:	300		0	300	−100	200	0	200	100
No. of changes	Change in ϕ_1'	R_1'		Change in ϕ_2'	R_2'	Change in ϕ_3'	R_3'	Change in ϕ_4'	R_4'
1		−25		−25	0		0		75
2		−25			20		20	20	−5
3	−6	−1			14		14		−5
4		3		4	−2		14		−1
5		7			−2	4	−2		3
6	2	−1			0		0		3
7		−1			1		1	1	−1
Final values	$\phi_1' = 296$	$R_1' = -1$		$\phi_2' = 279$	$R_2' = 1$	$\phi_3' = 204$	$R_3' = 1$	$\phi_4' = 221$	$R_4' = -1$

After the seventh step the residues were all reduced to ± 1. Any further integral changes would increase at least one of these residues, and hence the relaxation is terminated. The final values of the ϕ's are obtained by adding up the changes recorded for each step. The value of the ϕ's at the four points, to about three significant figures, should therefore be these ϕ'''s divided by 100; hence

$$\phi_1 = 2.96 \qquad \phi_2 = 2.79 \qquad \phi_3 = 2.04 \qquad \text{and} \qquad \phi_4 = 2.21$$

That these values are, in fact, correct to three significant figures can be seen by comparing them with the exact values,

$$\phi_1 = \tfrac{71}{24} = 2.958 \qquad \phi_2 = \tfrac{67}{24} = 2.792$$
$$\phi_3 = \tfrac{49}{24} = 2.042 \qquad \phi_4 = \tfrac{53}{24} = 2.208$$

13.18. Under- and Overrelaxation. As seen from the relaxation pattern of Fig. 13.7, the effect of changing ϕ, at a given point, by x is to reduce the residue at that point by $4x$ and to increase the residue at all four neighboring points by x. This can be looked upon as a dispersion of that part of the residue given by $4x$ among the four surrounding points. Since Laplace's difference equation need not be satisfied at the boundary points, one does not speak of the residue at boundary points. Thus, any residue that is dispersed to the boundary can be considered completely absorbed.

The straightforward relaxation process consists, then, of eliminating the largest residue by distributing this residue equally to the four surrounding points. The residues at each of these surrounding points will be increased or decreased in absolute magnitude depending, respectively, on whether its original residue is of like or opposite sign to this dispersed residue. After the largest residue has been dispersed, leaving a very small, perhaps even a zero, residue, the largest residue remaining is dispersed in the same way.

It is clear from this that it is usually of no importance to disperse the entire residue at a point since the subsequent dispersion of the residue at its neighboring points will distribute a portion of their residues back to this original point. This means that instead of taking pains to obtain an $x = \frac{1}{4}R_s$, which, as seen from Eq. (13.62), reduces the residue exactly to zero, one may just as well take for x any round number which, when multiplied by 4, accounts for most of the residue R_s. It is even better practice to anticipate the results of subsequent dispersion of the residues at the neighboring points by leaving a nonzero residue R_s' that will cancel with the residue driven back to this point.

If, more specifically, the residues over a portion of our region are, by and large, all of the same sign, one should choose x larger than $\frac{1}{4}R_s$ so that the new residue R_s' left at s is of opposite sign to R_s. This R_s' will then be of the right sign to be reduced in magnitude when the residues at the surrounding points are dispersed. Southwell refers to such a procedure as *overrelaxation*. How much larger x should be than $\frac{1}{4}R_s$ is left to one's experience. A poor guess, however, is not at all serious, since it only means that it will take longer to reduce the residues.

A rough rule for the choice of x when all the surrounding residues are of the same sign may be obtained by assuming that these residues are small compared with R_s. If this is the case, after the dispersion of the residue at s, these residues will be approximately x, while the new residue at s is $R_s - 4x$. Neglecting the effect on these residues of relaxing other points in the region and considering only the subsequent dispersion of these four residues surrounding the point s, one sees that the final residue at s is given roughly by $(R_s - 4x) + 4(x/4) = R_s - 3x$ since a residue of $\frac{1}{4}x$ is obtained from each of these four neighboring points. Setting this final residue at s equal to zero, we have the rough rule that

$$x \cong \frac{R_s}{3} \tag{13.63}$$

Since the assumptions made in deriving this estimate are very poorly approximated in most cases, one should use this expression only as a suggested value to be modified by one's experience.

If the residues immediately surrounding the point s being improved are opposite in sign to R_s, one needs to take account of the fact that the subsequent reduction of these residues will tend to cancel part of R_s. Therefore one chooses $x < \frac{1}{4}R_s$ so that $R_s' = R_s - 4x$ is of the same sign as R_s. That is, one distributes only part of the residue R_s. This process is termed *underrelaxation*, and its purpose is the same as that in over-relaxation, namely, to speed the reduction of residues to zero by taking account of the interaction of the residue reductions at neighboring points.

Example 2. Use the process of underrelaxation to shorten the solution of the problem in Example 1.

Since the residue at point 2 is -100 and at point 4 is 100, it is better on the initial change of ϕ_2 to leave a residue of -20 at point 2 to cancel the 20 that will be dispersed to this point when ϕ_4 is changed by 20. The way in which this technique cuts down the number of steps to six is illustrated in Work Sheet II, which should be compared with Work Sheet I. The gain for this example is very small because of the small number of points. This latter fact ensures rapid convergence for even the straight relaxation procedure. Clearly the most rapid convergence possible takes four steps, since there are four points whose values must be changed.

WORK SHEET II

(Boundary points omitted)

13.19. Relationship between the Liebmann Process and the Relaxation Method.

In the Liebmann process one replaces the value at a point by the average of its four neighbors. As can be seen by reference to Eq. (13.61), this reduces the residue at the point to zero. Therefore, in the relaxation terminology, Liebmann's averaging process is just a means of eliminating the residue at a given point.

The essential difference between the two methods, however, lies in the following two aspects:

1. In the Liebmann method the points are improved in a prescribed order independent of the values guessed for ϕ_s. In the relaxation method one corrects the point with the largest residue first and then proceeds to the point having the largest remaining residue, etc. The latter method, by correcting the ϕ_s in the region in which the values are the most inconsistent, clearly converges more rapidly.

2. The Liebmann process is fixed, in that one always reduces the residue to zero, while in Southwell's method one may choose to make only a

partial reduction in the residue or, on the other hand, to reduce the residue past zero so that its sign reverses. That is, one may under- or overrelax, as described in Sec. 13.18. This flexibility in the relaxation method permits one to anticipate the effect on a given residue R_s, which is to be reduced by a change in ϕ_s, that will arise from subsequent correction of neighboring points. It also makes it possible for one to deal largely with round numbers, which are more easily managed in pencil-and-paper calculations.

The flexibility of Southwell's relaxation method makes it far superior to other methods for hand computation. However, it is just this flexibility in the method that makes it extremely difficult to adapt to an automatic electronic computer. For instance, the advantage at each step of the calculation of selecting for improvement that point having the largest residue is very important in speeding the convergence. Moreover, it is not too difficult in hand computing to look back over as many as several hundred residues and pick out the largest residue or at least one that is very nearly the largest. For an automatic calculator, however, this process involves a large number of comparisons, each of which requires time. Unless some revolutionary method is devised for computing machines to select the largest residue R_s in a time that is not much longer than that needed to improve ϕ_s, once it is selected, the gain in convergence rate of Southwell's method over Liebmann's will not be realizable in automatic computations.

13.20. Block Relaxation.† The changing of the value of ϕ at any point simply disperses a part of the residue at that point to the four neighboring points. It is clear, therefore, that in a problem in which one has chosen the initial ϕ's in such a way that the sum of the residues is rather small, because of their being of both positive and negative signs, the residues will tend to cancel out as one proceeds with the relaxation. This means that for such a choice of the ϕ's the convergence will be much more rapid than if most of the residues are of the same sign. In the latter case, the residues will not cancel appreciably, but will have to be dispersed to the boundaries.

One may make the sum of the residues over a large block of points approximately zero without too much trouble by changing the value of all ϕ's within the block by a constant amount x. If N lines of the net connect points within the block to points lying outside the block, the effect of this block alteration is to change the sum of residues within the block by $-Nx$. Clearly x should be chosen so that

$$Nx \cong \Sigma R_s \tag{13.64}$$

† See R. V. Southwell, "Relaxation Methods in Theoretical Physics," pp. 55–58, Oxford University Press, New York, 1946.

where the summation is over all points of the block. Southwell refers to this process as *block relaxation.*

The residues are changed for only those points within the block having neighbors lying outside and for these outside neighbors. The

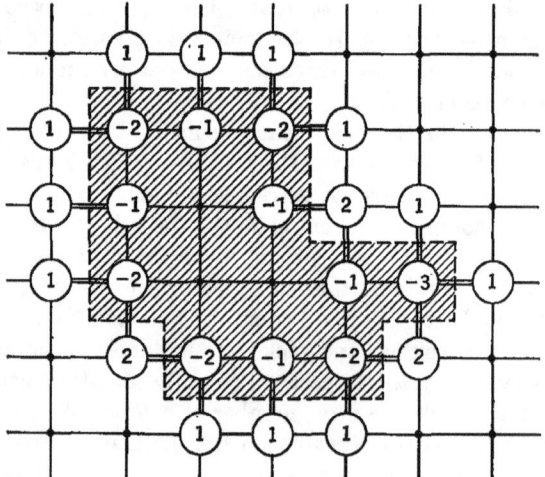

Fig. 13.8. Block relaxation of 14 points and the resulting relaxation pattern for the residues.

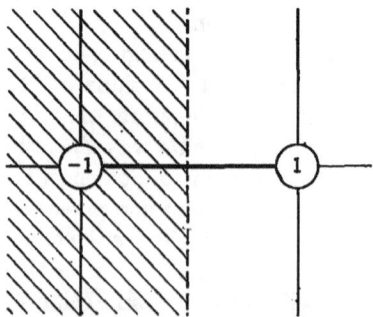

Fig. 13.9. Elementary relaxation pattern. The shaded region represents the region occupied by the block of points being relaxed.

relaxation pattern (which when multiplied by x gives the change in the residues) for a block of 14 points is shown in Fig. 13.8. Observe that this pattern is made up simply of superpositions of the elementary pattern shown in Fig. 13.9.

Example 3. Use block relaxation to reduce the sum of the residues within the dashed line of Fig. 13.10 to zero. The values are given on the boundary, but the values written to the left of the interior points of the lattice were chosen arbitrarily.

The sum of the residues within the block is $35 + 5 + 15 + 10 + 5 = 70$. There are 10 segments, or lines, of the net cut by the dashed line bounding the block of

5 points. Therefore, by Eq. (13.64), $10x = 70$, or $x = 7$, and hence all the points within the block are to be increased by 7 as shown in Fig. 13.10. The changes in the residues are also given in this figure. They are based on a relaxation pattern, resembling that of Fig. 13.8, formed by superposing the elementary relaxation pattern of Fig. 13.9 for each of the 10 segments cut.

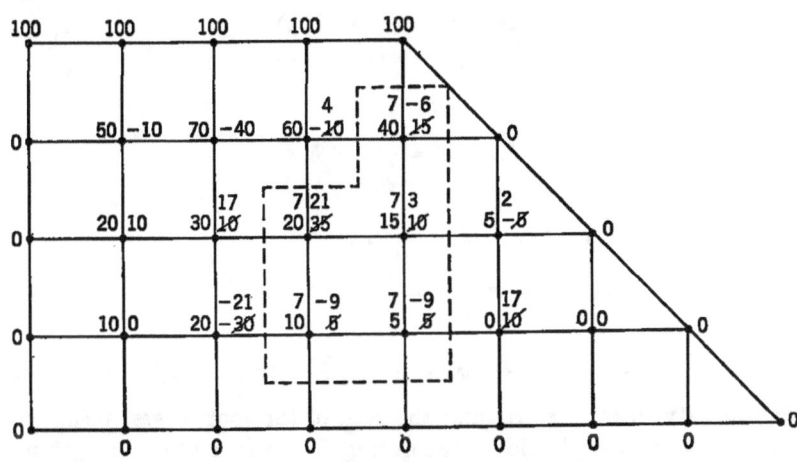

Fig. 13.10. Illustration of block relaxation.

13.21. Group Relaxation.† Southwell differentiates between a *block* relaxation and a *group* relaxation. The latter he uses to designate a more general treatment of a group of points than described above. Thus, if one conceives of a surface whose distance from the xy plane of the lattice at each lattice point is equal to ϕ, then a block relaxation is a straight displacement of a section of this surface, up or down as the case may be. On the other hand, a group relaxation permits rotation of this surface about one or more axes parallel to the xy plane and even distortions of the surface itself.

POISSON'S EQUATION

13.22. Liebmann's Method Applied to Poisson's Equation. Poisson's equation in two dimensions can be written

$$\nabla^2\phi \equiv \frac{\partial^2\phi}{\partial x^2} + \frac{\partial^2\phi}{\partial y^2} = f(x,y) \tag{13.65}$$

where $\phi = \phi(x,y)$. As seen from Chap. 12 or from the discussion of Laplace's equation, the finite-difference equation corresponding to Eq. (13.65) is indicated symbolically by

$$\Delta_h^2\phi_s \equiv \frac{1}{h^2}\left\{1 \quad \begin{matrix} 1 \\ -4 \\ 1 \end{matrix} \quad 1\right\}\phi_s = f_s \tag{13.66}$$

† See *ibid.*, p. 58.

where h is the net spacing and ϕ_s and f_s are the values of ϕ and f at a general net point s.

It is convenient to introduce, as in the theory of bivariate interpolation, the two operators E_x and E_y, which are defined by

$$\begin{aligned} E_x\phi(x,y) &= \phi(x+h,\,y) \\ E_y\phi(x,y) &= \phi(x,\,y+h) \end{aligned} \tag{13.67}$$

Their inverses satisfy the equations

$$\begin{aligned} E_x^{-1}\phi(x,y) &= \phi(x-h,\,y) \\ E_y^{-1}\phi(x,y) &= \phi(x,\,y-h) \end{aligned} \tag{13.68}$$

Using these operators, Eq. (13.66) can be written

$$(E_x + E_y + E_x^{-1} + E_y^{-1} - 4)\phi_s = h^2 f_s \tag{13.69}$$

Solving for ϕ_s,

$$\phi_s = \tfrac{1}{4}(E_x\phi_s + E_y\phi_s + E_x^{-1}\phi_s + E_y^{-1}\phi_s) - \tfrac{1}{4}h^2 f_s \tag{13.70}$$

The quantity in brackets is just the sum of the four closest neighbors of ϕ_s. Therefore ϕ_s should be the average of its four closest neighbors, just as in the case of Laplace's equation, minus, however, a term depending on the inhomogeneous function $f(x,y)$ and the square of the interval h.

Equation (13.70) can be used to improve arbitrarily assigned values of ϕ in exactly the same way that Eq. (13.16) was used in solving Laplace's equation. Employed in this way, Eq. (13.70) is termed the Liebmann improvement formula for Poisson's equation.

13.23. The Shortley-Weller Method Applied to Poisson's Equation.[†] Since Eq. (13.69) holds for all interior points of the lattice, it holds for $\phi_t = E_x^a E_y^b \phi_s$ provided a and b are integers and are so chosen that the point t lies within the lattice. That is,

$$(E_x + E_x^{-1} + E_y + E_y^{-1} - 4)(E_x^a E_y^b)\phi_s = h^2(E_x^a E_y^b)f_s \tag{13.71}$$

where clearly

$$E_x^a E_y^b f_s = f_t \tag{13.72}$$

One may add any number of equations of the form of Eq. (13.71); therefore

$$\begin{aligned} (E_x + E_x^{-1} &+ E_y + E_y^{-1} - 4)[E_x E_y + E_x^{-1} E_y + E_x E_y^{-1} + E_x^{-1} E_y^{-1} \\ &+ 2(E_x + E_x^{-1} + E_y + E_y^{-1}) + 6]\phi_s \\ &= h^2[E_x E_y + E_x^{-1} E_y + E_x E_y^{-1} + E_x^{-1} E_y^{-1} \\ &+ 2(E_x + E_x^{-1} + E_y + E_y^{-1}) + 6]f_s \end{aligned} \tag{13.73}$$

It is convenient to introduce the symbolic pattern notation of Eq. (13.66)

† See R. Weller, G. H. Shortley, and B. Fried, The Solution of Torsion Problems by Numerical Integration of Poisson's Equation, *J. Appl. Phys.*, 11: 283–290 (1940).

at this step. We already have that

$$E_x + E_x^{-1} + E_y + E_y^{-1} - 4 = \left\{\begin{matrix} & 1 & \\ 1 & -4 & 1 \\ & 1 & \end{matrix}\right\} \equiv P_1 \quad (13.74)$$

In complete analogy with this,

$$E_x E_y + E_x^{-1} E_y + E_x E_y^{-1} + E_x^{-1} E_y^{-1} + 2(E_x + E_x^{-1} + E_y + E_y^{-1}) + 6$$

$$= \left\{\begin{matrix} 1 & 2 & 1 \\ 2 & 6 & 2 \\ 1 & 2 & 1 \end{matrix}\right\} \equiv P_2 \quad (13.75)$$

Equation (13.73) becomes, therefore, in this symbolic pattern notation

$$\left\{\begin{matrix} & 1 & \\ 1 & -4 & 1 \\ & 1 & \end{matrix}\right\} \left\{\begin{matrix} 1 & 2 & 1 \\ 2 & 6 & 2 \\ 1 & 2 & 1 \end{matrix}\right\} \phi_s = h^2 \left\{\begin{matrix} 1 & 2 & 1 \\ 2 & 6 & 2 \\ 1 & 2 & 1 \end{matrix}\right\} f_s \quad (13.76)$$

The meaning of the pattern product of Eq. (13.76) for x increasing as we go downward and y increasing to the right is determined by the expressions for the two patterns in terms of the E's. Thus, since

E_x shifts the pattern P_2 one step down,

E_x^{-1} shifts the pattern P_2 one step up,

E_y shifts the pattern P_2 one step to the right,

E_y^{-1} shifts the pattern P_2 one step to the left, and

-4 multiplies the pattern P_2 by -4,

the product $P_1 P_2$ is the sum of these five patterns obtained from P_2. Carrying this out, one obtains as a result

$$P_1 P_2 = \left\{\begin{matrix} & 1 & \\ 1 & -4 & 1 \\ & 1 & \end{matrix}\right\} \left\{\begin{matrix} 1 & 2 & 1 \\ 2 & 6 & 2 \\ 1 & 2 & 1 \end{matrix}\right\}$$

$$= \left\{\begin{matrix} & 1 & 2 & 1 & \\ 1 & 0 & 0 & 0 & 1 \\ 2 & 0 & -16 & 0 & 2 \\ 1 & 0 & 0 & 0 & 1 \\ & 1 & 2 & 1 & \end{matrix}\right\} \quad (13.77)$$

Since the expressions for P_1 and P_2 in terms of the E's are commutative, one may multiply these symbolic patterns in the reverse order to that of (13.77) and obtain the same result.

Equation (13.76) can now be written

$$\left\{\begin{matrix} & 1 & 2 & 1 & \\ 1 & 0 & 0 & 0 & 1 \\ 2 & 0 & -16 & 0 & 2 \\ 1 & 0 & 0 & 0 & 1 \\ & 1 & 2 & 1 & \end{matrix}\right\} \phi_s = h^2 \left\{\begin{matrix} 1 & 2 & 1 \\ 2 & 6 & 2 \\ 1 & 2 & 1 \end{matrix}\right\} f_s \quad (13.78)$$

The left-hand side of this equation is just the composite-difference equation (13.12) developed for Laplace's equation, while the right-hand side is the inhomogeneous expression to be associated with it.

Equation (13.78) may be used to express the value of ϕ at the center of a block of 9 points in terms of the f's within and the values of ϕ at points surrounding the block. Applied to the block of 9 points shown in Fig. 13.11, one obtains an improvement formula for the ϕ at the center point (there designated by a), namely,

$$a' = \tfrac{1}{16}\left[\sum_{i=1}^{8} d_i + 2\sum_{i=1}^{4} e_i - h^2\left(6f_a + \sum_{i=1}^{4} f_{b_i} + 2\sum_{i=1}^{4} f_{c_i}\right)\right] \quad (13.79)$$

Equation (13.79) thus constitutes the first step in the 9-point block-improvement process of Shortley and Weller. The next step is to express b_i in terms of its diagonal neighbors and the value of f_{b_i} (the value of f at b_i). Since the spacing between diagonal points is $\sqrt{2}\,h$, the Poisson difference equation replacing Eq. (13.66) is

$$\frac{1}{2h^2}\left\{\begin{matrix} 1 & & 1 \\ & -4 & \\ 1 & & 1 \end{matrix}\right\}\phi_s = f_s \quad (13.80)$$

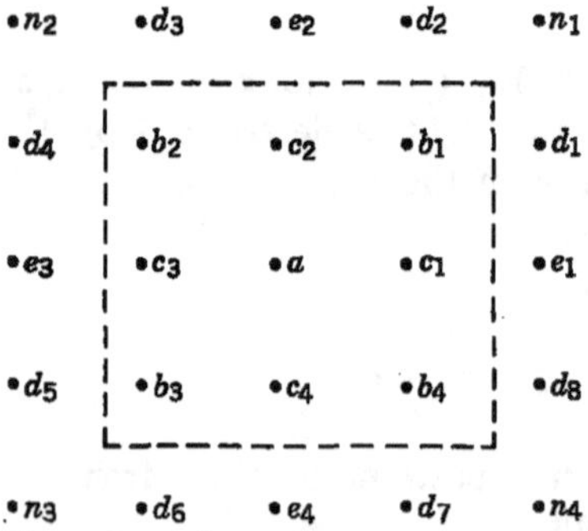

Applied to Fig. 13.11, Eq. (13.80) yields the following improvement formulas for the b's:

$$\begin{aligned} b_1' &= \tfrac{1}{4}(a' + e_1 + e_2 + n_1 - 2h^2 f_{b_1}) \\ b_2' &= \tfrac{1}{4}(a' + e_2 + e_3 + n_2 - 2h^2 f_{b_2}) \quad \text{etc.} \end{aligned} \quad (13.81)$$

Having the b's, we can obtain the improvement formula for the c's by employing Eq. (13.66); thus

$$\begin{aligned} c_1' &= \tfrac{1}{4}(a' + b_4' + b_1' + e_1 - h^2 f_{c_1}) \\ c_2' &= \tfrac{1}{4}(a' + b_1' + b_2' + e_2 - h^2 f_{c_2}) \quad \text{etc.} \end{aligned} \quad (13.82)$$

Notice that these improvement formulas are just those used for solving Laplace's equation with, however, certain additional terms depending linearly on the f's within the block. These latter terms can be evaluated once and for all for each point improvement formula; therefore the iterative process is essentially the same as for Laplace's equation.

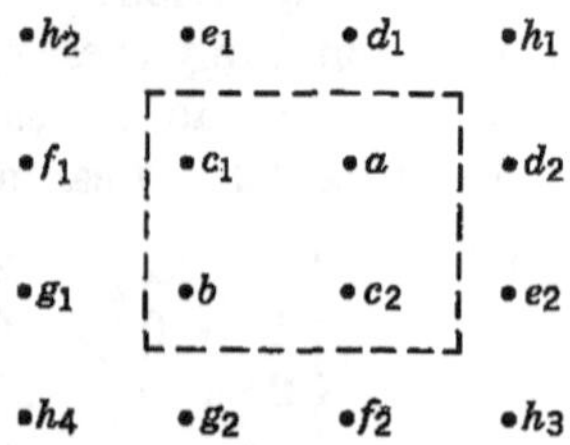

In a similar way one may obtain improvement formulas for the 4-point and for the

FIG. 13.11. Lettering of points in a 9-point block.

FIG. 13.12. Lettering of points in a 4-point block.

25-point blocks.   The resulting formulas for the 4-point block in terms of Fig. 13.12 are

$$a' = \tfrac{1}{24}[7(d_1 + d_2) + 2(e_1 + e_2 + f_1 + f_2) + (g_1 + g_2)$$
$$\qquad\qquad - h^2(7f_a + f_b + 2f_{c_1} + 2f_{c_2})]$$
$$b' = \tfrac{1}{4}[a' + f_1 + f_2 + h_4 - 2h^2f_b] \qquad\qquad (13.83)$$
$$c_1' = \tfrac{1}{4}[a' + b' + e_1 + f_1 - h^2f_{c_1}]$$
$$c_2' = \tfrac{1}{4}[a' + b' + e_2 + f_2 - h^2f_{c_2}]$$

13.24. Use of the Difference Function and Extrapolation.   In Sec. 13.15 we observed several advantages of treating the difference function, i.e., the difference between the first improved function and the original function.   This device is applicable to the above treatments of Poisson's equation.   Moreover, the improvement formulas applicable to this difference function are obtained from those above by setting the f's equal to zero; i.e., they are identical with those for Laplace's equation.†   Thus the f's are needed only in obtaining the difference function.

Since the improvement formulas applied to the difference function are the same as for Laplace's equation, the whole discussion relating to the relative rate of convergence of the Liebmann process and the 4-, 9-, and 25-point block processes of Shortley and Weller applies without amendment.   In particular, the extrapolation process can be applied to Poisson's equation without modification.

13.25. The Relaxation Method Applied to Poisson's Equation.   The residue for Poisson's equation is defined by the equation

$$R_s \equiv (E_x + E_x^{-1} + E_y + E_y^{-1} - 4)\phi_s - h^2f_s \qquad (13.84)$$

If the residue at every point is reduced to zero, the Poisson difference equation (13.69) is satisfied.   Clearly a change in ϕ_s of x produces a new residue

$$R_s' = R_s - 4x$$

just as in the Laplace difference equation.   Therefore the same relaxation pattern, that given in Fig. 13.7, holds for both cases.

This means that the only distinction between the relaxation method for Laplace's equation and for Poisson's equation arises in the initial calculation of the residues.   Since the way in which these residues are dispersed is exactly the same in both cases, the techniques of overrelaxation, underrelaxation, block relaxation, and group relaxation described previously are applicable directly to Poisson's equation.

† *Ibid.*

OTHER ELLIPTIC EQUATIONS

13.26. Biharmonic Equation. As a somewhat more complicated differential equation of the elliptic type, consider the biharmonic equation

$$\nabla^4\phi \equiv \frac{\partial^4\phi}{\partial x^4} + 2\frac{\partial^4\phi}{\partial x^2\,\partial y^2} + \frac{\partial^4\phi}{\partial y^4} = 0 \qquad (13.85)$$

This is the differential equation satisfied, for example, by the deflection of a plate if the magnitude of the deflection is small.

Since Eq. (13.85) can be written in the form

$$\nabla^2(\nabla^2\phi) = \left(\frac{\partial^2}{\partial x^2} + \frac{\partial^2}{\partial y^2}\right)\left(\frac{\partial^2}{\partial x^2} + \frac{\partial^2}{\partial y^2}\right)\phi = 0 \qquad (13.86)$$

the corresponding finite-difference equation relating to a square lattice is expressed symbolically as

$$\frac{1}{h^4}\left\{1 \quad \begin{matrix}1\\-4\\1\end{matrix} \quad 1\right\}\left\{1 \quad \begin{matrix}1\\-4\\1\end{matrix} \quad 1\right\}\phi_s = 0 \qquad (13.87)$$

where s is any interior point of the lattice except those adjacent to the boundary points. If we make use of the idea of products of patterns developed above, this equation becomes

$$\left\{\begin{matrix} & & 1 & & \\ & 2 & -8 & 2 & \\ 1 & -8 & 20 & -8 & 1 \\ & 2 & -8 & 2 & \\ & & 1 & & \end{matrix}\right\}\phi_s = 0 \qquad (13.88)$$

One may employ the relaxation method to solve this equation, in which case the residue at s is defined as

$$R_s \equiv \left\{\begin{matrix} & & 1 & & \\ & 2 & -8 & 2 & \\ 1 & -8 & 20 & -8 & 1 \\ & 2 & -8 & 2 & \\ & & 1 & & \end{matrix}\right\}\phi_s \qquad (13.89)$$

A change in ϕ_s of x causes a change in the residue R_s of $20x$ and a change in the residues of the four closest neighbors of s of $-8x$. The total effect on the residues of a change in ϕ_s is indicated by the relaxation pattern of Fig. 13.13. This pattern, when multiplied by x, the change in ϕ_s, yields the change in the residues at s and surrounding points.

Since the difference equation (13.88) applies only to those interior points which are not adjacent to the boundary points, it is clear that the boundary conditions must provide sufficient information to fix the outer two layers of points, i.e., the boundary points and those immediately adjacent to them. Thus, one might specify ϕ and its normal derivative on the closed boundary.

The relaxation process, as before, consists at each step in changing the ϕ associated with the largest residue in such a way as to reduce the residues to zero, or nearly so. When all the residues are zero, within a small tolerance, the ϕ's satisfy approximately the biharmonic difference equation (13.88), and hence for small enough net spacing h they approximate the numerical values of the solution of the differential equation (13.85) at these lattice points.

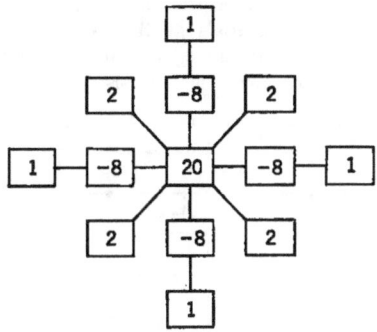

Fig. 13.13. Relaxation pattern for the biharmonic equation.

Although the relaxation is more complicated for the biharmonic difference equation than for the Laplace difference equation, the processes of overrelaxation, underrelaxation, block relaxation, etc., are obviously applicable to the biharmonic equation.

PROBLEMS

1. Obtain the weights shown in Eqs. (13.12) and (13.14) from the basic difference equation of (13.2).
2. Apply the Liebmann process to the problem given in Fig. 13.1 using zero as the initial value of all interior points. Carry out results to two decimal places. Check the results against the values given in Table 13.1.
3. Obtain the formula (13.17) for an unequal-arm star using the method of Sec. 7.4 to express the second derivatives in the Laplacian in terms of unequally spaced ordinates.
4. Determine what initial values would be obtained in the problem of Fig. 13.1 using the method of Frocht and Levens.
5. Reduce the size of the net in Prob. 2 to one-half in the region of the maximum, and find the new solution for this mixed-sized net.
6. Determine the rate of convergence of the improvement method of Kormes when applied to a rectangular area by finding the lowest characteristic value λ_1 associated with the improvement matrix T. HINT: Take for the lowest mode $\phi_{ij} = \sin{(\pi i/m)} \sin{(\pi j/n)}$.
7. Find the explicit expression for the Liebmann improvement matrix applied to a rectangular region containing six internal points. Show that it depends on the method of scanning.
8. A hollow, cylindrical electrical conductor of square cross section having 1 cm inside dimensions has running through it a cylindrical center conductor of circular cross

section $\frac{1}{2}$ cm in diameter. Find the potential distribution in the region between these conductors when a potential of 100 volts relative to the outer conductor is applied to the center conductor. Laplace's equation is thus satisfied in the region between the conductors. Apply Laplace's difference equation to the cross section of this region, using a square net with a spacing $h = \frac{1}{8}$ cm between net points. Use the relaxation method, and make full use of the symmetry present. When the potentials at the net points are obtained, employ linear interpolation to sketch in the equipotential surfaces corresponding to 25, 50, and 75 volts.

9. Write down the block-relaxation pattern for the following blocks of points:

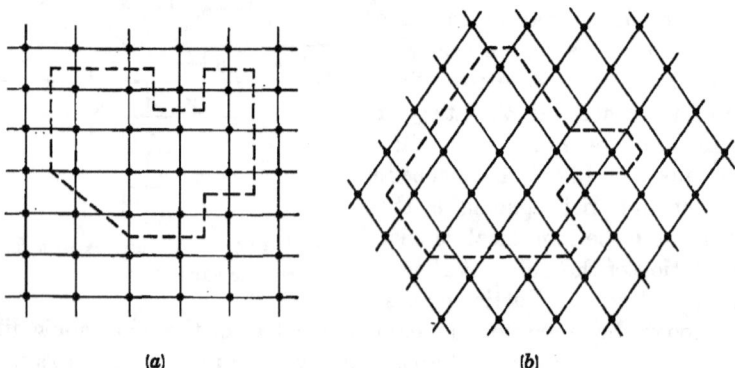

(a)   (b)

10. (a) Solve Poisson's equation (13.65) for a square region when the source function $f(x,y)$ is equal to a constant. Take nine interior points, and use the Shortley-Weller method. (b) Show how extrapolation can be used to speed up the calculations. (Assume that ϕ is zero on the boundary.)

11. Assuming that ϕ satisfies Laplace's difference equation, and making use of the pattern equation

$$\left\{ \begin{matrix} & 1 & \\ 1 & -4 & 1 \\ & 1 & \end{matrix} \right\} \left\{ \begin{matrix} & 1 & \\ 1 & 4 & 1 \\ & 1 & \end{matrix} \right\} = \left\{ \begin{matrix} & & 1 & & \\ & 2 & 0 & 2 & \\ 1 & 0 & -12 & 0 & 1 \\ & 2 & 0 & 2 & \\ & & 1 & & \end{matrix} \right\}$$

determine the potential at the five interior points of the following lattice:

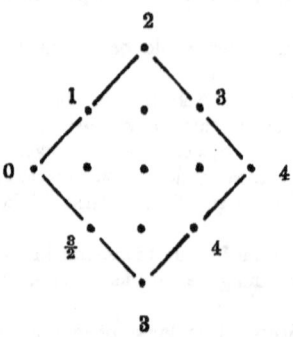

The numbers give the potential at the boundary points of the lattice.

CHAPTER 14

SOLUTION OF PARTIAL DIFFERENTIAL EQUATIONS
OF THE PARABOLIC AND HYPERBOLIC TYPES

In this chapter we shall study the partial difference equations corresponding to differential equations of both the parabolic and the hyperbolic types. The basis of classification was given in Chap. 12, where it was pointed out that each of the three types has associated with it one or more characteristic types of boundary conditions. Thus, the elliptic equation is associated with a Dirichlet or with a Neumann boundary condition, in which the value of the dependent variable, or its normal derivative, is specified on a closed surface. Such boundary conditions, however, are unsuited for the parabolic and the hyperbolic equations.

14.1. Heat-flow Equations. As an example of an equation of the parabolic type, let us again take the *heat-flow*, or *diffusion*, equation

$$\nabla^2 \phi - \frac{1}{a} \frac{\partial \phi}{\partial t} = 0 \tag{14.1}$$

For simplicity, assume that $\phi = \phi(x,t)$; then one obtains the one-dimensional heat-flow equation

$$\frac{\partial^2 \phi}{\partial x^2} - \frac{1}{a} \frac{\partial \phi}{\partial t} = 0 \tag{14.2}$$

By taking a rectangular lattice in the xt plane in which the lattice points are at $x_i = x_0 + ih$ and $t_j = t_0 + jp$, where i and j are integers, and expressing the derivatives in terms of differences, one obtains the corresponding difference equation. Letting

$$\phi_i^j = \phi(x_0 + ih, t_0 + jp) \tag{14.3}$$

one has (see Chap. 12)

$$\frac{\partial^2}{\partial x^2} \phi(x_i, t_j) = \frac{1}{h^2} (\phi_{i+1}^j - 2\phi_i^j + \phi_{i-1}^j) \tag{14.4}$$

and

$$\frac{\partial}{\partial t} \phi(x_i, t_j) = \frac{1}{p} (\phi_i^{j+1} - \phi_i^j) \tag{14.5}$$

Thus, a difference equation that approximates the differential equation

(14.2) is

$$\frac{1}{h^2}(\phi_{i+1}^j - 2\phi_i^j + \phi_{i-1}^j) = \frac{1}{ap}(\phi_i^{j+1} - \phi_i^j)$$

or, solving for ϕ_i^{j+1},

$$\phi_i^{j+1} = (1 - 2m)\phi_i^j + m(\phi_{i+1}^j + \phi_{i-1}^j) \qquad (14.6)$$

where

$$m = \frac{ap}{h^2} \qquad (14.7)$$

Here m may be varied by taking different time intervals p.

This equation is to be compared to the two-dimensional heat-flow difference equation (12.52). In the latter a choice of $m = \frac{1}{4}$ is suggested, since the first term then drops out; however, this choice of m in Eq. (14.6) does not cause the first term to vanish but leads to the equation

$$\phi_i^{j+1} = \frac{1}{4}(\phi_{i+1}^j + 2\phi_i^j + \phi_{i-1}^j) \qquad (14.8)$$

This is just what one would deduce directly from Eq. (12.52). Thus, the choice of m leading to the vanishing of the first term in Eq. (14.6) or (12.52) should not be considered as particularly significant. It will be shown later, however, that the stability of the difference equation does depend on the choice of m.

14.2. Boundary Conditions for Parabolic Equations. Typical boundary conditions for the one-dimensional heat-flow equation (14.2) are a specification of the temperatures (which may be made to vary with time) on two faces of a slab of material and the initial temperature distribution throughout the slab. This amounts to specifying the three following functions:

$$\begin{aligned} \phi(0,t) &= f_1(t) & t \geq 0 \\ \phi(L,t) &= f_2(t) & t \geq 0 \\ \phi(x,0) &= g(x) & 0 \leq x \leq L \end{aligned} \qquad (14.9)$$

The faces of the slab are located at $x = 0$ and $x = L$, and the temperature distribution is given at $t = 0$.

If one chooses a rectangular net as described by Eq. (14.3), letting $x_0 = 0$, $t_0 = 0$, and takes $h = L/M$, where M is an integer, the lattice points ϕ_0^j and ϕ_M^j represent the temperature of the two surfaces of the slab (see Fig. 14.1). The initial temperature distribution throughout the slab is then given by the ϕ_i^0, which lie along the x axis. Thus one substitutes for the boundary conditions (14.9) the conditions

$$\begin{aligned} \phi_0^j &= f_1(jp) & j \geq 0 \\ \phi_M^j &= f_2(jp) & j \geq 0 \\ \phi_i^0 &= g(ih) & 0 \leq i \leq M \end{aligned} \qquad (14.10)$$

applicable directly to the lattice.

Starting with the known values of ϕ_i^0, one may use the difference equation (14.6), with j set equal to zero, to determine ϕ_i^1 for $i = 1, 2,$ ..., $M - 1$. Since ϕ_0^1 and ϕ_M^1 are specified by the boundary conditions, one then has all the ϕ_i^1. From these, by letting $j = 1$ in Eq. (14.6), one may likewise determine the ϕ_i^2. Proceeding in this way, one is able to determine the ϕ_i^j on successive rows of the lattice of Fig. 14.1. This amounts to finding the temperature throughout the slab at successive times separated by the time interval p.

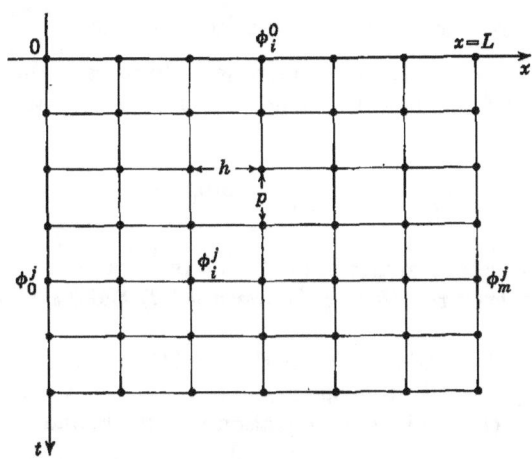

FIG. 14.1. Rectangular net used in the discussion of both the heat-flow equation and the wave equation.

There remains the question of how well the function ϕ_i^j of the discrete variable i for a fixed value of j represents the function $\phi(x,t)$ of the continuous variable x for $t = jp$. One would expect, generally, that the finer the net, the more accurate the representation and that as the net spacing approaches zero, the solution ϕ_i^j of the difference equation (14.6) would approach the solution $\phi(x,t)$ of the differential equation (14.2). It turns out, however, that the latter convergence depends on the choice of the parameter m in Eq. (14.6). Note that in keeping m fixed, as one does in such a problem, the difference equation (14.6) remains the same as the net size is reduced, but the net spacing p in the time direction reduces much more rapidly than h. Thus, if h is reduced to $\frac{1}{2}$, p must be reduced to $\frac{1}{4}$; therefore the individual rectangular nets become more and more elongated with the short dimension in the time direction.

Besides the question of the theoretical improvement of the solution as the net size is reduced, there is the question of the *stability* of the solution. This refers to the effect of round-off errors or incidental errors in the calculation of the ϕ_i^j for successive rows on all later values. If these errors tend to grow as one proceeds with the calculation, the solution is said to be unstable. It will be shown that the stability of the solution depends on the choice of m.

Instead of specifying the temperature on the faces of the slab, one could specify the rate at which heat is lost at these faces. One might, for example, assume at $x = 0$ a heat loss proportional to the difference in temperature between the face at $x = 0$ and some ambient temperature T_0. Since this heat must be conducted to the surface at the same rate it is lost, the heat loss per unit area at the face $x = 0$ is $\kappa \left[\dfrac{\partial \phi}{\partial x}\right]_{x=0}$, where κ is the thermal conductivity of the material. In place of the first boundary condition of Eqs. (14.9) we would then use the condition

$$\left[\frac{\partial}{\partial x}\,\phi(x,t)\right]_{x=0} = c[\phi(0,t) - T_0] \qquad (14.11)$$

where c is a constant giving the ratio of the heat loss per unit area, for a unit temperature difference between $\phi(0,t)$ and T_0, to the thermal conductivity κ.

If we replace the left-hand side of Eq. (14.11) by the equivalent difference ratio $(\phi_1^j - \phi_0^j)/h$, the boundary condition may be transformed for the purpose of the difference equation into the boundary condition

$$\phi_0^j = \frac{1}{1 + ch}\,(\phi_1^j + chT_0) \qquad (14.12)$$

All the ϕ_i^{j-1} for a given j being known, the difference equation (14.6) (as seen above) yields the values of ϕ_i^j for only the interior points $i = 1, 2, \ldots, M - 1$. The boundary condition of (14.12) is therefore needed to determine ϕ_0^j, which is needed, in turn, in calculating the next row.

Note that giving the boundary condition on all four sides of the lattice in Fig. 14.1, as would be done for an elliptic equation, would overspecify the solution. One could then use the values on three sides, as shown above, to obtain a solution, which would consequently determine the values on the fourth side. Thus, the specification of the solution on the positive-time side either would be superfluous, or, even more likely, it would lead to a contradiction.

14.3. Convergence of the Solution of the Difference Equation. As the net in Fig. 14.1 is made finer and finer by increasing $M = L/h$, the values ϕ_i^j at the lattice points should approach closer and closer to the value of the solution $\phi(x,t)$ of the differential equation (14.2) at the corresponding locations $x = ih$ and $y = jp$. Convergence requires, in fact, that the difference should go to zero as $M \to \infty$. We shall investigate such convergence for the boundary conditions

$$\phi(0,t) = \phi(L,t) = 0 \qquad t \geq 0$$
$$\phi(x,0) = g(x) \qquad 0 \leq x \leq L \tag{14.13}$$

which correspond to Eqs. (14.9) for the special case $f_1(t) = f_2(t) = 0$.

It is convenient to introduce a function $\phi^{(M)}(x,t)$ satisfying the difference equation

$$\phi^{(M)}(x, t + p) = (1 - 2m)\phi^{(M)}(x,t)$$
$$+ m[\phi^{(M)}(x + h, t) + \phi^{(M)}(x - h, t)] = (1 - m\delta_x^2)\phi^{(M)}(x,t) \tag{14.14}$$

where m is defined by Eq. (14.7) and δ_x^2 is the second central difference with respect to x. We require that $\phi^{(M)}(x,t)$ satisfy the same boundary conditions (14.13) as $\phi(x,t)$. Since for the lattice points $x = ih$, $y = jp$, where i and j are integers, the difference equation (14.14) reduces to Eq. (14.6) and the boundary conditions (14.9) reduce to the conditions (14.10), one has

$$\phi_i^j = \phi^{(M)}(ix,jh) \tag{14.15}$$

Convergence can, therefore, be formulated as the convergence of the function $\phi^{(M)}(x,t)$ to $\phi(x,t)$ as M approaches infinity.

First, we obtain the solution of the differential equation (14.2). For a given value of t the function $\phi(x,t)$ may be expanded as a function of x into a Fourier series

$$\phi(x,t) = \sum_{n=1}^{\infty} b_n(t) \sin \frac{n\pi x}{L} \tag{14.16}$$

whose coefficients depend on t. Substituting this into the differential equation (14.2), one has

$$- \sum_{n=1}^{\infty} \left[\left(\frac{n\pi}{L}\right)^2 b_n(t) + \frac{1}{a} b_n'(t) \right] \sin \frac{n\pi x}{L} = 0$$

which requires that

$$b_n'(t) = - \frac{an^2\pi^2}{L^2} b_n(t)$$

Thus,
$$b_n(t) = B_n e^{-an^2\pi^2 t/L^2}$$

where B_n is a constant, and hence from Eq. (14.16)

$$\phi(x,t) = \sum_{n=1}^{\infty} B_n e^{-an^2\pi^2 t/L^2} \sin \frac{n\pi x}{L} \tag{14.17}$$

The first boundary conditions of Eqs. (14.13) are immediately satisfied, so one need only satisfy the condition $\phi(x,0) = g(x)$. Since by Eq. (14.17) $\phi(x,0) = \Sigma_{n=1}^{\infty} B_n \sin (n\pi x/L)$, this can be done by letting

$$B_n = \frac{2}{L} \int_0^L g(x) \sin \frac{n\pi x}{L} dx \qquad (14.18)$$

To obtain a solution of the difference equation (14.14), we again take a fixed value t and expand $\phi^{(M)}(x,t)$ in a Fourier series; thus

$$\phi^{(M)}(x,t) = \sum_{n=1}^{\infty} b_n^{(M)}(t) \sin \frac{n\pi x}{L} \qquad (14.19)$$

Since the second central difference of $\sin (n\pi x/L)$ is

$$\delta^2 \sin \frac{n\pi x}{L} \equiv \sin \frac{n\pi(x + h)}{L} - 2 \sin \frac{n\pi x}{L} + \sin \frac{n\pi(x - h)}{L}$$

$$= 2 \left(\cos \frac{n\pi h}{L} - 1 \right) \sin \frac{n\pi x}{L}$$

$$= -4 \sin^2 \frac{n\pi h}{2L} \sin \frac{n\pi x}{L} \qquad (14.20)$$

one has

$$\delta_x^2 \phi^{(M)}(x,t) = \sum_{n=1}^{\infty} b_n^{(M)}(t) \delta_x^2 \sin \frac{n\pi x}{L}$$

$$= -4 \sum_{n=1}^{\infty} \sin^2 \frac{n\pi h}{2L} b_n^{(M)}(t) \sin \frac{n\pi x}{L}$$

therefore substitution of $\phi^{(M)}(x,t)$ from Eq. (14.19) into Eq. (14.14) gives rise to the equation

$$\sum_{n=1}^{\infty} \left[b_n^{(M)}(t + p) - \left(1 - 4m \sin^2 \frac{n\pi h}{2L} \right) b_n^{(M)}(t) \right] \sin \frac{n\pi x}{L} = 0 \qquad (14.21)$$

This requires, since $h = L/M$, that

$$b_n^{(M)}(t + p) = \left(1 - 4m \sin^2 \frac{n\pi}{2M} \right) b_n^{(M)}(t) \qquad (14.22)$$

To obtain a solution of Eq. (14.22) assume that

$$b_n^{(M)}(t) = B_n^{(M)} e^{\alpha_n t} \qquad (14.23)$$

where $B_n^{(M)}$ is a constant; then Eq. (14.22) is satisfied if

$$e^{\alpha_n(t+p)} = \left(1 - 4m \sin^2 \frac{n\pi}{2M}\right) e^{\alpha_n t} \tag{14.24}$$

or

$$\alpha_n = \frac{1}{p} \ln \left(1 - 4m \sin^2 \frac{n\pi}{2M}\right) \tag{14.25}$$

The constant $B_n^{(M)}$ in Eq. (14.23) could be a periodic function of period p. Since this generalization does not affect the values at the lattice points, it will be omitted. This places an additional condition on $\phi^{(M)}(x,t)$ but does not restrict the ϕ_i^j.

Substituting from Eq. (14.23) into Eq. (14.19), one has

$$\phi^{(M)}(x,t) = \sum_{n=1}^{\infty} B_n^{(M)} e^{\alpha_n t} \sin \frac{n\pi x}{L}$$

Since for $t = 0$

$$\phi^{(M)}(x,0) = g(x) = \sum_{n=1}^{\infty} B_n^{(M)} \sin \frac{n\pi x}{L}$$

it is clear that $B_n^{(M)} = B_n$, as given by Eq. (14.18), so that

$$\phi^{(M)}(x,t) = \sum_{n=1}^{\infty} B_n e^{\alpha_n t} \sin \frac{n\pi x}{L} \tag{14.26}$$

This solution differs from that of Eq. (14.17) for the differential equation only in the coefficients of t in the exponentials.

Any given α_n (n fixed) approaches $-an^2\pi^2/L^2$, as required, as M approaches infinity. This can be seen from Eq. (14.25), for as $M \to \infty$, $\sin(n\pi/2M) \to n\pi/2M$. Moreover, for small values of x, the $\ln(1 - x) \to -x$, and therefore as $M \to \infty$,

$$\alpha_n \to \frac{1}{p} \ln \left(1 - m \frac{n^2\pi^2}{M^2}\right) \to -\frac{m}{p} \frac{n^2\pi^2}{M^2}$$

From Eq. (14.7) $m/p = a/h^2$, and since $L = Mh$,

$$\alpha_n \to -\frac{a}{h^2} \frac{n^2\pi^2}{M^2} = -a \frac{n^2\pi^2}{L^2} \tag{14.27}$$

as M approaches infinity.

Thus, each individual term of the series in Eq. (14.26) for $\phi^{(M)}(x,t)$ as the net size is reduced approaches the corresponding term of $\phi(x,t)$ in Eq. (14.17). If the initial distribution given by $g(x)$ is expressible in a finite Fourier series so that only a finite number of terms occurs in these equations, this convergence of the individual terms ensures convergence of $\phi^{(M)}(x,t)$, the solution of the difference equation, to the solution $\phi(x,t)$ of the differential equation. In general, however, convergence of the individual terms is not sufficient.

As seen from Eq. (14.24), each increase of p in the time multiplies the exponential $e^{\alpha_n t}$ by the factor $1 - 4m \sin^2 (n\pi/2M)$. If $m < \frac{1}{2}$, this factor is always less than 1 in absolute value for all n and M, and hence $e^{\alpha_n t}$ decreases as t increases. If, however, $m > \frac{1}{2}$, then for n an odd multiple of M the factor is $1 - 4m < -1$. Thus $|e^{\alpha_n t}|$ will increase exponentially with time and cause trouble. No matter how large M, therefore, for $m > \frac{1}{2}$ some of the terms in Eq. (14.19) will grow exponentially with time, instead of decreasing exponentially, as in Eq. (14.17). Convergence of $\phi^{(M)}(x,t)$ to $\phi(x,t)$ should, therefore, be expected to hold for $m > \frac{1}{2}$ only for special restrictions on the function $g(x)$.

Hildebrand† has shown that if $g(x)$ is continuous in the interval 0 to L except for a finite number of finite jumps and is of bounded variation, and if m is fixed with $0 < m < \frac{1}{2}$, then, for any value of x in the interval and for any positive value of t,

$$\lim_{M \to \infty} \phi^{(M)}(x,t) = \phi(x,t)$$

14.4. Stability of the Solution. In Sec. 14.3 it was shown that for $m < \frac{1}{2}$ and the boundary conditions of Eqs. (14.13) each term in the solution of the difference equation decreased exponentially with time. This means that an error in one of the values of ϕ_i^j due, say, to round-off will alter the values of the ϕ_i^j in subsequent rows but that these effects will die out as one proceeds. If, however, $m > \frac{1}{2}$, some of the terms in the solution will increase exponentially with time. Thus, even if an exact solution of the difference equation does converge to the differential equation for particular choices of $g(x)$, some round-off errors will tend to be amplified exponentially as one proceeds. Therefore, the solution is said to be unstable for $m > \frac{1}{2}$.

A simple way of understanding the dependence of the instability on m is to go back to the difference equation (14.6) and assume that in the jth row of Fig. 14.1 we have errors of constant magnitude ϵ which alternate in sign. Then, by that equation, the errors in the $(j + 1)$th row are of magnitude $|1 - 4m|\epsilon$ and also alternate in sign. Thus, if $m > \frac{1}{2}$, the errors grow from row to row, being multiplied each time by $4m - 1$.

14.5. The Wave Equation. Consider the one-dimensional wave equation (12.10)

$$\frac{\partial^2 \phi}{\partial x^2} - \frac{1}{v^2} \frac{\partial^2 \phi}{\partial t^2} = 0 \tag{14.28}$$

† See F. B. Hildebrand, On the Convergence of Numerical Solutions of the Heat-flow Equation, *J. Math. Phys.*, **31**: 35–41 (1952).

satisfied, for example, by a vibrating string when the displacement is small. Here v designates the velocity of propagation along the string. For boundary conditions we shall take the following:

$$
\begin{aligned}
\phi(0,t) &= f_1(t) \\
\phi(L,t) &= f_2(t) \\
\phi(x,0) &= g_1(x) \\
\phi_t(x,0) &= g_2(x)
\end{aligned}
\qquad (14.29)
$$

where $\phi_t(x,t) \equiv \partial\phi/\partial t$. Note that the first three of these are the same conditions imposed on the heat-flow equation [see Eqs. (14.9)]. The addition of the fourth condition, however, is fundamental in that the differential equation contains the second time derivative rather than the first. In both Eqs. (14.9) and (14.29) the first two boundary conditions will not occur if the entire range of x is covered by the differential equation. Boundary conditions such as the above, giving at some time $t = 0$ the value of the dependent variable together with its first time derivative throughout some special region, are said to be of the *Cauchy* type.

We again use the rectangular net of Fig. 14.1 and the same designation ϕ_i^j for the value of $\phi(ih,jp)$, but ϕ now should be thought of as some kind of a displacement rather than as a temperature. The second derivates of Eq. (14.28) can then be replaced by the second differences divided by the square of the net spacing [see, for example, Eq. (12.48)]. Thus, the difference equation corresponding to Eq. (14.28) is

$$
\frac{1}{h^2}(\phi_{i+1}^j - 2\phi_i^j + \phi_{i-1}^j) = \frac{1}{v^2 p^2}(\phi_i^{j+1} - 2\phi_i^j + \phi_i^{j-1})
$$

which, solved for ϕ_i^{j+1}, may be written

$$
\phi_i^{j+1} = 2(1 - m^2)\phi_i^j + m^2(\phi_{i+1}^j + \phi_{i-1}^j) - \phi_i^{j+1} \qquad (14.30)
$$

where
$$
m = \frac{vp}{h} \qquad (14.31)
$$

It is convenient to choose the ratio of the mesh spacings p and h so that $m = 1$. The difference equation (14.30) then reduces to

$$
\phi_i^{j+1} = \phi_{i+1}^j + \phi_{i-1}^j - \phi_i^{j-1} \qquad (14.32)
$$

Using the pattern notation of previous chapters, we may write this difference equation in the form

$$
\left\{
\begin{matrix}
 & -1 & \\
1 & 0 & 1 \\
 & -1 &
\end{matrix}
\right\} \phi_i^j = 0 \qquad (14.33)
$$

14.6. Boundary Conditions for the Difference Equation. The fourth boundary condition of Eqs. (14.29) may be written

$$\frac{1}{p}(\phi_i^1 - \phi_i^0) = g_2(ih) \tag{14.34}$$

Since the first three boundary conditions are the same as for the heat-flow equation [see Eqs. (14.9) and (14.10)], the boundary conditions for the difference equation may be written

$$\begin{aligned}
\phi_0^j &= f_1(jp) & j &\geq 0 \\
\phi_M^j &= f_2(jp) & j &\geq 0 \\
\phi_i^0 &= g_1(ih) & 0 &\leq i \leq M \\
\phi_i^1 &= g_1(ih) + pg_2(ih) & 0 &\leq i \leq M
\end{aligned} \tag{14.35}$$

Here, as before, $h = L/M$, where M is an integer. The last equation of Eqs. (14.35) is obtained from Eq. (14.34) by replacing ϕ_i^0 by $g_1(ih)$.

The boundary conditions (14.35) specify the values of the ϕ_i^j on the first two rows and on the first and last columns. In the heat-flow equation the second row is not specified, and for Laplace's equation the last row is specified in place of the second row—of course, in this case, t would represent not time but another space coordinate. These remarks illustrate the type of boundary condition appropriate to each of the three types of difference equations, hyperbolic, parabolic, and elliptic (for a definition of this classification see Chap. 12).

14.7. Solution of the Difference Equation. The solution of the difference equation (14.32) together with its boundary conditions (14.35) is obtained by starting with the known values in the first two rows and working forward in time by means of the difference equation. Thus the procedure here, as with a parabolic equation, represents a *marching problem*,† in which values are obtained in succession from the knowledge of previous values without the need of successive improvements.

Starting with a knowledge of the values of ϕ_i^j in the first two rows, obtained from the last two equations of Eqs. (14.35), one may, by taking $j = 2$ in Eq. (14.32), obtain the values ϕ_i^3, $i = 1, 2, \ldots, M$ for all the interior points of the third row. The two end values ϕ_0^3 and ϕ_M^3 are given by the first two boundary conditions. Having all the ϕ_i^3, we may set $j = 3$ in the difference equation (14.32) and obtain the ϕ_i^4. Thus, all subsequent rows may be found from a knowledge of the first two and from the first two boundary conditions.

There remains the question of how closely the solution of the difference equation obtained in a practical application of the method represents the true values of the differential equation at the net points. This

† This terminology was introduced by L. F. Richardson, *Trans. Roy. Soc. (London)*, **226A**: 300 (1927).

depends on the *truncation error* introduced by replacing the differential equation by the difference equation and on the *round-off error* introduced by the need to use finite decimal fractions.   It is important to know, for example, how the truncation error depends on the interval h and whether it goes to zero as $h \to 0$.

14.8. Convergence for Hyperbolic Equations.   We shall study the convergence of the solution of the difference equation (14.32) to the solution

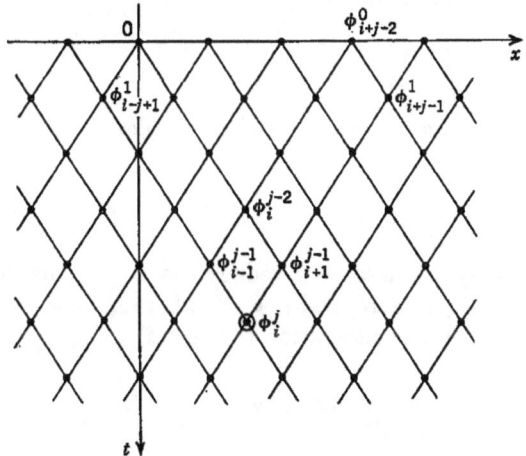

Fig. 14.2. Dependence of ϕ_i^j on the initial values ϕ_s^0 and ϕ_s^1.

of the one-dimensional wave equation (14.28) for the boundary conditions

$$\phi(x,0) = g_1(x)$$
$$\phi_t(x,0) = g_2(x) \tag{14.36}$$

which are to hold for all values of x.   The corresponding conditions on the ϕ_i^j are then [see Eq. (14.35)]

$$\phi_i^0 = g_1(ih)$$
$$\phi_s^1 = g_1(ih) + pg_2(ih) \tag{14.37}$$

for all integer values of i both positive and negative.

The difference equation (14.32) may be written

$$\phi_i^j - \phi_{i-1}^{j-1} = \phi_{i+1}^{j-1} - \phi_i^{j-2} \tag{14.38}$$

which states that the difference between corner values is the same for opposite sides of each rhombic net shown in Fig. 14.2.   This net connects only half the points, the other points being on a similar net.   Since these differences are constant, one deduces at once that

$$\phi_i^j - \phi_{i-1}^{j-1} = \phi_{i+j-1}^1 - \phi_{i+j-2}^0 \tag{14.39}$$

Letting $j = s$ and $i = s + k$ in Eq. (14.39) and summing s from 2 to j, one has

$$\sum_{s=2}^{j} \phi_{s+k}^{s} - \sum_{s=2}^{j} \phi_{s+k-1}^{s-1} = \sum_{s=2}^{j} \phi_{2s+k-1}^{1} - \sum_{s=2}^{j} \phi_{2s+k-2}^{0}$$

Since all but two terms on the left cancel, and since $j + k = i$,

$$\phi_i^j - \phi_{k+1}^1 = \sum_{s=1}^{j-1} (\phi_{2s+k+1}^1 - \phi_{2s+k}^0)$$

Thus, from Eqs. (14.37)

$$\phi_i^j = \phi(ih, jp) = \sum_{s=0}^{j-1} [g_1((2s + k + 1)h) + pg_2((2s + k + 1)h)]$$

$$- \sum_{s=0}^{j-1} g_1((2s + k)h) + g_1(kh) \quad (14.40)$$

Let $x = ih$, $t = jp$ be fixed, but reduce the size of the net by letting $h \to h/N$, $p \to p/N$, $i \to iN$, and $j \to jN$, where $N \to \infty$.

From the definition of integration (Riemann)

$$\lim_{h \to 0} h \sum_{s=0}^{n-1} f(x_0 + sh) = \int_{x_0}^{x_n} f(x) \, dx \quad (14.41)$$

Thus, since $k = i - j$ and $p = h/v$, $kh = x - vt$ and $(2j + k)h = x + vt$, and hence

$$\lim_{h \to 0} \frac{h}{v} \sum_{s=0}^{j-1} g_2((2s + k + 1)h) = \frac{1}{2v} \int_{x-vt}^{x+vt} g_2(x) \, dx \quad (14.42)$$

Likewise

$$\lim_{h \to 0} \sum_{s=0}^{j-1} [g_1((2s + k + 1)h) - g_1((2s + k)h)] = \lim_{h \to 0} h \sum_{s=0}^{j-1} g_1'((2s + k)h)$$

$$= \frac{1}{2} \int_{x-vt}^{x+vt} g_1'(x) \, dx = \frac{1}{2}[g_1(x + vt) - g_1(x - vt)] \quad (14.43)$$

Therefore, since $g_1(kh) = g_1(x - vt)$, on letting $h \to 0$ in Eq. (14.40) one obtains the limiting equations

$$\phi(x,t) = \frac{1}{2}[g_1(x + vt) + g_1(x - vt)] + \frac{1}{2v} \int_{x-vt}^{x+vt} g_2(x) \, dx \quad (14.44)$$

That this limiting form [Eq. (14.44)] of the solution of the difference equation (14.32) as $h \to 0$ satisfies the differential equation (14.28) can be seen at once by the fact that $\phi(x,t)$ is a sum of functions of either the $x - vt$ or the $x + vt$. It is quickly determined that $\phi(x,t)$ in Eq.

(14.44) satisfies the boundary conditions (14.36). Thus, the convergence of the finite-difference solution is established for the above simple case. While it does not guarantee convergence for a more general case, it does indicate in rather simple terms the general nature of a convergence proof.

The lines of constant $x - vt$ and of constant $x + vt$ are called *characteristics* of the differential equation, and play an important role in the theory of hyperbolic differential equations. The choice of $m = vp/h = 1$ makes the lattice lines of Fig. 14.2 characteristics of the differential equation. For this choice of m the value of ϕ_i^j is determined by just those ϕ_i^0 and ϕ_i^1 lying inside the diagonal lattice lines through lattice point (i,j). This is exactly similar to the solution $\phi(x,t)$ of the differential equation [see Eq. (14.44)], which solution is determined from just those values of $g_1(x)$ and $g_2(x)$ lying between the two characteristics through the point (x,t).

The points of the rectangular lattice not on the diagonal lattice of Fig. 14.2 lie on a similar lattice, and there is no direct coupling between them except through the boundary conditions. One could, of course, eliminate one lattice, but there is an advantage in keeping both of them, since it permits one to judge the size of the accumulated round-off error in terms of the discrepancy between the ϕ_i^j for the two nets.

Courant, Friedrichs, and Lewy† have analyzed the difference equation (14.30) for $m \neq 1$ and have shown that convergence of the solution occurs only for $m \leq 1$. For $m > 1$ only the part of the range of initial values between the two characteristics through the point under study will influence the value of ϕ. This is in contrast to the differential equation, which requires that ϕ be dependent on all initial values between the characteristics. A change in values outside the diagonal lattice lines through the point (i,j) will not influence ϕ_{ij}^j, even for $h \to 0$, while it will change the corresponding value of $\phi(x,t)$ for the differential equation. This means that the difference solution will not ordinarily converge for $m > 1$.

14.9. Truncation Error. As stated above, the truncation error expresses the difference between the solution of the difference equation and the differential equation for a finite h.

By Taylor's expansion

$$\int_{x_s}^{x_s+h} f(x)\ dx = hf(x_s) + \tfrac{1}{2}h^2 f'(\xi_s)$$

where $x_s = x_0 + sh$ and $x_s < \xi_s < x_s + h$. Summing this equation for $s = 0, 1, 2, \ldots , n - 1$,

$$\int_{x_0}^{x_n} f(x)\ dx = h \sum_{s=0}^{n-1} f(x_0 + sh) + \tfrac{1}{2}h^2 \sum_{s=0}^{n-1} f'(\xi_s) \qquad (14.45)$$

† R. Courant, K. Friedrichs, and H. Lewy, Über die partiellen Differenzengleichungen der mathematischen Physik, *Math. Ann.*, **100**: 32–74 (1928).

This equation replaces Eq. (14.41). Thus, for finite h Eqs. (14.42) to (14.44) will have additional terms in h^2 corresponding to the last term in Eq. (14.45). Therefore the truncation error is of the order of h^2. This means that this error varies inversely as the number of points in the lattice.

14.10. Propagation of Round-off Errors. Suppose the round-off error in ϕ_i^j is 1 in some decimal position; then through the difference equation (14.32) an error of 1 will appear in ϕ_{i+1}^{j+1} and ϕ_{i-1}^{j+1}. These errors will then produce errors in the $(j+2)$th row; thus the effect of the original

```
0 │ 0   0   0 ₓ 0   0  ①  0   0   0   0   0   0
  │
0 │ 0   0   0   0   1   0   1   0   0   0   0   0
  │
0 │ 0   0   0   1   0   1   0   1   0   0   0   0
  │
0 │ 0   0   1   0   1   0   1   0   1   0   0   0
  ↓t
0   0   1   0   1   0   1   0   1   0   1   0   0

0   1   0   1   0   1   0   1   0   1   0   1   0

1   0   1   0   1   0   1   0   1   0   1   0   1
```

FIG. 14.3. The effect of an error at one lattice point on subsequent values.

error will be passed on from row to row as shown in Fig. 14.3. Since the difference equation is linear, the total effect of the errors in each of the values is obtained by adding up the effects of each error. Thus, our analysis shows the finite-difference solution to be essentially stable in that the individual errors grow only linearly and not exponentially.

The error in any given value ϕ_i^j is due to errors inherited from those ϕ_i^k in earlier rows $(k < j)$ that lie in the triangular area between the diagonal lines through the (i,j) lattice point. The weights to be given these previous errors are all either 1 or 0, and are represented exactly by the triangle of numbers in Fig. 14.3 reversed in time. Thus, the maximum error possible in any value due to round-off of previous values is just the sum of the unit weights in the triangle. This sum obviously increases quadratically with j, the number of steps in the computation. On the average, however, the error should increase only as the square root of this sum, i.e., it should increase only linearly with j.

PROBLEMS

1. A slab of material 10 cm thick having a conductivity $\kappa = 0.04$ cal/cm-deg, heat capacity $c = 0.25$ cal/g, and density $\rho = 2$ g/cm³ is insulated on one side and is held at a temperature of T_0°C on the other. If the material is initially at a temperature of $T_0 + 100$ deg, how long will it take for the temperature at the center of the slab to be reduced to $T_0 + 10$ deg? Take $h = 1$ cm and the modulus equal to one-half.

2. A long, square bar is insulated on three sides, while the fourth side is held at 80°C. If the temperature of the bar is originally at 20°C, how does the temperature throughout the bar vary with the time? Use a square lattice with 9 interior points and take 10 time intervals.

3. Find the temperature fluctuation in a copper sphere 1 m in radius subjected at its surface to a sinusoidal temperature change of 10°C peak to peak. The period of the temperature cycle is 1 hour. Take the net spacing in the r direction to be 10 cm.

4. Find the behavior of a uniform string of length L clamped at both ends and under a tension T. Let the mass per unit length be m, and assume that the string is given initially a triangular displacement. The initial velocity is taken to be zero, and the maximum displacement is taken to be at the center of the string. Take for the spacial interval of the net $h = L/10$.

5. Use the same technique as used in Fig. 14.3 for the wave equation to determine for the one-dimensional heat-flow equation the effect of an error in one of the values of the temperature at a net point. Take first a modulus m less than $\frac{1}{2}$ and then one greater than $\frac{1}{2}$.

6. Set up the heat-flow equation in cylindrical coordinates assuming symmetry of temperature about the z axis. Then replace the differential equation by an appropriate difference equation.

7. A function $\phi(x,t)$ satisfies the differential equation

$$\phi_{xx} - \phi_{tt} = 0$$

with boundary conditions

$$\left. \begin{array}{l} \phi(0,t) = 0 \\ \phi_x(1,t) = 0 \end{array} \right\} \quad \text{for all } t \geq 0$$

and initial conditions

$$\left. \begin{array}{l} \phi(x,0) = x(2 - x) \\ \phi_t(x,0) = 0 \end{array} \right\} \quad \text{for } 0 \leq x \leq 1$$

Set up the appropriate difference equation, and using a square lattice with spacing in the x and t directions of $h = 0.2$, find $\phi(x,t)$ in the region $0 \leq x \leq 1$ and $0 \leq t \leq 2$. HINT: Fit the boundary condition $\phi_x(1,t) = 0$ by assuming a $\phi(1.2,t)$ that is equal to $\phi(0.8,t)$.

8. (a) Determine the appropriate difference equation for the wave equation in two spacial dimensions. (b) A rectangular drumhead is clamped at the edges and given the initial displacement indicated in the net below. What will be its subsequent behavior?

-1	-1	0	1	1
$\cdot$	$\cdot$	$\cdot$	$\cdot$	$\cdot$
-2	-2	0	2	2
$\cdot$	$\cdot$	$\cdot$	$\cdot$	$\cdot$
-1	-1	0	1	1
$\cdot$	$\cdot$	$\cdot$	$\cdot$	$\cdot$

displacement zero on the boundary

9. Solve Prob. 7 with the differential equation replaced by

$$\phi_{xx} + 0.2\phi_t - \phi_{tt} = 0$$

10. A function $\phi(x,y,z,t)$ satisfies the differential equation

$$\phi_{xx} + \phi_{yy} + \phi_{zz} - \phi_{tt} = 0$$

in the unit cube $0 \leq x \leq 1$, $0 \leq y \leq 1$, and $0 \leq z \leq 1$ and satisfies the boundary condition that ϕ vanishes on the surfaces of the cubical region. Let $\phi = 1$ inside the cube, at $t = 0$. How will ϕ throughout the cube vary with time? (a) Solve the problem by a finite-difference technique using 27 interior points in the cube. (b) Solve the differential equation directly using the method of separation of variables. How does this solution compare for practical purposes with the finite-difference technique of part (a)?

CHAPTER 15

INTEGRAL EQUATIONS

An integral equation is an equation involving the unknown function under the integral sign. Although most physical problems are usually analyzed in terms of differential equations, they often may be expressed alternatively as integral equations, and some even find their most natural expression in this form. The very interesting connections between differential and integral equations and between the Green's function and the kernel of an integral equation are not discussed here but may be found in several references.†

15.1. Classification and Terminology. Consider the integral equation

$$\phi(x) = f(x) + \lambda \int_a^b K(x,y)\phi(y)\,dy \qquad (15.1)$$

involving the unknown function $\phi(x)$ and the known functions $f(x)$ and $K(x,y)$. The function $K(x,y)$ is termed the *kernel* of the integral equation and plays the leading role in our discussion. The limits a and b will usually be assumed finite and $K(x,y)$ bounded in the square region R, for which $a \le x \le b$, $a \le y \le b$. If either of these conditions does not hold, the integral equation is said to be *singular*. The parameter λ in Eq. (15.1) could, of course, be combined with the kernel $K(x,y)$, but it will be found useful in determining the relative importance of the integral in the equation. Thus, if λ is very small, $\phi(x)$ will be approximately equal to $f(x)$.

The integral equation (15.1) is of the *Fredholm type* since it involves constant limits a and b and of the *second kind* since the unknown occurs outside as well as inside the integral. It is *nonhomogeneous* because of the occurrence of the known function $f(x)$. Such an equation generally has a solution, and we shall start our discussion of methods with equations of this form.

The homogeneous equation corresponding to Eq. (15.1) is

$$\phi(x) = \lambda \int_a^b K(x,y)\phi(y)\,dy \qquad (15.2)$$

† See, for instance, William V. Lovitt, "Linear Integral Equations," Chap. 4 Dover Publications, New York, 1950.

and this equation will usually possess a nontrivial solution [$\phi(x)$ not identically zero] for only certain *characteristic values* of λ. Associated with these characteristic values λ_i ($i = 1, 2, 3, \ldots$) will be the solutions $\phi_i(x)$, termed the *characteristic functions* or *eigenfunctions* of the integral equation. In the problem of a vibrating string the λ_i determine the natural vibration frequencies and $\phi_i(x)$ the corresponding modes of vibration—expressed, for example, by the relative lateral displacement as a function of distance along the string.

An integral of the *first kind*

$$f(x) = \int_a^b K(x,y)\phi(y) \, dy \qquad (15.3)$$

need not involve the parameter λ, since this parameter may be included in the unknown function $\phi(x)$. Equation (15.3) may be looked upon as defining a certain transform of the function $\phi(x)$. Since $f(x)$ is assumed given, the problem becomes one of finding the inverse transform giving $\phi(x)$ in terms of $f(x)$. Only for special kernels, however, can the solution be written in the form

$$\phi(x) = \int_a^b k(x,y)f(x) \, dx$$

where $k(x,y)$ is a kernel said to be reciprocal to $K(x,y)$.

If the limits of integration in any of the above integral equations are taken to be a and x instead of a and b, the integral equation is said to be of the *Volterra type;* thus the integral equation

$$\phi(x) = f(x) + \lambda \int_a^x K(x,y)\phi(y) \, dy \qquad (15.4)$$

is a *nonhomogeneous* integral equation of the *second kind* of the *Volterra type*.

All of the above integral equation are *linear*, meaning that only the first power of the unknown function $\phi(x)$ enters into the equation. One may also study *nonlinear* equations of the form

$$\phi(x) = f(x) + \lambda \int_a^b g(y,\phi(y)) \, dy \qquad (15.5)$$

where the function $g(y,\phi)$ is known.

In contrast to the great changes occurring in the theory of differential equations in going from ordinary differential equations of one independent variable to partial differential equations of more than one independent variable, no comparable changes occur in the theory of integral equations. Thus, one may readily adapt the methods of this chapter to integral equations of more than one variable, such as the equation

$$\phi(x,y) = f(x,y) + \lambda \iint_R K(x,y;\xi,\eta) \, d\xi \, d\eta \qquad (15.6)$$

where R is any region of the xy plane over which the functions $f(x,y)$ and $\phi(x,y)$ are defined.

15.2. Method of Successive Approximations. As stated earlier, if λ is small enough, an approximate solution of the integral equation

$$\phi(x) = f(x) + \lambda \int_a^b K(x,y)\phi(y) \, dy \qquad (15.7)$$

is $\phi_1(x) = f(x)$. Using this approximation, one may correct for the integral and obtain a better approximation

$$\phi_2(x) = f(x) + \lambda \int_a^b K(x,y)\phi_1(y) \, dy$$

Repetition of this process leads to successive approximations $\phi_s(x)$, each defined in terms of the preceding approximation by the equation

$$\phi_s(x) = f(x) + \lambda \int_a^b K(x,y)\phi_{s-1}(y) \, dy \qquad (15.8)$$

This procedure is seen to be very similar to Picard's method of integrating a differential equation (see Sec. 8.4).

Let
$$\epsilon_s(x) = \phi(x) - \phi_s(x) \qquad (15.9)$$

then, as seen by subtracting Eq. (15.8) from Eq. (15.7), one has that

$$\epsilon_s(x) = \lambda \int_a^b K(x,y)\epsilon_{s-1}(y) \, dy \qquad (15.10)$$

Suppose one is dealing with a nonsingular integral equation, so that a and b are finite and $K(x,y)$ bounded, i.e.,

$$|K(x,y)| \leq M \qquad (15.11)$$

and let N_s be the maximum value of $|\epsilon_s(x)|$ in the interval $a \leq x \leq b$. Then, by taking the absolute value of $\epsilon_s(x)$ in Eq. (15.10) and substituting the value of x that makes it a maximum, one has

$$N_s = |\epsilon_s(\bar{x})| \leq \lambda \int_a^b |K(\bar{x},y)| \, |\epsilon_{s-1}(y)| \, dy \leq \lambda M(b - a)N_{s-1}$$

Thus, N_s will approach zero as $s \to \infty$ if†

$$\lambda M(b - a) < 1 \qquad (15.12)$$

The analytical method described above suffers from the fact that the successive integrations required may be difficult, or impossible, to per-

---

† This, of course, is not the case if $N_1 = \infty$. It is readily shown, however, from Eq. (15.7) that

$$N_1 = |\phi(\bar{x}) - f(\bar{x})| \leq \lambda M \int_a^b |\phi(y)| \, dy$$

and therefore N_1 must be finite if $|\phi(x)|$ is integrable.

form. One can, however, usually employ numerical integration and proceed as described below.

Let the interval $a \leq x \leq b$ be divided into n equal parts by the points

$$x_i = a + ih \qquad i = 0, 1, 2, \ldots, n$$

and let

$$f_i = f(x_i)$$
$$\phi_i = \phi(x_i) \tag{15.13}$$
$$K_{ij} = K(x_i, y_j)$$

then from Eq. (15.7)

$$\phi_i = f_i + \lambda \int_a^b K(x_i, y) \phi(y)\, dy$$

$$\cong f_i + \lambda h \sum_{j=0}^{n} A_j K_{ij} \phi_j$$

where the weights A_j will depend on the particular quadrature formula used (see Chap. 7). One is lead, therefore, to the set of linear equations

$$\phi_i = f_i + \lambda \sum_{j=0}^{n} \Re_{ij} \phi_j \qquad i = 0, 1, 2, \ldots, n \tag{15.14}$$

where $\Re_{ij} = hA_j K_{ij}$, as a finite-difference approximation to the integral equation (15.7).

A solution of Eq. (15.14) can be found by successive approximation using the iteration equations†

$$\phi_i^s = f_i + \lambda \sum_{j=0}^{n} \Re_{ij} \phi_j^{s-1} \tag{15.15}$$

provided the absolute value of the largest characteristic value of the matrix $\Re_{ij}$ is less than $|1/\lambda|$ (see Sec. 13.13). This will certainly be the case if

$$\left| \lambda \sum_{j=0}^{n} \Re_{ij} \right| < 1 \tag{15.16}$$

Since for any quadrature formula used $h\Sigma_j A_j = b - a$, inequality (15.16) is the finite-difference equivalent of inequality (15.12).

15.3. Liouville-Neumann Series. A modification of the above method of successive approximation permits one to change the inhomogeneous term $f(x)$ and still make use of a major part of the previous computation. One merely carries out formally the integration required. Thus, starting again with $\phi_1(x) = f(x)$ one obtains from Eq. (15.8) for the next approximation

$$\phi_2(x) = f(x) + \lambda \int_a^b K(x, y) f(y)\, dy$$

† In these equations ϕ_i^s represents the sth iteration for ϕ_i obtained after s applications of the set of equations in (15.15), starting with $\phi_i^0 = 0$.

From this same iteration equation

$$\phi_3(x) = f(x) + \lambda \int_a^b K(x,y) \left[f(y) + \lambda \int_a^b K(y,z)f(z)\, dz \right] dy$$

$$= f(x) + \lambda \int_a^b K(x,y)f(y)\, dy + \lambda^2 \int_a^b K_2(x,y)f(y)\, dy$$

where $\qquad K_2(x,y) = \int_a^b K(x,t_1)K(t_1,y)\, dt_1$

Repeated iterations using Eq. (15.8) lead to the general equation

$$\phi_s(x) = f(x) + \sum_{r=1}^{s-1} \lambda^r \int_a^b K_r(x,y)f(y)\, dy \qquad (15.17)$$

where $K_r(x,y)$ is termed the *rth iterated kernel* and is defined by the equation

$$K_r(x,y) = \int_a^b K(x,t)K_{r-1}(t,y)\, dt$$

$$= \int_a^b \cdots \int_a^b K(x,t_1)K(t_1,t_2)K(t_2,t_3) \cdots K(t_{r-1},y)\, dt_1\, dt_2 \cdots dt_{r-1} \qquad (15.18)$$

and $K_1(x,y) = K(x,y)$. In the limit of $s \to \infty$ in Eq. (15.17) one has

$$\phi(x) = f(x) - \lambda \int_a^b k(x,y;\lambda)f(y)\, dy \qquad (15.19)$$

where $k(x,y;\lambda)$ is called the *reciprocal kernel* and is defined by

$$k(x,y;\lambda) = - \sum_{r=1}^{\infty} \lambda^{r-1}K_r(x,y) \qquad (15.20)$$

provided this series converges. Note that by transposing the integral in Eq. (15.19) we obtain an integral equation of the same form as Eq. (15.7) but with $\phi(x)$ and $f(x)$ interchanged and with $k(x,y;\lambda)$ replacing $K(x,y)$.

It is readily seen that the finite-difference analogue of Eq. (15.19) obtained by repeated application of Eq. (15.15) is

$$\phi_i = f_i - \lambda \sum_{j=0}^{n} \mathfrak{k}_{ij}(\lambda)f_j \qquad (15.21)$$

where $\qquad \mathfrak{k}_{ij}(\lambda) = - \sum_{r=1}^{\infty} \lambda^{r-1}(\mathfrak{K}^r)_{ij} \qquad (15.22)$

and $\qquad (\mathfrak{K}^1)_{ij} = \mathfrak{K}_{ij}$

$$(\mathfrak{K}^r)_{ij} = \sum_{k=0}^{n} \mathfrak{K}_{ik}(\mathfrak{K}^{r-1})_{kj} \qquad (15.23)$$

Note that $\mathfrak{K}^r$ is truly the rth power of the matrix $\mathfrak{K}$.

15.4. Volterra Equation of the Second Kind. Taking the first difference of the integral equation†

$$\phi(x) = f(x) + \int_a^x K(x,y)\phi(y)\,dy \tag{15.24}$$

one has

$$\phi(x) - \phi(x - h) = f(x) - f(x - h) + \int_{x-h}^x K(x,y)\phi(y)\,dy$$

If, now, one sets $x = x_i = a + ih$ in this equation and applies a quadrature formula to evaluate the integral, one has (assuming $i > 0$)

$$\phi_i - \phi_{i-1} = f_i - f_{i-1} + h\sum_{j=0}^i A_{ij}K_{ij}\phi_j \tag{15.25}$$

where the subscript notation is that defined in Eq. (15.13) and the A_{ij} are the weights required by the particular quadrature formula used.

The ϕ_s are evaluated in order. From Eq. (15.24) $\phi(a) = f(a)$ and therefore $\phi_0 = f_0$. Since Eq. (15.25) can be written

$$\phi_i = \frac{1}{1 - \Re_{ii}}\left[f_i - f_{i-1} + (1 + \Re_{i,i-1})\phi_{i-1} + \sum_{j=0}^{i-2} \Re_{ij}\phi_j\right] \tag{15.26}$$

where $\Re_{ij} = hA_{ij}K_{ij}$, ϕ_1 is found from the equation

$$\phi_1 = \frac{1}{1 - \Re_{11}}[f_1 - f_0 + (1 + \Re_{10})\phi_0] = \frac{f_1 + \Re_{10}f_0}{1 - \Re_{11}}$$

By the use of Eq. (15.26) one can then find ϕ_2, ϕ_3, etc.

For simple quadrature formulas the A_{ij} are zero when j is appreciably smaller than i, and therefore $\Re_{ij}$ has nonzero elements only near the diagonal. This makes the recurrence formula (15.26) for the ϕ_i depend only on a few of the earlier values. Some simplification also occurs for open-type formulas, since then (see Chap. 7) $A_{ii} = 0$, and $\Re_{ii}$ can be dropped in Eq. (15.26).

15.5. Solution of the Finite-difference Approximation. As was shown in Sec. 15.2, the Fredholm equation of the second kind

$$\phi(x) = f(x) + \lambda\int_a^b K(x,y)\phi(y)\,dy$$

can be approximated by the set of linear equations

$$\phi_i = f_i + \lambda\sum_{j=0}^n \Re_{ij}\phi_j \qquad i = 0, 1, 2, \ldots, n \tag{15.27}$$

† It is convenient, for the purposes of this discussion, to include λ in the kernel $K(x,y)$.

where $\Re_{ij} = hA_jK_{ij}$, A_j being the weights assigned by quadrature formula used, with ϕ_i, f_i, and K_{ij} defined as in Eq. (15.13). Now any of the methods described in Chap. 10 can be used to solve these equations, and the resulting $\phi_i = \phi(a + ih)$ constitute in tabular form an approximation to the unknown function $\phi(x)$. This approximation becomes better the smaller the interval $h = (b - a)/n$ between successive arguments, but, of course, the number of equations to be solved becomes larger.

A very convenient method for hand computation is the method of relaxation described in Chap. 13.

15.6. Fredholm Theory. One may, of course, express the solution of the set of linear equations formally by means of determinants. In the Fredholm theory one finds the limiting form of this solution as the interval h approaches zero. Only a bare outline of the theory will be given here.†

Cramér's rule applied to Eqs. (15.27) may be written

$$\phi_s = \frac{1}{\Delta} \begin{vmatrix} 1 - \lambda\Re_{00} & -\lambda\Re_{01} & \cdots & -\lambda\Re_{0,s-1} & f_0 & -\lambda\Re_{0,s+1} & \cdots & -\lambda\Re_{0n} \\ -\lambda\Re_{10} & 1 - \lambda\Re_{11} & \cdots & -\lambda\Re_{1,s-1} & f_1 & -\lambda\Re_{1,s+1} & \cdots & -\lambda\Re_{1n} \\ -\lambda\Re_{20} & -\lambda\Re_{21} & \cdots & 1 - \lambda\Re_{2,s-1} & f_2 & -\lambda\Re_{2,s+1} & \cdots & -\lambda\Re_{2n} \\ \cdots & \cdots & \cdots & \cdots & \cdots & \cdots & \cdots & \cdots \\ -\lambda\Re_{n0} & -\lambda\Re_{n1} & \cdots & -\lambda\Re_{n,s-1} & f_n & -\lambda\Re_{n,s+1} & \cdots & 1 - \lambda\Re_{nn} \end{vmatrix}$$

(15.28)

where

$$\Delta = \begin{vmatrix} 1 - \lambda\Re_{00} & -\lambda\Re_{01} & -\lambda\Re_{02} & \cdots & -\lambda\Re_{0n} \\ -\lambda\Re_{10} & 1 - \lambda\Re_{11} & -\lambda\Re_{12} & \cdots & -\lambda\Re_{1n} \\ -\lambda\Re_{20} & -\lambda\Re_{21} & 1 - \lambda\Re_{22} & \cdots & -\lambda\Re_{2n} \\ \cdots & \cdots & \cdots & \cdots & \cdots \\ -\lambda\Re_{n0} & -\lambda\Re_{n1} & -\lambda\Re_{n2} & \cdots & 1 - \lambda\Re_{nn} \end{vmatrix}$$

(15.29)

provided $\Delta \neq 0$. Expanding the determinant in Eq. (15.28) by elements of the column involving the f_k, one has

$$\phi_s = \frac{1}{\Delta} \sum_{r=0}^{n} f_r \Delta_{rs}$$

(15.30)

where Δ_{rs} is the cofactor of the element f_r.

One can expand the determinants involved in Eq. (15.30) in powers of λ; thus,

$$\Delta = 1 - \lambda \sum_i \Re_{ii} + \frac{\lambda^2}{2!} \sum_{i,j} \begin{vmatrix} \Re_{ii} & \Re_{ij} \\ \Re_{ji} & \Re_{jj} \end{vmatrix} - \frac{\lambda^3}{3!} \sum_{i,j,k} \begin{vmatrix} \Re_{ii} & \Re_{ij} & \Re_{ik} \\ \Re_{ji} & \Re_{jj} & \Re_{jk} \\ \Re_{ki} & \Re_{kj} & \Re_{kk} \end{vmatrix} + \cdots$$

$$+ (-\lambda)^{n+1} \begin{vmatrix} \Re_{00} & \Re_{01} & \cdots & \Re_{0n} \\ \Re_{10} & \Re_{11} & \cdots & \Re_{1n} \\ \cdots & \cdots & \cdots & \cdots \\ \Re_{n0} & \Re_{n1} & \cdots & \Re_{nn} \end{vmatrix}$$

(15.31)

† See Lovitt, *op. cit.*, chap. 3.

$$\Delta_{ss} = 1 - \lambda \sum_i{}' \mathfrak{K}_{ii} + \frac{\lambda^2}{2!} \sum_{i,j}{}' \begin{vmatrix} \mathfrak{K}_{ii} & \mathfrak{K}_{ij} \\ \mathfrak{K}_{ji} & \mathfrak{K}_{jj} \end{vmatrix}$$

$$- \frac{\lambda^3}{3!} \sum_{i,j,k}{}' \begin{vmatrix} \mathfrak{K}_{ii} & \mathfrak{K}_{ij} & \mathfrak{K}_{ik} \\ \mathfrak{K}_{ji} & \mathfrak{K}_{jj} & \mathfrak{K}_{jk} \\ \mathfrak{K}_{ki} & \mathfrak{K}_{kj} & \mathfrak{K}_{kk} \end{vmatrix} + \cdots \quad (15.32)$$

where the apostrophe means that the summation indexes do not assume the same value as s, while for $r \neq s$

$$\Delta_{rs} = \lambda \mathfrak{K}_{sr} - \lambda^2 \sum_i \begin{vmatrix} \mathfrak{K}_{sr} & \mathfrak{K}_{si} \\ \mathfrak{K}_{ir} & \mathfrak{K}_{ii} \end{vmatrix} + \frac{\lambda^3}{2!} \sum_{i,j} \begin{vmatrix} \mathfrak{K}_{sr} & \mathfrak{K}_{si} & \mathfrak{K}_{sj} \\ \mathfrak{K}_{ir} & \mathfrak{K}_{ii} & \mathfrak{K}_{ij} \\ \mathfrak{K}_{jr} & \mathfrak{K}_{ji} & \mathfrak{K}_{jj} \end{vmatrix} + \cdots \quad (15.33)$$

In the limit of $h \to 0$, remembering that $\mathfrak{K}_{ij} = hA_j K_{ij}$ and that in the limit the A_j can be taken equal to 1, the summations over i, j, and k become multiple integrals. We shall take the corresponding variables of integration to be t_1, t_2, and t_3. The fixed indexes r and s will be replaced by $x = a + sh$ and $y = a + rh$. The limit of Δ and Δ_{rs} as $h \to 0$ are designated by $D(\lambda)$ and $hD(x,y;\lambda)$ and are given by

$$D(\lambda) = 1 - \lambda \int_a^b K(t,t)\, dt + \frac{\lambda^2}{2!} \iint_a^b \begin{vmatrix} K(t_1,t_1) & K(t_1,t_2) \\ K(t_2,t_1) & K(t_2,t_2) \end{vmatrix} dt_1\, dt_2$$

$$- \frac{\lambda^3}{3!} \iiint_a^b \begin{vmatrix} K(t_1,t_1) & \cdots & K(t_1,t_3) \\ \cdots \cdots \cdots \cdots \cdots \\ K(t_3,t_1) & \cdots & K(t_3,t_3) \end{vmatrix} dt_1\, dt_2\, dt_3 + \cdots \quad (15.34)$$

and

$$D(x,y;\lambda) = \lambda K(x,y) - \lambda^2 \int_a^b \begin{vmatrix} K(x,y) & K(x,t) \\ K(t,y) & K(t,t) \end{vmatrix} dt$$

$$+ \frac{\lambda^3}{2!} \iint_a^b \begin{vmatrix} K(x,y) & K(x,t_1) & K(x,t_2) \\ K(t_1,y) & K(t_1,t_1) & K(t_1,t_2) \\ K(t_2,y) & K(t_2,t_1) & K(t_2,t_2) \end{vmatrix} dt_1\, dt_2 + \cdots \quad (15.35)$$

The limit of Δ_{ss} as $h \to 0$ is the same as that for Δ, namely, $D(\lambda)$.

The solution of the set of equations as given by Eq. (15.30) may be written

$$\phi_s = \frac{\Delta_{ss}}{\Delta} f_s + \sum_r{}' \frac{\Delta_{rs} f_r}{\Delta}$$

which in the limit becomes

$$\phi(x) = f(x) + \int_a^b \frac{D(x,y;\lambda) f(y)}{D(\lambda)}\, dy \quad (15.36)$$

Thus, the inhomogeneous equation

$$\phi(x) = f(x) + \lambda \int_a^b K(x,y) \phi(y)\, dy \quad (15.37)$$

has the solution given by Eq. (15.36), provided λ is not a zero of the *Fredholm determinant* $D(\lambda)$. On the other hand, if we are treating a homogeneous equation, one for which $f(x) \equiv 0$, a nontrivial solution exists only for those λ that make $D(\lambda) = 0$. This is exactly parallel to the case of a finite set of linear equations, such as given by Eq. (15.27).

It is convenient to express $D(x,y;\lambda)$ and $D(\lambda)$ merely as unspecified power series in λ

$$D(x,y;\lambda) = \sum_{k=0}^{\infty} (-1)^k \frac{\lambda^{k+1}}{k!} D_k(x,y) \qquad (15.38)$$

and

$$D(\lambda) = \sum_{k=0}^{\infty} (-1)^k \frac{\lambda^k}{k!} D_k \qquad (15.39)$$

and to look for suitable recurrence relations for the coefficients of λ. From these equations and Eqs. (15.34) and (15.35), it is seen that

$$D_k = \int_a^b D_{k-1}(x,x) \, dx \qquad (15.40)$$

Substituting from Eq. (15.36) into Eq. (15.37), we have

$$\int_a^b \left[\frac{D(x,y;\lambda)}{D(\lambda)} - \lambda K(x,y) - \lambda \int_a^b K(x,t) \frac{D(t,y;\lambda)}{D(\lambda)} \, dt \right] f(y) \, dy = 0$$

Since this is to hold for arbitrary $f(y)$, the relation

$$D(x,y;\lambda) - \lambda K(x,y) D(\lambda) = \lambda \int_a^b K(x,t) D(t,y;\lambda) \, dt \qquad (15.41)$$

must be satisfied. Substituting the series expansion in λ for $D(\lambda)$ and $D(x,y;\lambda)$ and equating the coefficients of like powers of λ, one obtains the recurrence equation

$$D_k(x,y) = D_k K(x,y) - k \int_a^b K(x,t) D_{k-1}(t,y) \, dt \qquad (15.42)$$

Use of this equation together with Eq. (15.40) permits one to calculate $D_k(x,y)$ from a knowledge of $D_{k-1}(x,y)$. Thus, starting with a knowledge that

$$D_0(x,y) = K(x,y)$$

one is able to determine the $D_k(x,y)$ and D_k in order and thus arrive at an expansion for $D(x,y;\lambda)$ and $D(\lambda)$.

15.7. Kernels of the Form $K(x,y) = \Sigma_i F_i(x) G_i(y)$. If the kernel of the integral equation

$$\phi(x) = f(x) + \lambda \int_a^b K(x,y) \phi(y) \, dy$$

is expressible in the form

$$K(x,y) = \sum_{i=1}^{m} F_i(x) G_i(y) \qquad (15.43)$$

the solution of the integral equation reduces to a set of m linear equations in m unknowns. Furthermore, even when the kernel is not of this form, one may approximate it by such a kernel and achieve an approximate solution to the integral equation.

Substituting $K(x,y)$ from Eq. (15.43) into the integral equation, one determines at once that

$$\phi(x) = f(x) + \lambda \sum_{i=1}^{m} C_i F_i(x) \tag{15.44}$$

where

$$C_i = \int_a^b G_i(y)\phi(y)\, dy \tag{15.45}$$

Thus, the form of the solution is known, and one need only determine the constants C_i. This can be done very simply by substituting $\phi(x)$ from Eq. (15.44) into Eq. (15.45). One obtains a set of linear equations

$$C_i = \lambda \sum_{j=1}^{m} A_{ij}C_j + B_i \qquad i = 1, 2, \ldots, m \tag{15.46}$$

for the C_i where

$$A_{ij} = \int_a^b G_i(y)F_j(y)\, dy \tag{15.47}$$

and

$$B_i = \int_a^b G_i(y)f(y)\, dy \tag{15.48}$$

The coefficients A_{ij} and B_j would ordinarily be obtained from these integrals by numerical integration. If λ were small, one could use Eqs. (15.46) in their present form as iteration equations for the C_i of the left-hand side. Otherwise, one might use the relaxation method or any of the methods of Chap. 10 to solve these equations.

If the kernel is not of the form given by Eq. (15.43), we may nevertheless use a finite double Fourier-series expansion, a finite Taylor's-series expansion in two variables, or an interpolation formula in two variables (see Chap. 11) to obtain an approximation of $K(x,y)$ of the appropriate form. Use of the Lagrangian interpolation formula (see Sec. 11.15), however, leads to essentially the same set of equations as obtained in Sec. 15.2.

15.8. Characteristic Value Equations. The homogeneous Fredholm equation of the second kind

$$\phi(x) = \lambda \int_a^b K(x,y)\phi(y)\, dy \tag{15.49}$$

as seen from Fredholm's theory, does not possess a solution unless λ is a root of the equation

$$D(\lambda) = 0 \tag{15.50}$$

We are interested, therefore, in the roots λ_k of Eq. (15.50) in the corresponding solutions $\phi_k(x)$ of the integral equation.

The technique for finding these characteristic values and functions is to use any of the methods described earlier for the inhomogeneous equation to replace the integral equation by a set of linear equations. These equations will, of course, be homogeneous, and the vanishing of the determinant of the coefficients of the unknowns will determine the λ_k. Only the ratio of the unknowns, say the ordinates ϕ_i, can be solved for. This corresponds to the obvious fact that if $\phi(x)$ is a solution of Eq. (15.49) for a particular value of λ, then $C\phi(x)$ is also a solution for the same value of λ. For this reason, it is customary to choose this constant of such a size that the characteristic functions $\phi_i(x)$ are normalized, i.e., they satisfy the equation

$$\int_a^b |\phi_i(x)|^2 \, dx = 1 \tag{15.51}$$

Most integral equations arising in physical problems are characterized by a *symmetric kernel*, i.e.,

$$K(y,x) = K(x,y) \tag{15.52}$$

Since the characteristic values λ_k and characteristic functions $\phi_k(x)$ associated with these kernels have very interesting properties, we shall limit our discussion to such kernels. We shall, furthermore, assume that the kernel is a real function of x and y.

It will first be shown that *for a real symmetric kernel all the characteristic values are real.* Let λ in Eq. (15.49) be any one of the characteristic values and $\phi(x)$ a real or complex solution of the integral equation for this λ. Multiplying both sides of the equation by $\phi^*(x)$, the complex conjugate of $\phi(x)$, and integrating from a to b, one obtains the equation

$$\int_a^b \phi^*(x)\phi(x) \, dx = \lambda \iint\limits_a^b \phi^*(x)K(x,y)\phi(y) \, dx \, dy \tag{15.53}$$

The double integral on the right is real since its complex conjugate is

$$\iint\limits_a^b \phi(x)K(x,y)\phi^*(y) \, dx \, dy$$

and if we interchange the variables of integration x and y and make use of Eq. (15.52), we obtain the original integral. From Eq. (15.53) we see that λ is the ratio of two real integrals and must, therefore, be real.

The characteristic functions can be restricted to real functions, since it follows quite simply that if $\phi(x) = u(x) + iv(x)$ is a complex characteristic function corresponding to the characteristic value λ, its real and its imaginary parts are also characteristic functions corresponding to the same λ. One merely has to substitute $\phi(x) = u(x) + iv(x)$ into Eq. (15.49) and equate real and imaginary parts.

For a real symmetrical kernel, the characteristic functions $\phi_i(x)$ and $\phi_j(x)$ corresponding to different characteristic values λ_i and λ_j are orthogonal in the interval $a \leq x \leq b$. For real functions $\phi_i(x)$ and $\phi_j(x)$ orthogonality requires that

$$\int_a^b \phi_i(x)\phi_j(x) \, dx = 0 \qquad \lambda_i \neq \lambda_j \qquad (15.54)$$

The proof of Eq. (15.54) consists first in writing down the equations satisfied by $\phi_i(x)$ and $\phi_j(x)$, namely,

$$\phi_i(x) = \lambda_i \int_a^b K(x,y)\phi_i(y) \, dy$$

$$\phi_j(x) = \lambda_j \int_a^b K(x,y)\phi_j(y) \, dy$$

We multiply the first equation by $\lambda_j\phi_j(x)$ and the second equation by $\lambda_i\phi_i(x)$ and integrate all terms with respect to x between the limits $x = a$ and $x = b$. Subtracting the resulting equations leads to the condition

$$(\lambda_j - \lambda_i) \int_a^b \phi_i(x)\phi_j(x) \, dx = \lambda_i\lambda_j \left[\iint\limits_a^b \phi_j(x)K(x,y)\phi_i(y) \, dx \, dy \right.$$

$$\left. - \iint\limits_a^b \phi_i(x)K(x,y)\phi_j(y) \, dx \, dy \right]$$

The two double integrals on the right-hand side of this equation are equal, as can be seen by interchanging the variables of integration x and y in the second integral and making use of the symmetry of $K(x,y)$. Therefore

$$(\lambda_j - \lambda_i) \int_a^b \phi_i(x)\phi_j(x) \, dx = 0$$

and, since $\lambda_i \neq \lambda_j$, Eq. (15.54) follows.

A further property of a symmetric kernel is that for such a kernel the characteristic equation (15.49) always has at least one characteristic value.

15.9. Volterra Equation of the First Kind. Assuming the integral equation

$$f(x) = \int_a^x K(x,y)\phi(y) \, dy \qquad (15.55)$$

can be differentiated with respect to x, one has

$$f'(x) = K(x,x)\phi(x) + \int_a^x \frac{\partial K(x,y)}{\partial x} \phi(y) \, dy \qquad (15.56)$$

If $K(x,x)$ does not vanish identically, we may divide this equation through by $K(x,x)$ and obtain a Volterra equation of the second kind, whose solution has been discussed in Sec. 15.4.

If, on the other hand, we use the technique of that section to write down the finite-difference equation corresponding to Eq. (15.25), we have (for $i > 0$)

$$f_i - f_{i-1} = h \sum_{j=0}^{i} A_{ij} K_{ij} \phi_j = \sum_{j=0}^{i} \Re_{ij} \phi_j$$

or

$$\phi_i = \frac{1}{\Re_{ii}} \left(f_i - f_{i-1} - \sum_{j=0}^{i-1} \Re_{ij} \phi_j \right) \qquad (15.57)$$

Setting $x = a$ in Eq. (15.55), we see that $f(a)$ must be zero for the integral equation to be consistent; therefore, f_0 should be set equal to zero. The first recurrence relation given by Eq. (15.57)—that for $i = 1$ —is therefore

$$\phi_1 = \frac{1}{\Re_{11}} (f_1 - \Re_{10} \phi_0) \qquad (15.58)$$

To obtain ϕ_0, we must make use of Eq. (15.56). Setting $x = a$ in this equation, we have

$$f'(a) = K(a,a) \phi(a)$$

which, in our finite-difference notation, may be written

$$\phi_0 = \frac{f'_0}{K_{00}} \qquad (15.59)$$

Having ϕ_0, we can then compute ϕ_1 from Eq. (15.58) and ϕ_2, ϕ_3, etc., in succession from Eq. (15.57).

15.10. Fredholm Equation of the First Kind. A Fredholm equation of the first kind

$$f(x) = \int_a^b K(x,y) \phi(y) \, dy \qquad (15.60)$$

will not generally have a solution, as can be seen by assuming that the kernel is of the form (see Sec. 15.7)

$$K(x,y) = \sum_{i=1}^{m} F_i(x) G_i(y) \qquad (15.61)$$

Substituting this $K(x,y)$ in Eq. (15.60), we have

$$f(x) = \sum_{i=1}^{m} C_i F_i(x) \qquad (15.62)$$

where

$$C_i = \int_a^b G_i(y) \phi(y) \, dy \qquad (15.63)$$

From Eq. (15.62), it follows that a solution will not exist unless $f(x)$ can be expanded in terms of the $F_i(x)$.

If $f(x)$ is expansible in the $F_i(x)$ and $\phi_1(x)$ is a particular solution of the integral equation, this solution is not unique, since one can add to $\phi_1(x)$ any function that is orthogonal to all the $G_i(y)$.

A general method of solving Eq. (15.60) is to expand all the functions involved in terms of a complete set of normal and orthogonal functions $\psi_i(x)$; thus let

$$f(x) = \sum_{i=1}^{\infty} b_i \psi_i(x)$$

$$K(x,y) = \sum_{i,j=1}^{\infty} a_{ij}\psi_i(x)\psi_j(y) \qquad (15.64)$$

$$\phi(x) = \sum_{i=1}^{\infty} c_i \psi_i(x)$$

In actual computation, one would ordinarily use only a fairly small number of $\psi_i(x)$ to approximate $f(x)$ and $K(x,y)$. One could, for instance, use a finite Fourier expansion for this purpose. The coefficients b_i and a_{ij} could then be obtained by standard methods from the values of $f(x)$ and $K(x,y)$ at certain suitable points.

To keep our method quite general we shall determine the b_i and a_{ij} by making use only of the orthogonal properties of the $\psi_i(x)$. Thus, multiplying the first equation in (15.64) by $\psi_r(x)$ and integrating from $x = a$ to $x = b$, we have

$$b_r = \int_a^b f(x)\psi_r(x)\,dx \qquad (15.65)$$

By multiplying the second equation by $\psi_r(x)\psi_s(y)$ and integrating both x and y between the limits of a and b we also have

$$a_{rs} = \iint_a^b \psi_r(x)K(x,y)\psi_s(y)\,dx\,dy \qquad (15.66)$$

Our remaining problem is to determine the coefficients c_i in terms of the b_i and a_{ij}. Substituting from Eqs. (15.64) into the integral equation, one finds that

$$\sum_{i=1}^{\infty} b_i \psi_i(x) = \sum_{i,j=1}^{\infty} a_{ij}\psi_i(x)\int_a^b \psi_j(y)\sum_{k=1}^{\infty} c_k \psi_k(y)\,dy = \sum_{i,j=1}^{\infty} a_{ij}c_j \psi_i(x)$$

and therefore

$$\sum_{j=1}^{\infty} a_{ij}c_j = b_i \qquad i = 1, 2, 3, \ldots \qquad (15.67)$$

Our integral equation is thus reduced to the solution of an infinite set of linear equations. In practice, as stated above, one would usually be content with an approximation involving only a finite number of $\psi_i(x)$, and the equations above are then replaced by the finite set

$$\sum_{j=1}^{m} a_{ij}c_j = b_i \qquad i = 1, 2, \ldots, m \qquad (15.68)$$

The necessary and sufficient condition that these equations have a solution is that the *rank* of the matrix (a_{ij}) be the same as the rank of the *augmented matrix*†

$$\begin{bmatrix} a_{11} & a_{12} & \cdots & a_{1m} & b_1 \\ a_{21} & a_{22} & \cdots & a_{2m} & b_2 \\ \cdot & \cdot & \cdots & \cdot & \cdot \\ a_{m1} & a_{m2} & \cdots & a_{mm} & b_m \end{bmatrix}$$

A matrix is said to be of rank r if it contains at least one r-rowed determinant which is not zero while all determinants it contains of order higher than r are zero. One obvious case in which a solution would not exist arises if $b_k \neq 0$ but all the a_{kj} $(j = 1, 2, \ldots, m)$ vanish.

One can, of course, apply the procedure of Sec. 15.2 to replace the integral equation (15.60) by the set of finite-difference equations

$$\sum_{j=0}^{n} A_j K_{ij}\phi_j = f_i \qquad i = 0, 1, 2, \ldots, n \qquad (15.69)$$

where again the A_j are the weights appropriate to the quadrature formula used to approximate the integral in Eqs. (15.60) and f_{ij}, ϕ_j, and K_{ij} are defined as in Eqs. (15.13). The difficulty of this approach is that it is not readily determined whether the integral equation possesses a solution, since the consistency of the equations (15.69) will sometimes be dependent upon the choice of n.

PROBLEMS

1. Show that for a singular integral equation the method of successive approximations will converge to the true solution in the sense that $\int_a^b |\phi(x) - \phi_s(x)|\, dx \to 0$ as $s \to \infty$ provided $\lambda M < 1$ where M is the maximum value of $\int_a^b |K(x,y)|\, dx$.

2. Write down the recurrence relation of Eq. (15.26) for the special case in which one uses the trapezoidal rule for the quadrature formula.

3. (a) Prove that the reciprocal kernel $k(x,y;\lambda)$ satisfies the integral equation

$$K(x,y) + k(x,y;\lambda) = \lambda \int_a^b K(x,t)k(t,y;\lambda)\, dt$$

(b) Prove that $k(x,y;\lambda)$ and $K(x,y)$ can be interchanged in this equation.

† See M. Bôcher, "Introduction to Higher Algebra," p. 46, The Macmillan Company, New York, 1907.

4. (a) Determine the solution of the integral equation

$$\phi(x) = x + \lambda \int_0^1 (x + y)\phi(y)\, dy$$

using Fredholm's method. (b) Determine the solution using the method of Sec. 15.7.

5. (a) Show that for a kernel $K(x,y) = F(x)G(y)$ there can be no more than one characteristic value and one characteristic function, and find these. (b) Show that if $F(x)$ and $G(x)$ are orthogonal, there are no characteristic values.

6. Assuming that $F(x)$ and $G(x)$ are normalized and mutually orthogonal in the interval $a \leq x \leq b$, what are the characteristic values and characteristic functions for a kernel $K(x,y) = F(x) + G(y)$?

7. (a) Use a power-series expansion in x to find a general solution of the integral equation

$$x^2 + 2 = \int_0^1 (x^2 + y^2)\phi(y)\, dy$$

(b) Reduce the integral equation to a set of linear equations by the method of Sec. 15.2. Let $n = 2$, and use Simpson's rule. Solve these equations, and compare these numerical values with the solution obtained in part (a).

8. (a) Solve the integral equation

$$\phi(x) = x^2 + 3 + 2 \int_0^1 (3x^2 + 3y^2 - 2y)\phi(y)\, dy$$

(b) Find the values of λ for which there exist nontrivial solutions of the integral equation

$$\phi(x) = \lambda \int_0^1 (3x^2 + 3y^2 - 2y)\phi(y)\, dy$$

9. Find a numerical solution of the integral equation

$$\phi(x) = x + \int_0^1 \phi(y) \sin xy\, dy$$

using a finite-difference approximation. Take an interval of $h = 0.2$, and use Simpson's rule.

10. Using a finite-difference technique and an interval of $h = 0.2$, find the solution of the integral equation

$$x^2 = \int_0^x \frac{x - y}{x + y} \phi(y)\, dy$$

over the range $0 \leq x \leq 1$.

11. Find the solution of the integral equation

$$e^x - 1 = \int_0^x e^{xy}\phi(y)\, dy$$

by differentiating the equation and employing a finite-difference method to the resulting equation. Take an interval of $h = 0.2$, and obtain the solution over the range $0 \leq x \leq 1$.

ESTIMATION OF ERROR IN
NUMERICAL COMPUTATION

A.1. Kinds of Errors. A solution to a problem in applied mathematics may be in error because of any of the following sources of errors:

1. Approximate nature of the physical law or principle upon which the solution is based.

2. *Truncation errors* caused by the use of a closed form, such as the first few terms of an infinite series, to express a quantity defined by a limiting process.

3. *Round-off errors* in numerical computation arising from the need to use finite decimal numbers in the computation.

4. *Errors in computation* or *mistakes* arising from faulty handling of the rules of arithmetic or from improper transcribing of the numbers used.

The first of these is not part of the subject matter of this book, and is hence ignored. Truncation errors have been treated in the text as a particular subject was developed, and hence are not discussed here.

While errors in computation are theoretically avoidable by taking sufficient pains, from a more realistic view they always have a certain finite probability of occurrence at each step of the calculation. Adequate checks throughout the computation are then seen to be necessary to guard against such mistakes' going undetected. As a practical matter a sizable proportion of the calculations can profitably be spent on such checks to avoid incorrect values for the final results or a lengthy search for the mistakes at the end of the computation.

The rest of this appendix is concerned exclusively with methods of determining the accumulation of round-off errors in numerical computation.

A.2. General Procedure for Obtaining an Error Estimate. We shall treat first a few very simple computations in order to develop the type of error analysis suitable for most problems and then later develop general formulas for the maximum possible error in various numerical procedures. These formulas permit an estimate to be made of the rate of growth of errors as the computation proceeds without a detailed step-by-step analysis; thus one may determine rather quickly whether a certain way of carrying out the computation is *stable*, i.e., whether the growth of error is within bounds.

All numerical computations can conveniently be broken down into the elementary operations of addition, subtraction, multiplication, and division. Therefore a study of the round-off errors in these operations is sufficient for an over-all analysis of error in a given computation.

Addition. Suppose one wishes to add two numbers N_1 and N_2 that are likely to be in error, either because of the approximate nature of the computation used in their determination or because of the need to represent them by a finite decimal fraction. Thus 3.1416 may take the place of π and 1.414 the place of $\sqrt{2}$. How does one determine the upper limit of the error in the sum from the maximum possible errors that can be present in the numbers N_1 and N_2? Let the true values of these numbers be $\tilde{N}_1$ and $\tilde{N}_2$; then

$$\tilde{N}_1 = N_1 + \epsilon_1$$
$$\tilde{N}_2 = N_2 + \epsilon_2 \qquad \text{(A.1)}$$

where ϵ_1 and ϵ_2 are the errors in these numbers.† Thus the sum of the true numbers, $\tilde{N}_3 = \tilde{N}_1 + \tilde{N}_2$, is given by

$$\tilde{N}_3 = N_3 + \epsilon_3$$
where
$$N_3 = N_1 + N_2$$
and
$$\epsilon_3 = \epsilon_1 + \epsilon_2 \qquad \text{(A.2)}$$

Thus the errors simply add.

Since ϵ_1 and ϵ_2 are usually not known, Eq. (A.2) cannot be used directly. One presumably knows, however, that

$$|\epsilon_1| \leq e_1 \qquad |\epsilon_2| \leq e_2$$

where the *maximum errors* e_1 and e_2 are positive numbers giving upper limits for the maximum absolute value of the errors in N_1 and N_2. From Eq. (A.2), therefore,

$$|\epsilon_3| \leq |\epsilon_1| + |\epsilon_2| \leq e_1 + e_2$$

and thus the maximum error in N_3 is given by

$$e_3 = e_1 + e_2 \qquad \text{(A.3)}$$

Subtraction. It is clear that if one subtracts N_1 and N_2 one has in place of the Eq. (A.2) the equation

$$\epsilon_3 = \epsilon_1 - \epsilon_2$$

---

† This definition of the errors is opposite in sign to that usually given. It is in keeping, however, with the general procedure in numerical analysis of expressing the error by the addition to an approximate formula of a so-called error term.

but since ϵ_1 and ϵ_2 may be opposite in sign, one gets the same estimate of maximum error e_3 in N_3 as that given by Eq. (A.3). Thus there arises the following simple rule:

If two given numbers are added or subtracted, the maximum error in the sum or difference is just the sum of the maximum errors in the two given numbers.

Multiplication. Suppose one multiplies the two numbers given in (A.1). One finds for the product

$$\tilde{N}_3 = \tilde{N}_1\tilde{N}_2 = N_1N_2 + N_1\epsilon_2 + N_2\epsilon_1 + \epsilon_1\epsilon_2$$

Since ϵ_1 and ϵ_2 are presumably small compared to the numbers N_1 and N_2, the term $\epsilon_1\epsilon_2$ may be neglected. One then has

$$N_3 = N_1N_2$$
$$\epsilon_3 = N_1\epsilon_2 + N_2\epsilon_1$$

From the latter equation

$$|\epsilon_3| \leq |N_1|\,|\epsilon_2| + |N_2|\,|\epsilon_1| \leq |N_1|e_2 + |N_2|e_1$$

so that the maximum error e_3 in the product is given by

$$e_3 = |N_2|e_1 + |N_1|e_2 \tag{A.4}$$

Dividing Eq. (A.4) by $|N_3| = |N_1|\,|N_2|$, one has

$$\frac{e_3}{|N_3|} = \frac{e_1}{|N_1|} + \frac{e_2}{|N_2|} \tag{A.5}$$

It is convenient, therefore, to define the *maximum relative error* in a number N by

$$r = \frac{e}{|N|} \tag{A.6}$$

then Eq. (A.5) becomes simply

$$r_3 = r_1 + r_2 \tag{A.7}$$

Division. If two numbers represented by N_1 and N_2 are divided, the quotient is

$$\tilde{N}_3 = \frac{N_1 + \epsilon_1}{N_2 + \epsilon_2} = \frac{(N_1 + \epsilon_1)(N_2 - \epsilon_2)}{N_2^2 - \epsilon_2^2}$$

Again neglecting squares of the errors,

$$\tilde{N}_3 = \frac{1}{N_2^2}(N_1N_2 + N_2\epsilon_1 - N_1\epsilon_2)$$

or

$$N_3 = \frac{N_1}{N_2}$$
$$\epsilon_3 = \frac{\epsilon_1}{N_2} - \frac{N_1\epsilon_2}{N_2^2} \tag{A.8}$$

By taking the absolute value of both sides of this equation one obtains, in the same manner as above, the equation

$$e_3 = \frac{e_1}{|N_2|} + \frac{|N_1|e_2}{|N_2|^2} \tag{A.9}$$

This equation divided through by $|N_3| = |N_1|/|N_2|$ reduces to Eq. (A.5). Thus the following simple rule covers both multiplication and division:

If two given numbers are multiplied or divided, the resulting number has a maximum relative error given by the sum of the relative errors of the two given numbers.

Example 1. From the tables of Jahnke and Emde

$$\begin{aligned} J_0(0.0) &= 1.0000 \\ J_0(0.1) &= 0.9975 \\ J_0(0.2) &= 0.9900 \\ J_0(0.3) &= 0.9766 \\ J_0(0.4) &= 0.9604 \end{aligned} \tag{A.10}$$

Use Lagrange's interpolation formula to determine $J_0(0.241)$, and compute the maximum round-off error in the calculations, assuming a table of Lagrangian interpolation coefficients (see Chap. 5) is available accurate to five decimal places.

The formula for $J_0(0.241)$ is

$$J_0(0.241) = A_{-2}J_0(0.0) + A_{-1}J_0(0.1) + A_0J_0(0.2) + A_1J_0(0.3) + A_2J_0(0.4) \tag{A.11}$$

where

$$\begin{aligned} A_{-2} &= 0.02260 \\ A_{-1} &= -0.15449 \\ A_0 &= 0.79694 \\ A_1 &= 0.36920 \\ A_2 &= -0.03425 \end{aligned} \tag{A.12}$$

The value of $J_0(0.241)$ is therefore

$$J_0(0.241) = 0.98503 \tag{A.13}$$

where an extra figure has been retained deliberately.

To determine the maximum error in the calculation we note that all the numbers in (A.10) have the same maximum error 5×10^{-5} and a maximum relative error of about the same amount [see Eq. (A.6)]. Thus for the purposes of error analysis we may treat all the J_0 in (A.11) as equal and determine the error in finding

$$(A_{-2} + A_{-1} + A_0 + A_1 + A_2)J_0 \tag{A.14}$$

where J_0 is assumed to have a relative error of 5×10^{-5}. Since the maximum absolute error† in the A_k is 5×10^{-6}, the maximum error in the sum of the A_k by (A.3) is 2.5×10^{-5}. The sum of the A_k is 1; therefore the maximum relative error in the sum of the A_k is also 2.5×10^{-5}. Finally, therefore, as seen from (A.14), we have

† Here as elsewhere the term *maximum error* is used to denote an upper limit for the absolute value of an error that may be present in a number. Sometimes the term *maximum absolute error* is used for the same quantity to emphasize that absolute rather than relative error is being considered. For simplicity, however, we shall omit the word "maximum" where no misunderstanding is likely to occur.

the product of two numbers with relative errors, respectively, of 2.5×10^{-5} and 5×10^{-5}. Thus the maximum relative error in $J_0(0.241)$ as given by (A.13) is 7.5×10^{-5}. By Eq. (A.6) the absolute error is therefore

$$e \cong (0.98)(7.5 \times 10^{-5}) \cong 7 \times 10^{-5}$$

thus, the number given in (A.13) may be wrong by as much as 7 in the last place. If one now rounds this number

$$J_0(0.241) = 0.9850$$

one introduces an added error of 3×10^{-5}, which may be in the same direction as that above; hence, the number of 0.9850 may be in error by as much as 10^{-4}, and the last figure may thus be in error by 1 owing to round-off error alone.

A.3. Stability of a Problem. A problem in numerical analysis can quite generally be formulated as a problem of finding one or more numbers y_i, $i = 1, 2, \ldots, m$, in terms of given numbers x_j, $j = 1, 2, \ldots, n$. Thus in a general sense the problem may be identified with a set of functions

$$y_i = f_i(x_1, x_2, \ldots, x_n) \qquad i = 1, 2, \ldots, m \qquad (A.15)$$

If one varies the given numbers x_j by δx_j, the numerical answer y_i will vary by

$$\delta y_i = \sum_{j=1}^{n} \frac{\partial f_i}{\partial x_j} \delta x_j \qquad (A.16)$$

If the coefficients $\partial f_i / \partial x_j$ are not too large, uncertainties δx_j in the input numbers x_j will cause only reasonably small changes in the answers δy_i, and the problem is said to be *stable*. On the other hand, if one or more of these coefficients become very large, any uncertainty in the input numbers changes the answers so greatly that the answers y_i are almost, or completely, meaningless. The problem is then said to be *unstable*. Such problems can be treated only in so far as it remains possible to increase the accuracy of the given numbers x_j. When this cannot be done, they must either be abandoned or treated statistically.

A.4. Stability of a Solution. The stability treated above relates to the problem as it is formulated and not to the particular way in which the calculations are made. There exist, however, many stable problems that have unstable numerical solutions. This arises from the fact that a solution usually involves auxiliary intermediate numbers calculated from the given x_j, which must then serve as input numbers for subsequent steps in the calculation. Now if the computation of the final numbers from the intermediate numbers is unstable, the method of solution will fail because of the round-off error in the intermediate numbers.

Such a solution may be made stable by increasing the number of figures retained in the intermediate results. Thus one of the big advantages of a computing machine with more decimal (or binary) digits is that it may permit the use of a solution that is otherwise unstable. Such a solution is often simpler than any of those having greater stability. It may often, therefore, be advisable to retain several extra figures in all intermediate computations and to round the final figures as indicated by an analysis of the over-all error in the answers.

It is convenient at times to think of the round-off errors as being comparable to noise in an electronic system. Thus there are errors in the given figures that can be regarded as constituting the input noise; these errors may be amplified (or attenuated) in the process of computing the output numbers y_i. The relative errors in these outputs constitute the "noise-to-signal" ratio inherent in the problem. However, a particular method of calculation may introduce additional noise at each stage of the calculation owing to the need of rounding the intermediate numbers. The more figures retained in a computation, the more "noise-free" the computation is, and the longer a chain of calculations can be extended before the output is obscured by the noise.

A.5. Loss of Significant Figures in a Computation. The *number of significant* figures in a number N expressed in decimal notation refers to the number of digits, starting at the left with the first nonzero digit and proceeding to the right, that are assumed to be correct. More precisely, the last figure counted usually is required to be merely approximately correct. In so far as the number of significant figures in a computed value has precise meaning, it can be set equal to the quantity

$$\log_{10} \frac{|N|}{e} = - \log_{10} r \qquad (A.17)$$

rounded off to the nearest integer. Thus if one represents π by 3.14, the error e is 0.0016, so that

$$\log_{10} \frac{3.14}{0.0016} \cong \log_{10} 2{,}000 = 3.3$$

and the number of significant figures is therefore 3. The loss of a significant figure is, by (A.17), equivalent to multiplying the relative error by 10.

Consider a step in the computation involving the determination of a single number N_T from a set of numbers N_k, $k = 1, 2, \ldots, n$. One may take the true numbers to be these numbers plus error terms ϵ_T and ϵ_k and set

$$N_T + \epsilon_T = f(N_1 + \epsilon_1, N_2 + \epsilon_2, \ldots, N_n + \epsilon_n) \qquad (A.18)$$

Assuming that the numbers in the computation still retain several significant figures, the errors are small, and therefore the squares of the errors may generally be neglected. From a Taylor's-series expansion of the right-hand side, therefore,

$$N_T + \epsilon_T = f(N_1, N_2, \ldots, N_n) + \sum_{i=1}^{n} \frac{\partial f}{\partial N_i} \epsilon_i$$

so that

$$N_T = f(N_1, N_2, \ldots, N_n) \tag{A.19}$$

and

$$\epsilon_T = \sum_{i=1}^{n} \frac{\partial f}{\partial N_i} \epsilon_i \tag{A.20}$$

From the above equation (see Sec. A.2)

$$|\epsilon_T| \le \sum_{i=1}^{n} \left| \frac{\partial f}{\partial N_i} \right| |\epsilon_i| \le \sum_{i=1}^{n} \left| \frac{\partial f}{\partial N_i} \right| e_i$$

Since e_T is the maximum value that ϵ_T may have,

$$e_T = \sum_{i=1}^{n} \left| \frac{\partial f}{\partial N_i} \right| e_i \tag{A.21}$$

is the maximum absolute error in N_T. The maximum relative error r_T by (A.6) is

$$r_T = \frac{1}{|N_T|} \sum_{i=1}^{n} |N_i| \left| \frac{\partial f}{\partial N_i} \right| r_i \tag{A.22}$$

Letting a weighted average of the r_i be defined by

$$\bar{r} = \frac{\displaystyle\sum_{i=1}^{n} \left| N_i \frac{\partial f}{\partial N_i} \right| r_i}{\displaystyle\sum_{i=1}^{n} \left| N_i \frac{\partial f}{\partial N_i} \right|} \tag{A.23}$$

one may write Eq. (A.22) in the form

$$r_T = \lambda \sigma \bar{r} \tag{A.24}$$

where

$$\lambda = \left| \frac{\displaystyle\sum_{i=1}^{n} N_i \frac{\partial f}{\partial N_i}}{f} \right| \tag{A.25}$$

and
$$\sigma = \frac{\sum_{i=1}^{n} \left| N_i \frac{\partial f}{\partial N_i} \right|}{\left| \sum_{i=1}^{n} N_i \frac{\partial f}{\partial N_i} \right|} \tag{A.26}$$

Thus the average relative error $\bar{r}$ of the inputs is multiplied by an amplification factor

$$A = \lambda \sigma = \frac{1}{|f|} \sum_{i=1}^{n} \left| N_i \frac{\partial f}{\partial N_i} \right| \tag{A.27}$$

This factor may, of course, be greater or less than 1. The meaning of the factors λ and σ will be seen from the special cases treated below.

A.6. Summation. Suppose one sums a set of numbers N_i to obtain N_T; then

$$f(N_1, N_2, \ldots, N_n) = \sum_{i=1}^{n} N_i \tag{A.28}$$

and $\partial f/\partial N_j = 1$. Thus

$$\lambda = \left| \frac{\sum_{i=1}^{n} N_i}{f} \right| = 1 \tag{A.29}$$

and
$$\sigma = \frac{\sum_{i=1}^{n} |N_i|}{\left| \sum_{i=1}^{n} N_i \right|} \geq 1 \tag{A.30}$$

In summation, therefore, the relative error is multiplied by σ, the ratio of the sum of the absolute values of the numbers over the absolute value of the sum.

If all the numbers are positive, $\sigma = 1$, and there is no amplification of the relative error. On the other hand, if the positive numbers nearly cancel the negative numbers, σ is large, and the computation tends to be unstable. It is usually the occurrence of large values of σ in some step of a numerical computation that produces instability.

A.7. Products. Suppose N_T is the product of a set of numbers N_i; then in (A.25) and (A.26)

$$f(N_1, N_2, \ldots, N_n) = \prod_{i=1}^{n} N_i \tag{A.31}$$

so that

$$\sum_{i=1}^{n} N_i \frac{\partial f}{\partial N_i} = \sum_{i=1}^{n} \left(\prod_{j=1}^{n} N_j \right) = nf$$

Thus
$$\lambda = n \quad \text{and} \quad \sigma = 1 \qquad \text{(A.32)}$$

and, therefore, the weighted average $\bar{r}$ of the relative errors is multiplied simply by n, the number of numbers being multiplied. From (A.23) this weighted average is just

$$\bar{r} = \frac{1}{n} \sum_{i=1}^{n} r_i \qquad \text{(A.33)}$$

the ordinary average of the r_i.

The maximum relative errors by (A.24) and (A.31) are thus seen to grow proportional to the number of factors involved in a product. The evaluation of high-order determinants directly would thus be an unstable process both because of the number of factors involved in the individual terms and because of the partial cancellation of the terms that is due to differences in signs. The latter, as we have seen, leads to a large value of σ.

A.8. Powers and Roots. If all the numbers N_i in (A.31) are the same, one has

$$N_T = f(N) = N^n \qquad \text{(A.34)}$$

and since all the relative errors r_i are the same in (A.33), $\bar{r} = r$. Since $A = \lambda \sigma = n$, as before,

$$r_T = nr \qquad \text{(A.35)}$$

The nth root N may be written as the $(1/n)$th power of the number; thus

$$N_T = \sqrt[n]{N} = N^{1/n} \qquad \text{(A.36)}$$

and by (A.35)

$$r_T = \frac{1}{n} r \qquad \text{(A.37)}$$

A.9. Logarithms and Exponentials. For

$$N_T = f(N) = \ln N$$

one has

$$N \frac{\partial f}{\partial N} = 1$$

so that by (A.27)

$$A = \frac{1}{|f|} = \frac{1}{|\ln N|} \qquad \text{(A.38)}$$

Thus the relative error in the logarithm of a number N is small if the N is large or if it is near zero. On the other hand, the relative error is greatly magnified if N is near 1.

If

$$N_T = e^N$$

then

$$A = \frac{|Ne^N|}{|e^N|} = |N| \qquad (A.39)$$

and the relative error increases by a factor $|N|$ over that for N.

A.10. Rounding Off Numbers. If a number is rounded off at the mth decimal place in the usual manner, a change in the number of $5 \times 10^{-(m+1)}$ may occur. This change may either increase or decrease the absolute error in the number. However, for the purpose of an error analysis which computes the maximum error that may arise one must add the error to that already present. This means that any rounding of intermediate numbers should decrease the reliability of the final result. Therefore the only reason for rounding intermediate numbers is to reduce the work involved in carrying along additional figures. A compromise is to carry along a few extra figures and round off beyond this point. The number of extra figures must be increased as the length of the computation increases to keep these intermediate rounding-off errors small.

The rounding off of the final answers also adds to the maximum absolute error, but it serves a useful function in indicating how many of the final figures are thought to be correct.

A.11. Application of Statistics to Error Analysis. The error analysis given above applies to the maximum absolute error or maximum relative error made in a given computation. This is the customary way of treating round-off errors in ordinary computations of moderate length. The advent of electronic calculators, which permit computations to be performed involving thousands and even millions of additions and multiplications, has required more realistic estimates of the errors to be expected based on the high probability of a great deal of canceling of the positive and negative errors.

Thus, if a thousand numbers accurate to three significant figures are multiplied, their individual maximum relative errors may be greater than 0.1 per cent, and the maximum relative error in the product is 1,000 times as large. The maximum relative error in the answer could thus be 100 per cent. The root mean square of the errors averaged over all possibilities for the combination of the positive and negative individual errors, however, can be obtained by multiplying the individual error of 0.1 per cent by $\sqrt{1,000}$. Thus the final result is very likely (68 per cent of the time) to be in error by less than 3.2 per cent. It will be extremely unlikely to be in error by more than 10 per cent.

The analysis of error along these lines would require a rather careful development of some of the basic concepts of statistics, and hence its development is omitted for lack of space.

APPENDIX B

REFERENCES

The following references have extensive bibliographies on numerical analysis:

Bennett, A. A., W. E. Milne, and H. Bateman: Numerical Integration of Differential Equations, Report of the committee on numerical integration, *Bull. Natl. Research Council, U.S. No.* 92, 1933.

Grinter, L. E. (ed.): "Numerical Methods of Analysis in Engineering," The Macmillan Company, New York, 1949.

Hildebrand, F. B.: "Introduction to Numerical Analysis," McGraw-Hill Book Company, Inc., 1956.

Householder, Alston S.: "Principles of Numerical Analysis," McGraw-Hill Book Company, Inc., New York, 1953.

Milne, W. E.: "Numerical Solution of Differential Equations," John Wiley & Sons, Inc., New York, 1953.

Nörlund, N. E.: "Vorlesungen über Differenzenrechnung," Springer-Verlag OHG, Berlin, 1924.

Paige, L. J., and O. Taussky (ed.): "Simultaneous Linear Equations and the Determination of Eigenvalues," National Bureau of Standards, Applied Mathematics Series 29, 1953.

Staff of the Computation Laboratory: "A Manual of Operation for the Automatic Sequence Controlled Calculator," Harvard University Press, Cambridge, Mass., 1946.

REFERENCES

INDEX